D0225553

BIOLOGICAL
NMR
SPECTROSCOPY

BIOLOGICAL NMR SPECTROSCOPY

Edited by

John L. Markley

University of Wisconsin, Madison

Stanley J. Opella

University of Pennsylvania

New York Oxford
Oxford University Press
1997

Oxford University Press

Oxford New York
Athens Auckland Bangkok Bogota Bombay Buenos Aires
Calcutta Cape Town Dar es Salaam Delhi Florence Hong Kong
Istanbul Karachi Kuala Lumpur Madras Madrid Melbourne
Mexico City Nairobi Paris Singapore Taipei Tokyo Toronto

and associated companies in
Berlin Ibadan

Published by Oxford University Press, Inc.
198 Madison Avenue, New York, New York 10016

Library of Congress Cataloging-in-Publication Data
Biological NMR spectroscopy / edited by John L. Markley, Stanley J. Opella.
p. cm.
Includes bibliographical references and index.
ISBN 0-19-509468-9
1. Nuclear magnetic resonance spectroscopy. 2. Biomolecules—Analysis.
I. Markley, John L. II. Opella, Stanley J.
QP519.9.N83B56 1997
574.19'285—dc20 96-12586

9 8 7 6 5 4 3 2 1

Printed in the United States of America
on acid-free paper

Table of Contents

Introduction

The founders of the field of biological nuclear magnetic resonance (NMR) spectroscopy had a clear vision of its future potential. The 65th birthday of one of these pioneers, Professor Oleg Jardetzky, provided an occasion for bringing together the small group of scientists whose research defined the field at its inception, Jardetzky and his mentors, Linus Pauling and William N. Lipscomb, and Mildred Cohn and Robert G. Shulman. Another key figure in the early days of the field, William D. Phillips who was expected to participate in the Symposium and this book died shortly before the Symposium held at Stanford University in March, 1994. The historical section of this volume (Section I) conveys the excitement of the beginnings of biological NMR spectroscopy, when every experiment was new and potentially important. Section II, which constitutes the bulk of this Volume, provides a contemporary overview of the legacy of these early experiments.

It is nothing short of astonishing to be able to say that biological NMR spectroscopy has fulfilled the dreams of its founders. Its potential as a method for determining structures and describing the dynamics of proteins and nucleic acids in aqueous solution has been realized in recent years, and the approach has become an essential part of structural biology. These capabilities, which are practically routine now, reflect the rapid pace of technological invention and the scientific context of the period, 1957 to the present. Invention refers to the continuous stream of advances in instrumentation and magnets, computers and software, pulse sequences, and experimental design for NMR studies of proteins and other biopolymers. Context refers to the advances in biology. Even the most casual reader of newspaper headlines is aware that we live in the age of molecular biology, since it is now recognized that all aspects of life reflect the linear sequence of bases in DNA that store biological information and, in turn, specify the amino acid sequences of the proteins that express biological functions. However, we also realize that biology is fundamentally both three-dimensional and dynamic. And here is where structural biology will come to play the dominant role in describing the basic principles of biology and furthering biomedical and biotechnological applications. The limits of what can be obtained from sequences alone are already at hand. The limits of what can be obtained from structural biology, in general, and biological NMR spectroscopy, in particular, have not even been approached. The frontiers of protein complexes, membrane proteins, and carbohydrates, as well as more detailed descriptions of the interplay between structure and dynamics in protein function beckon.

Oleg Jardetzky is one of the pillars of biological NMR spectroscopy. He identified at its earliest stages the importance of resolving and assigning resonances from individual sites in a protein. This enabled the unique ability of

NMR spectroscopy to differentiate among chemically identical groups within the environment of a folded protein to be fully exploited. Perhaps the major experimental tool at our disposal is the use of isotopic labeling; what is so obviously a routine laboratory method now was a totally foreign undertaking in the beginning. The same can be said for signal averaging for sensitivity enhancement. These are the procedures that led to resonances being used as monitors of the structure, dynamics, and chemistry of protein groups. The latter got its start with the titration of individual histidine residues in proteins. All of these things, and many others mentioned in the historical section of the volume, were simply not present anywhere in science in 1957. The ideas and technology could not be borrowed from any other field, they had to be invented through the process of basic research.

1995 marks the 50th anniversary of the discovery of the nuclear magnetic resonance phenomenon, recognized by the 1952 Nobel Prize to Felix Bloch and Edward M. Purcell. Biological NMR spectroscopy is generally recognized as starting in 1957 with the publication of the first NMR spectrum of a protein followed shortly thereafter by its interpretation in terms of the constituent amino acids. The second Nobel Prize in NMR was awarded to Richard Ernst in 1991, and he has contributed the Foreword for this volume. Then the story as presented in this book goes into the hands of Oleg Jardetzky, Mildred Cohn, Bob Shulman, and Joe Ackerman, who provides a moving tribute to Bill Phillips and his contributions to the field.

Readers of this book do not need to be reminded of the importance of structural biology in the scheme of science or of the importance of NMR spectroscopy to structural biology. However, we hope that this book will serve to remind all of us of the willingness to participate in unconventional research and the high level of innovation required of the founders in order to establish the field of biological NMR spectroscopy. This can be seen directly in the contributions in the historical section. It can also be seen in the successful applications of biomolecular and biomedical NMR spectroscopy described in the scientific contributions that make up the majority of the book. It has been an exciting adventure from the tentative beginnings to these latest developments in biological NMR spectroscopy.

Acknowledgments

This book is a direct outgrowth of the Symposium held at Stanford. We are grateful to all of the financial sponsors of the meeting as well as all of the participants. We especially appreciate the generous contributions from the principal sponsors, Bruker Instruments, Glaxo Research Institute, and Oxford University Press. Essential support from Varian Associates, Magnex Scientific, Martek Biosciences Corporation, Otsuka Electronics, Ajinomoto Company, Inc., Escom Science Publishers BV, IBM, Cambridge Isotope Laboratories, Inc., Tecmag, Inc., Programmed Test Sources, Inc., Molecular Simulations Inc., Intermagnetics General Corporation, Isotec Inc., and Dr. Harold Amos is also greatly appreciated.

We thank the authors of all of the manuscripts for their timely submissions. We especially thank Richard Ernst for the Foreword, the founders, Oleg Jardetzky, Mildred Cohn, and Bob Shulman, for their personal reminisces, and Joe Ackerman for the tribute to Bill Phillips.

We warmly thank Robin Holbrook whose assistance in organizing the meeting was indispensable and acknowledge Russ Altman's help with local arrangements. We especially want to thank Jennifer Wang and Linda Matarazzo Cherkassky for preparing the book in camera ready form. This took enormous patience and sophistication in software, in light of the diversity of all aspects of the contributions.

And finally we thank Oleg Jardetzky for making all of this possible.

John L. Markley
Madison, USA

Stanley J. Opella
Philadelphia, USA

Foreword

My heartiest congratulations and thanks to you, dear Oleg, on the occasion of 65 intense years of invaluable contributions to science and to mankind! What would be NMR today without your foresight and your ingenuity? You foresaw the importance of biomedical NMR before many of the later contributors to the field were even born. You made attempts to this direction as one of the very first scientists and you continued to fertilize the field with ideas, concepts, critics, and valuable applications ever since the late fifties.

Indeed, the first contributions of Professor Oleg Jardetzky to biological NMR date back to 1956, when the technological development was not yet ready to successfully solve by NMR relevant biological questions. But Oleg Jardetzky never ceased to pursue his dream, and he contributed himself significantly to the advancement of NMR technology. For example signal averaging: It was the first successful attempt by Oleg Jardetzky and others to circumvent the incredibly low sensitivity of NMR and helped enormously to tackle biological systems. Probably Oleg Jardetzky was the first who recognized as early as 1965 the importance of Fourier transform spectroscopy. I still remember his enthusiastic support of this concept, in which I had difficulties myself to believe, sitting at a lake in New England on a hot summer afternoon during a Gordon Conference and discussing about the future prospects of pulse techniques. He has also picked up rapidly the two-dimensional NMR techniques for structural studies and solved many structural and functional questions of biomedicine ever since.

His early seeds have become in the mean time monumental trees which turned out to be indispensable for our understanding of biological processes. Equally valuable as his original contributions are his often rather polemic and critical discussion contributions. They are always stimulating and keep the discussion lively.

This book summarizing the 65th Birthday Symposium demonstrates vividly in how many ways you, dear Oleg, have contributed to biological NMR. It certainly creates expectations for your future productivity which I am sure you will even surpass. I hope you will continue to enjoy to actively contribute to science for many fruitful years to come. Best wishes!

R.R. Ernst
Zurich, Switzerland

The Founders Medal. Presented to Oleg Jardetzky, Mildred Cohn, and Robert G. Shulman at the XVIth International Conference on Magnetic Resonance in Biological Systems, Veldhoven, The Netherlands, August 1994. "For Outstanding Contributions to Biological Magnetic Resonance"

Section 1: History of Biological NMR Spectroscopy

1

Simple Insights from the Beginnings of Magnetic Resonance in Molecular Biology

Oleg Jardetzky

Stanford Magnetic Resonance Laboratory
Stanford University
Stanford, CA 94305 USA

Birthday symposia inevitably provide an opportunity for reflection. Noting that greater minds than mine have offered an apology for their life (St. Augustine, 1853 edition; St. Thomas Aquinas; John Henry cardinal Newman, 1864), I shall attempt to answer the question: What have been the lasting contributions of my generation - the generation that began its work before Richard Ernst's epoch making development of 2D NMR, and the equally momentous development of high field spectrometers, pioneered by Harry Weaver at Varian, Rex Richards at Oxford and Günther Laukien at Bruker, revolutionized the technology and put biological applications within everyone's reach? I offer these insights in the spirit that to fully understand a subject one must understand its history.

The essence of scientific endeavor is to see something no one has seen before - or understand something no one had understood before. If there had been such a contribution, it was to understand what biological questions could be asked by NMR and to develop prototype experiments showing how. Difficult as it is to imagine this today when such understanding is taken for granted, the now obvious just wasn't obvious then. Quite the contrary: well considered expert opinion of the day held the undertaking to be of very dubious merit. Linus Pauling, with whom it was my great fortune to spend my postdoctoral year, was never much interested in nuclear magnetic resonance (and did not think much of its promise for biological applications, as he clearly pointed out at this symposium). But, Linus Pauling firmly believed in giving the young the freedom to explore, and so the first crude interpretation of a protein NMR spectrum, taken a few weeks earlier by Martin Saunders, Arnold Wishnia and J. G. Kirkwood at Yale, was based on the first amino acid and peptide spectra we had taken at Caltech. When I got my first faculty job at Harvard, and wanted to apply for an NMR spectrometer, it was not quite as easy. I was called by the department chairman into a conference with the Dean and both tried to convince me that such a high risk request from a totally unknown young man would never be funded and I could destroy my academic career and damage the reputation of the University. My response was to take the plane to Washington the next morning and present the proposal - "NMR in Molecular Biology" - to the then director of the NSF Molecular Biology program, William V. Consolazio. With the assurance of his support and of a broadly based peer review, the proposal was submitted, reviewed and funded, and the first NMR laboratory dedicated to biological research was founded at the Harvard Medical School in 1959.

To develop the now commonplace understanding of what NMR could do in molecular biology required thought and required experimentation, and the basic insights we now take for granted grew gradually after alternatives were carefully ruled out. It would not reflect reality to credit any one individual with having done all that was needed to develop this understanding, but we were a small group. There was an occasional chemist and an occasional physicist who did an occasional experiment of potential biological relevance, but those who persevered in exploring the potential systematically were initially Mildred Cohn, Bob Shulman, and I, along with our students, joined a few years later by Bill Phillips. Being a small group, we worked very differently from the now modern scientist, who has to protect his intellectual property and market himself, lest he remain unnoticed and unfunded and therefore unable to go on. Without a qualm we shared our guesses, our daydreams and our doubts, our experimental designs and results - far in advance of publication - and when the work came to overlap, as did Bill Phillips' and ours in later years, we often encouraged each other to do the same or similar experiments, so the results could be compared and the generality of the conclusions tested. Much of what we did we did not bother to publish, because it was not on the critical path of mapping the landscape. We argued and sometimes irritated each other by sharply pointed criticism - but there was room enough for everyone, so there was no need to compete. In a sense, it was a different phase - a different kind - of science. It was exploratory rather than exploitive science. It was not aiming to exploit existing and invent bigger and better techniques to implement ideas obvious to everyone, but to develop the basic framework of ideas to determine what could and should be done. There are areas of science where this is still necessary, but NMR is now a mature technology and is not one of them. The value of its current massive data generation phase is obvious and likely to be lasting. If anything of enduring value survives from the early phase, it is a set of a few simple, but fundamental insights, most now taken for granted, but originally wrung from nature with a primitive technology and some uncertainty of interpretation.

The high resolution NMR spectrum of a protein contains information on its secondary and tertiary structure (1961-1969).

The first glimpse of this insight came from a comparison of our protein spectra in D_2O to those in trifluoroacetic acid, first published by Frank Bovey and G. V. D. Tiers. In our first review of biological applications of NMR we therefore could categorize the applications by the type of information obtained, as "(a) the determination of primary chemical structure, (b) study of conformation or secondary and tertiary structure, (c) the study of rate processes and molecular motion, and (d) detection of interactions between molecules" (Jardetzky & Jardetzky, 1962, p. 354) and noted "....the broad lines generally observed in protein spectra may result in part from restricted motion and in part from inexact superposition of individual amino acid lines...much more extensive correlations are necessary before detailed interpretation becomes convincing" (*ibid.*, p. 363). By 1965 Bill Phillips' and our experiments, still largely unpublished, prompted the prediction: "....it will be possible to use NMR as a specific method for the study of tertiary structure" (Jardetzky, 1965, p. 3), although duly noting that "the principal obstacle to unequivocal interpretation - the extremely poor resolution resulting from the overlap of a very large number of broad peaks - has not been overcome." The point was clinched by the clear demonstration in the experiments of McDonald and Phillips published in 1967-1969 (McDonald & Phillips, 1967; 1969) and Cohen and Jardetzky (1968) that the spectrum of the completely denatured protein was to a good approximation the sum of the spectra of the constituent amino acids and much less complex than the spectrum of the native protein. In our study, incomplete denaturation, with disulphide bridges still intact, could also be distinguished from the native fold and the completely denatured chain for the first time.

Proteins undergo internal motions and the high resolution NMR spectrum of a protein contains information on protein dynamics (1961-1969).

This was fairly obvious because native proteins had broad lines, while denatured proteins has sharp lines, but the real issue was whether globular proteins were as rigid as the early crystallographers thought them to be, or whether regional differences in mobility were reflected in the NMR spectra. This could be probed by studying the binding of small molecules to proteins. In 1961 we reported: "Selective broadening of absorption in the spectra of low molecular weight molecules can be used to determine the chemical groups preferentially stabilized by the formation of specific molecular complexes in solution."

(Jardetzky & Fischer, 1961, p. 46). In 1964/1965 we showed that the three-ring compound sulfaphenazole could bind to bovine serum albumin by *either* the sulfonamide *or* the substituent phenyl ring, the difference in the relaxation rates of the bound form at the two sites being a factor of 3-4 (Jardetzky & Wade-Jardetzky, 1965, p. 228). Discussing this result, already in 1964 it was pointed out that this discrepancy could result either from a difference in the number of nearest neighbors or from a difference in the local correlation times (Jardetzky, 1964, p. 516 and Table II). For use in the study of proteins, I therefore proposed a generalization (Eq. 16) of the relaxation equations formulated by Gutowsky and Woessner, "*with the significant difference that different correlation times are assumed for different pairs of nuclei and the internuclear distances are averaged over time,*" (*ibid.*, pp. 512 - 513; italics in the original) pointing out "If more than a single correlation time determines the relaxation rate of a given group or if there is a change of interproton distances, Eq. 16 must be solved in detail to obtain the desired information. In principle, this is possible by making relaxation measurements on a series of deuterium substituted analogs and thus obtaining the contribution of each individual proton to the relaxation rate of any other proton." (*ibid.*, pp. 517 - 518). This was, of course, before the invention of 2D NMR, and it was later developed by Kazuyuki Akasaka into a method for analyzing relaxation phenomena which he called DESERT (Akasaka *et al.*, 1975). Chemical shift averaging as an indication of conformational transitions of aromatic rings in proteins was first described in our spectra of selectively deuterated staphylococcal nuclease, almost in passing "...the existence of a conformational equilibrium involving a tyrosine residue seems fairly certain. The equilibrium is probably rapid on the NMR scale (see below), since a single tyrosine peak is observed for each residue." (Jardetzky, 1970, p. 120). Of greater interest and described in greater detail was the equilibrium affecting His 48 in ribonuclease, which was slow on the NMR time scale and represented the first detection of a conformational transition in native proteins by NMR (Meadows & Jardetzky, 1968). The clearest demonstration of segmental flexibility in proteins by NMR came a few years later in the study of tobacco mosaic virus prompted by the fact that crystallographers could not distinguish between static disorder and mobility in the RNA binding region (Jardetzky *et al.*, 1978).

The most important structural information obtainable by NMR comes from relaxation parameters (1961-1965).

The use of relaxation by paramagnetic ions to define distances in small molecules was demonstrated by Bob Shulman's study of the Mn-ATP complex (Sternlicht *et al.*, 1965a,b). The already cited 1964 review, before proposing the generalization of relaxation equations and pointing out that internuclear distances can be determined in proteins, summarized the understanding explicitly:

"Most of the existing chemical correlations take into account the information obtained from the measurement of chemical shifts and coupling constants. In contrast, relaxation studies have received relatively little attention, despite the well established fact that relaxation processes are extremely sensitive to variations in the molecular environment and therefore provide the potentially most informative measurements for the study of molecular interactions." (Jardetzky, 1964, p. 500).

The application of NMR to biological problems requires signal averaging to counteract the inherent low sensitivity (1962).

The first time that a computer (the CAT - computer of average transients) was attached to an NMR instrument, we reported in Nature:

"The unique potentialities of high resolution magnetic resonance as a method for obtaining detailed information on molecular structure and molecular interactions in solution are well recognized. However, the applicability of the method to the examination of interactions in biological systems has been hampered by its low sensitivity, requiring the use of comparatively concentrated solutions (~0.05 M or higher).

"We have now applied a technique which has allowed us to perform a series of crude but informative experiments on the binding of diphosphopyridine nucleotide (DPN) to the enzyme yeast alcohol dehydrogenase (ADH) in stoichiometric proportions. Our procedure has been to couple the output of a Varian model 4300B high-resolution spectrometer operating at 60 Mc/s to a 400 channel digital average response computer, the Mnemotron model 400 computer of average transients. The sweep of the computer was triggered through a Tektronix model 535 A oscilloscope from a marker placed in the sample tube." (Jardetzky *et al.*, 1963, p. 183).

This was our one and only contribution to NMR technology.

The complexity of protein NMR spectra is generally so high as to require isotopic spectral editing (1965).

This was first reported at the memorable meeting in Tokyo in 1965:

> "Detailed comparison of the chemical shifts in a large series of amino acids and their derivatives, obtained by Fujiwara and collaborators, ourselves as well as others has convinced us that the degree of resolution required for a direct, complete interpretation of protein NMR spectra is not to be expected even at the highest now attainable NMR frequencies of 300-400 Mc. The close similarity in the magnitudes of the relaxation times of amino acids in peptide linkage, precludes the alternative possibility of a detailed assignment of lines by selective saturation.
>
> "A complete interpretation of a protein NMR spectrum is therefore contingent on the preparation of partially deuterated analogs. The feasibility of this approach is suggested by the recent success of Katz and coworkers in isolating a completely deuterated enzyme. Given the possibility of observing the proton resonance spectrum of an individual amino acid in peptide linkage, against the background of a completely deuterated polypeptide chain, nuclear magnetic resonance becomes decisively superior to any existing method in the wealth of detailed structural information which it can provide about a protein in solution." (Jardetzky, 1965, pp. 1-2).

The idea was shared with J. J. Katz during a visit to Argonne, who was generous enough to provide deuterated amino acids for my initial deuteration experiments, carried out in Cambridge in 1965/66. The ultimate realization of the proposal was published in Science in 1968 (Markley *et al.*, 1968).

Contributions of secondary and tertiary structure to the NMR spectrum are at least in part separately identifiable (1967).

In the paper by Markley *et al.* (1967) on the helix-coil transition of polyamino acids we noted:

> "The peak corresponding to the proton on the α-carbon of the polypeptide backbone shifts upfield on helix formation. The magnitude of this shift, which is attributed to the magnetic anisotropy of the peptide bond, appears to be a sensitive measure of per cent helicity. The chemical shift on helix formation of the peak corresponding to the proton on the peptide nitrogen is determined by two opposing factors: (1) differences in hydrogen bonding; (2) magnetic anisotropy of the adjacent peptide bond. Side chain resonances are not shifted appreciably on helix formation." (Markley *et al.*, 1967, p. 25).

The paper by Nakamura and Jardetzky (1967) summarized the available information: "....the chemical shifts accompanying the incorporation of a given amino acid into a polypeptide chain can be systematized in terms of a few, relatively simple rules. The findings offer little hope for the use of NMR for the analysis of the primary sequence of large peptides, but underscore its usefulness in the study of secondary and tertiary structure." (Nakamura and Jardetzky, 1967, p. 2212). The conclusions reached in these early studies have been amply confirmed by the massive statistical analysis of chemical shifts in proteins by Wishart *et al.* (1991). Their importance however lies not so much in the information they contain on the origin of structural shifts, but in the already noted simple fact that without the secondary and tertiary structure giving rise to a separation of lines for identical amino acid residues, there would be no way of extracting structural information about proteins from NMR spectra.

High resolution NMR in principle allows a complete description of protein structure, dynamics and interactions (1970).

Summarizing the state of knowledge in the field in 1970, Gordon Roberts and I wrote about the possibilities offered by NMR:

> "1. The complete definition of the conformation of a protein in solution. It should be noted that this is in principle possible on the basis of high resolution NMR data alone. However it is an extremely difficult and laborious task and has thus far not been attempted.
>
> "2. The mechanisms of folding and unfolding of protein chains. Little detailed work on this question has been done. However, the high information content of a protein NMR spectrum should allow one to distinguish between a two-step and a multistep mechanism of denaturation and to describe precisely the sequence of structural changes.
>
> "3. Changes in conformation involving individual amino acid residues or entire regions of the polypeptide chain. Both conformational equilibria in the protein itself and conformational changes produced by ligand binding have been detected by NMR and will be discussed below. In addition to providing estimates of the rates of such changes, NMR makes it possible to define the amino acid residues involved." (Roberts and Jardetzky, 1970, p. 487).

Everything we have done since - perhaps even what the field as a whole has done since - was to exploit these basic insights: the first partial NMR structures of enzyme binding sites (Meadows *et al.*, 1969; Markley *et al.*, 1970), the first NMR detection of protein conformational changes (Meadows *et al.*, 1969; Nelson *et al.*, 1974), the detection of mobile segments in larger proteins (Jardetzky *et al.*, 1978), model-free analysis of relaxation data on proteins (King & Jardetzky, 1978), the first NMR determination of the three dimensional fold of a protein fragment, the lac-repressor headpiece, in solution (Ribeiro *et al.*, 1981 and Jardetzky, 1984), the critical analysis and development of methods for protein structure calculation (Lane & Jardetzky, 1987; Altman & Jardetzky, 1989; Zhao & Jardetzky, 1994), and finally the detailed NMR analysis of the structure and dynamics of a complete allosteric system (Zhao *et al.*, 1993; Zhang *et al.*, 1994; Gryk *et al.*, 1995). It was of course Richard Ernst's own and his students' monumental contribution to teach us how to exploit them best (Ernst *et al.*, 1987). And, it was Willie Gibbons' insight in 1975 that a combination of coupling constant and nuclear Overhauser effect measurements along the peptide backbone could be used as a scheme for the sequential assignment of protein resonances which made a systematic and complete interpretation of protein NMR spectra possible (Gibbons *et al.*, 1976). This approach was extended and refined by Kurt Wüthrich and Gerhard Wagner and their colleagues in Zurich into a practical procedure for assigning the spectra of small proteins. Publication of several protein structures determined by NMR followed, but it was the successful result of Robert Huber's challenge to Wüthrich to do a double-blind structure determination of tendamistat, a protein whose structure had not been determined previously, by x-ray diffraction and NMR in parallel (Billeter *et al.*, 1989; Braun *et al.*, 1989) that did much to convince the world that NMR was a useful structural tool.

In retrospect it is of course easy to say that all these general ideas and paradigms were obvious all along. *A priori* these ideas did not necessarily have to be true. Thirty years ago, there were those who thought that only primary structure should be reflected in the spectra, other influences being too weak. Others thought that relaxation parameters of macromolecules would forever remain uninterpretable, and thus of no use in the study of either structure or dynamics. There were those, as Amory Lovins, who since became a famous consultant on environmental affairs, who in 1965 proposed to Ed Purcell and to me to improve the sensitivity of NMR by a factor of at least 1000, using unstable vacuum tubes - a project he never had time to complete, but which would have rendered signal averaging obsolete. Then there were those who thought isotopic labeling of macromolecules would be impossible, and if not, far too expensive. There were many who thought that it was proof of scientific immaturity to even think of a protein structure determination by NMR. It was our extreme fortune that this thinking dominated neither the peer review panels of the NSF and NIH, nor the executive boards of Merck, Bell and DuPont, who underwrote some of our more expensive daydreams. Yet, we also had to learn that seeing something others did not see before does not always earn admiration - sometimes quite the contrary. As the possibility of protein structure determination by NMR was first discussed in public, at a CIBA Foundation symposium in 1970 (Porter and O'Connor, eds., 1970, p. 130), peer disapproval was recorded for all posterity:

> "F. M. Richards: Without reference to other techniques, how much information could you specify....? Are you able to specify distances and orientations to the other groups whose signals you can measure, or put certain limits on them?
> O. Jardetzky: ....In principle you can get this information if you measure relaxation times and the dependence of relaxation times on each neighbor in the vicinity.
> H. M. McConnell: That is an exaggeration!
> O. Jardetzky: It is (only) an exaggeration in the sense that it is a diabolically difficult undertaking. It is a very hard thing to do, but if you did have just two protons, you could make a good guess about the distance between them. Admittedly nobody has tried to do this."

Reflecting on how the now obvious was not always obvious, one is tempted to ask the question: what is it that is now not obvious that will be obvious later? The caveat referred to above and some of its consequences immediately come to mind, and we might consider them as additional insights:

Wherever there is conformational averaging, NMR solution structures calculated from averaged NMR parameters have no physical meaning.

For some time, the dominant creed has been that all proteins can be treated as rigid in solution. This prompted a veritable race to report more and more precise coordinates for NMR structures and to develop more and more sophisticated refinement methods for this purpose. Only recently is one beginning to see doubts on this point, and, with increasing frequency, references to my 1980 paper (Jardetzky, 1980) in which the nature and implications of conformational averaging were described - a paper which I thought for 17 years to be too self-evident and too trivial to publish. The paper was a summary of the comments I made at a 1963 Gordon conference in response to Bob Shulman's report of the structure of the Mn-ATP complex. They greatly angered him at the time, but they are true - and in the long run did not wreck our friendship.

This caveat and a host of other observations have given rise to a still more general insight about the role of NMR in Molecular Biology, which is still far from being generally accepted - in fact it is being zealously disputed by some -

but which undoubtedly will be, once the dust has settled. It was summarized in Gordon Roberts' and my book "NMR in Molecular Biology" in 1981:

> "Thus, NMR does not derive its importance as a method in molecular biology from being a technique in which one can routinely proceed from the measurement by straightforward calculation to a definite answer.... The importance of NMR rests rather on the fact that it provides a much greater wealth of different clues on questions of structure, dynamics and function than other methods. The individual clues may not always be unequivocal, but interpreted in the context of carefully designed chemical and biological, as well as spectroscopic, experiments, can lead to important discoveries" (Jardetzky & Roberts, 1981, p. 9).

Once this becomes generally understood, it will be appreciated that the primary role of NMR in Molecuar Biology is not as a - necessarily second rate - method for structure determination, but as a unique tool for the study of *dynamics* and *function*.

Epilogue

What we did *was* exploratory science. The essence of exploration is to try something, show that it can work by a prototype experiment and then go on to the next question and do the same, rather than exploit the first finding to its fullest, as is now commonly done. The freedom to pursue exploration rested on the free, permissive, unpressured, scientific climate which we had when we began, but which has all but disappeared in our lifetime.

In today's climate favoring utilitarian science, public accountability, narrowly conceived hypothesis testing, setting and achieving specific goals, mass production of data by proven methods, and cut-throat competition for resources and funds, Pauling's readiness to grant a postdoctoral fellow freedom to explore, Consolazio's visionary funding of a vague proposal with an uncertain outcome, the free bouncing back-and-forth of ideas in the early days of NMR without concern for priority and credit have become virtually unthinkable.

We may not have had an NMR in Molecular Biology today, perhaps not even 2D NMR and high field magnets, if a few of us had not had the opportunity to dream and play with it in the early days, groping in the dark and sorting out soluble and unsolved problems. After all, no lesser a group of NMR experts than the management of Varian declared in the mid-sixties that NMR had gone as far as it could go and stopped the development of high field magnets after pioneering it. Had everyone shared their views and the views of the few distinguished physicists who already in 1958 advocated abolishing the Gordon conferences on Magnetic Resonance, since nothing significant was left to be done in the field - what did happen could not have happened. If one looks closely, no technology has developed without the stepwise development of its use going hand in hand. It should not be forgotten that the giant steps of accomplished craftsmen cannot be taken if not preceded by the modest and halting - and yet seminal - steps of those who dream. To be sure, there is an inevitability to the course of science and at some time, somewhere, someone would have discovered that there was more to be done, and done it - but it would have been later rather than sooner and somewhere where the freedom to explore had not become extinct.

The freedom and the opportunity to daydream are important. The freedom and the opportunity to explore paths that others don't appreciate, or even disapprove of, are important. Taking the time to understand generalities and to develop a perspective is important. The freedom not to have to compete is important. The freedom to arouse - and survive - controversy is important.

Exploration has its dangers. Tackling the unknown can easily end in failure. Explorers rarely get their name attached to a specific contribution, because they jump from one thing to another before everyone sees their point and because it is easier to remember those who exploit an idea in tens or hundreds of different variations than those who originate it only once. Today this can be deadly in funding and tenure decisions. It is a tribute to those who set the climate at the time we were young that we did not unduly suffer for it.

Yet, it is not nostalgia for the "good old times" that prompts me to underscore this point. It is simply the recognition of a simple fact. If we forget that to find and shape an idea before it becomes common property is at least as important as its implementation after it does and that a perspective is at least as important as a specific result, we as a scientific community - perhaps we as a society - are on a self-defeating path. A society that increasingly thinks only in terms of directing and channeling craftsmanship in the pursuit of clearly visible goals is cutting itself off from the source of all innovation.

I was fortunate to have lived when and where I did. There is a feeling known only to those who have climbed a mountain at the break of dawn, long before it became crowded by afternoon hikers who must step on each other to move up the last inch. The exhilaration of even minor discovery in a field that still lies in the dark is something akin to it. It was a rare - now nearly unimaginable - privilege to have belonged to the generation that entered this branch of science - Molecular Biology - at its dawn. Perhaps it was even a greater privilege to have been part of a generation that understood and had not yet forgotten that science, as medicine, is not a trade, but a calling, that it is not about wealth, power, public attention and personal glory, not even about the good life and social standing, but above all, about the simple love of understanding the unknown.

References

Akasaka, K., Imoto, T., Shibata, S., and Hatano, H. (1975). *J. Magn. Reson.* **18**, 328.

Altman, R. B., and Jardetzky, O. (1989). in *Nuclear Magnetic Resonance, Part B: Structure and Mechanisms (Methods in Enzymology* 177), (N. J. Oppenheimer and T. L. James, eds.), Academic Press, New York, pp. 218.

Augustine, Saint, Bishop of Hippo (1853). Confessions of S. Augustine, (revised from a former translation by E. B. Pusey), John Henry Parker, Oxford, 363 pps.

Billeter, M., Kline, A. D., Braun, W., Huber, R., and Wüthrich, K. (1989). *J. Mol. Biol.* **206**, 677.

Braun, W., Epp, O., Wüthrich, K., and Huber, R. (1989). *J. Mol. Biol.* **206**, 669.

Cohen, J. S., and Jardetzky, O. (1968). *Proc. Nat. Acad. Sci. U.S.A.* **60**, 92.

Ernst, R. R., Bodenhausen, G., and Wokaun, A. (1987). *Principles of Nuclear Magnetic Resonance in One and Two Dimensions*, Clarendon Press, Oxford, 610 pps.

Gibbons, W. A., Crepaux, D., Delayre, J., Dunand, J.-J., Hajdukovic, G., and Wyssbrod, H. R. (1976). in Peptides: Chemistry, Structure and Biology, (R. Walter and J. Meienhofer, eds), Ann Arbor Sci., NY, p. 127.

Gryk, M. R., Finucane, M. D., Zheng, Z., and Jardetzky, O. (1995). *J. Mol. Biol.*, in press.

Jardetzky, O. (1964). Adv. Chem. Phys. VII, (J. Duchesne, ed.), Interscience, NY, 499.

Jardetzky, O. (1965). Proc. Int. Conf. Magnetic Resonance, Tokyo, Japan, N-3-14, 1.

Jardetzky, O. (1970). in Molecular Properties of Drug Receptors (Ciba Found. Symp.), (R. Porter and M. O'Connor, eds.), J. and A. Churchill, London, 113.

Jardetzky, O. (1980). *Biochim. Biophys. Acta* **621**, 227.

Jardetzky, O. (1984). in Progress in Bioorganic Chemistry and Molecular Biology, (Yu. A. Ovchinnikov, ed.), Elsevier Science Publishers B.V., Amsterdam, p. 55.

Jardetzky, O., and Fischer, J. J. (1961). Proc. 4th Int. Conf. on Med. Electronics, NY, p. 46.

Jardetzky, O., and Jardetzky, C. D. (1962). in Methods of Biochemical Analysis, IX, (D. Glick, ed.), Interscience, NY, 235.

Jardetzky, O., and Roberts, G. C. K. (1981). NMR in Molecular Biology, Academic Press, New York, NY, 681 pps.

Jardetzky, O., and Wade-Jardetzky, N. G. (1965). *Mol. Pharmacol.* **1**, 214.

Jardetzky, O., Akasaka, K., Vogel, D., Morris, S., and Holmes, K. C. (1978). *Nature* **273**, 564.

Jardetzky, O., Wade, N. G., and Fischer, J. J. (1963). *Nature* **197**, 183.

King, R., and Jardetzky, O. (1978), *Chem. Phys. Lett.* **55**, 15.

Lane, A. N., and Jardetzky, O. (1987). *Euro. J. Biochem.* **164**, 389.

Markley, J. L., Meadows, D. H., and Jardetzky, O. (1967). *J. Mol. Biol.* **27**, 25.

Markley, J. L., Putter, I., and Jardetzky, O. (1968). *Science* **161**, 1249.

Markley, J. L., Williams, M. N., and Jardetzky, O (1970). *Proc. Nat. Acad. Sci. U.S.A.* **65**, 645.

McDonald, C. C., and Phillips, W. D. (1967) *J. Am. Chem. Soc.* **89**, 6332.

McDonald, C. C., and Phillips, W. D. (1969). *J. Am. Chem. Soc.* **91**, 1513.

Meadows, D.H., and Jardetzky, O. (1968). *Proc. Nat. Acad. Sci. U.S.A.* **61**, 406.

Meadows, D. H., Roberts, G. C. K., and Jardetzky, O. (1969). *J. Mol. Biol.* **45**, 491.

Nakamura, A., and Jardetzky, O. (1967). *Proc. Nat. Acad. Sci. U.S.A.* **58**, 2212.

Nelson, D. J., Cozzone, P. J., and Jardetzky, O. (1974). in Molecular and Quantum Pharmacology, (E. Bergmann and B. Pullman, eds.), D. Reidel Publ. Col., Dordrecht-Holland, 501.

Newman, John Henry (1864). Apologia Pro Vita Sua, Longman, Green, Longman, Roberts, and Green, London, 127 pp.

Porter, R., and O'Connor, M. eds. (1970). Molecular Properties of Drug Receptors, (CIBA Foundation Symposium), J. and A. Churchill, London, 298 pps.

Ribeiro, A. A., Wemmer, D., Bray, R. P., and Jardetzky, O. (1981). *Biochem. Biophys. Res. Comm.* **99**, 668.

Roberts, G. C. K., and Jardetzky, O. (1970). in Advances in Protein Chemistry **24**, Academic Press, Inc., NY, 447.

Sternlicht, H., Shulman, R. G., and Anderson, E. W. (1965a). *J. Chem. Phys.* **43**, 3133.

Sternlicht, H., Shulman, R. G., and Anderson, E. W. (1965b). *J. Chem. Phys.* **43**, 4123.

Wishart, D. S., Sykes, B. D., and Richards, F. M. (1991). *J. Mol. Biol.* **222**, 311.

Wüthrich, K. (1989). *Science* **243**, 45.

Zhang, H., Zhao, D., Revington, M., Lee, W., Jia, X., Arrowsmith, C., and Jardetzky, O. (1994). *J. Mol. Biol.* **238**, 592.

Zhao, D., and Jardetzky, O. (1994). *J. Mol. Biol.* **239**, 601.

Zhao, D., Arrowsmith, C. H., Jia, X., and Jardetzky, O. (1993). *J. Mol. Biol.* **229**, 735.

2

Choice of Problems in the Early Days of Biological NMR Spectroscopy

M. Cohn

Department of Biochemistry and Biophysics
University of Pennsylvania
Philadelphia, PA 19104 USA

It was a mere ten years after the discovery of NMR that Oleg Jardetzky under the mentorship of the physical chemist John Wertz (Wertz and Jardetzky, 1956) began using ^{23}Na NMR with the aim of studying Na^+ transport in biological systems as suggested by William Lipscomb. Jardetzky found that Na^+ NMR provided a unique method for following the binding of Na^+ in weak complexes. Advantage was taken of the sensitivity of quadrupolar nuclei to their chemical environment as reflected in their relaxation rates which could be readily observed at a field of 7,030 gauss available at the time (Jardetzky and Wertz, 1956). From the very first, Jardetzky limited his choice to those problems that could be investigated uniquely or most effectively by NMR spectroscopy.

One of Jardetzky's principal goals was to elucidate, at least in part, the three-dimensional structures of biological macromolecules in aqueous solution, a distant goal in the late 1950's. He realized that before attempting to tackle the structure of these complex molecules, proteins and nucleic acids, by NMR it was essential to initially characterize the spectra of their components, amino acids and nucleosides. In 1957, he published a note in the Journal of Chemical Physics (Takeda and Jardetzky, 1957) on a few amino acids, not only reporting the chemical shifts of all the protons but also showing that in a dipeptide, for example, glycylglycine, the two CH_2 groups are non-equivalent. In 1958, he published an NMR paper, a systematic study of the proton NMR spectra of amino acids, in the Journal of Biological Chemistry (Jardetzky and Jardetzky, 1958), thus introducing many facets of NMR spectroscopy to the biochemical community. This seminal paper included: 1) the chemical shifts of the protons of 22 amino acids and their dependence on pH, concentration and ionic strength and 2) the effect of rate processes on the NMR spectrum as exemplified by the exchange of the guanidino protons of arginine with water. Increased structural information from peptide NMR spectroscopy attracted many investigators to this area of research. Nakamura and Jardetzky (1967) and Saunders, Wishnia and Kirkwood (1957) were among the first to specify systematically the effects of peptide bond formation and of primary structure on the proton spectra of amino acids, thus making it possible to distinguish these effects from changes due to secondary and tertiary structural features of proteins.

The first NMR spectrum of a protein, ribonuclease, consisting of four broad peaks, was reported in 1957 by Saunders, Wishnia, and Kirkwood (1957). Soon thereafter, Jardetzky and Jardetzky (1957) analyzed the magnitude of each peak of this low resolution spectrum demonstrating that the summation of the proton chemical shifts of the constituent amino acids of the protein determined by them (Jardetzky and Jardetzky, 1958) could account in first approximation for the observed protein peaks. Subsequently, Cohen and Jardetzky (1968) pointed out that only the completely denatured protein could be represented by the sum of the spectra of its constituent amino acids.

Following the study of amino acid spectra, an investigation in 1960 of the proton NMR of the components of nucleic acids (Jardetzky and Jardetzky, 1960), purines, pyrimidines and nucleosides, led to the discovery of base stacking which laid the basis of future investigators' interpretations of nucleic acid spectra. Conformational information could also be gleaned from the nucleoside spectra, since the α and β anomers could readily be distinguished. In further studies (Jardetzky, Pappas and Wade, 1963), the lifetimes of the base stacking and hydrogen-bonded interactions were determined

In the mid-1960s, Jardetzky really hit his stride and introduced many applications of NMR to biochemical problems. Although the ability to analyze protein spectra was very limited, Jardetzky pointed out that interactions of proteins with small molecules could be profitably studied by NMR. Thus he opened up the area of identifying the groups in the ligand molecule involved in the ligand-protein interaction by NMR spectroscopy. Such interaction was predicated on observing chemical shift changes and/or selective broadening of individual protons of the ligand upon binding, the former due to change in the environment of the observed nucleus and the latter due to a decrease in the motional freedom upon binding to protein. This approach was suggested by Jardetzky, Fischer and Pappas (1961) and implemented in a study of the binding of penicillin to bovine serum a few years later (Fischer and Jardetzky, 1965). Many other protein-ligand interactions including enzyme-inhibitor complexes were subsequently investigated by Jardetzky's group and other investigators in that decade (Roberts and Jardetzky, 1970).

Serious attempts to assign resonance peaks to individual amino acids began in the second half of the '60s initially with those proton resonances that fell outside the general envelope of the protein spectrum. In 1966, Bradbury and Scheraga (1966) observed three resolved C2 H histidine resonances in the spectrum of bovine pancreatic ribonuclease at 60 MHz. The next year, Jardetzky and his collaborators (Meadows *et al.*, 1967) with a 100 MHz spectrometer, were able to resolve the C2 H resonances of all four histidine residues of RNase as well as those of staphylococcus nuclease and lysozyme. For the purpose of resolving other aromatic residues and eventually assigning all the amino acids, in 1965 Jardetzky suggested (Jardetzky, 1965) selective deuteration to simplify the usual complex spectrum with many overlapping peaks. Protein could be isolated from organisms grown on deuterated amino acids with the exception of one or two protonated amino acids, thus eliminating most of the resonances in crowded regions of the spectrum. This ingenious strategy was realized experimentally by his group in 1968 (Markley *et al.*, 1968). Many variations of the strategy of isotopic substitution have proven useful for the determination of protein structure by NMR spectroscopy.

The next problem Jardetzky undertook to solve was the assignment of resonances to specific residues of a given amino acid in a protein for which the amino acid sequence is known. For example, in RNase, pH titrations of the histidine residues revealed four pK values, each associated with an individual histidine. In a landmark paper, Jardetzky and Scheraga and their coworkers (Meadows *et al.*, 1968) used several strategies to assign each of the four histidine residues characterized by their pKs to a specific histidine residue in the RNase sequence. These strategies included 1) minor chemical modification of a single known histidine residue 2) specific cleavage of the 20-21 peptide bond of RNase (recombination of fragments yielded an active enzyme) and 3) deuteration of the single histidine in the small fragment allowing it to be unequivocally assigned in the spectrum of the recombined complex. It was the first time the pKs of amino acids in a protein had been identified with specific residues in that protein. Further developments for assignment ensued in Jardetzky's group; for example, the comparison between wild type and mutant forms of staphylococcal nuclease (Figure 15, Roberts and Jardetzky, 1970) and the identification of the amino acid residues at the active site of enzymes deduced from the spectral changes accompany complex formation upon complexation with specific inhibitors (Jardetzky and Wade-Jardetzky, 1971)

Only a few highlights among Jardetzky's many contributions to NMR spectroscopy before 1970 have been included in this brief discussion. In retrospect, the application of NMR was in its infancy from 1956 to 1970; nevertheless the number of papers related to biochemical problems had grown from 1 to 800 during that period (Jardetzky and Wade-Jardetzky, 1971). Oleg Jardetzky had the vision and dedication to nurture this emerging technique's applicability to biochemical problems and by developing ingenious strategies kept it viable in the face of skepticism of both biochemists and NMR experts.

And he welcomed those who joined him in this arduous pursuit. What could be regarded at the time only as his undue optimism when expressed in a review in 1971 (Jardetzky and Wade-Jardetzky, 1971), may be regarded today as truly prophetic:

> "And thus today high-resolution NMR has emerged alongside X-ray diffraction as one of the two most powerful methods for the study of the structure and conformation of macromolecules, of molecular interactions, and of the time course of molecular processes. It is unique in the combination of high information content, sensitivity to both molecular dynamics and molecular geometry, and applicability to molecules in solution."

The maturation of NMR spectroscopy's ability to solve biochemical and biological problems as manifested by the papers presented at this symposium and the high regard in which this method is now held, owes much to Oleg Jardetzky's vision and dedication in the period before high magnetic field spectrometers, Fourier, transform and multidimensional techniques became available.

References

Bradbury, J.H., and Scheraga, H.A. (1966). *J. Am, Chem. Soc.* **88**, 4240.

Cohen, J.S., and Jardetzky, O. (1968). *Proc. Natl. Acad. Sci. U.S.A.* **60**, 92.

Fischer, J.J., and Jardetzky, O. (1965). *J. Am. Chem. Soc.* **87**, 3237.

Jardetzky, C.D., and Jardetzky, O. (1960). *J. Am. Chem. Soc.* **82**, 222.

Jardetzky, O. (1965). *Proc. Intern. Symp. Nucl. Magn. Reson.*, Tokyo, Abstr. N- 3.

Jardetzky, O., and Jardetzky, C.D. (1957). *J. Am. Chem. Soc.* **79**, 5322.

Jardetzky, O., and Jardetzky, C.D. (1958). *J. Biol. Chem.* **233**, 383.

Jardetzky, O., and Wade - Jardetzky, N.G. (1971). *Annu. Rev. Biochem.* **40**, 605.

Jardetzky, O., and Wertz, J.E. (1956). *Arch. Biochem. Biophys.* **65**, 569.

Jardetzky, O., Fischer, J.J., and Pappas, P. (1961). *Biochem. Pharmacol.* **8**, Abstr. 387.

Jardetzky, O., Pappas, P., and Wade, N. (1963). *J. Am. Chem. Soc.* **85**, 1657.

Markley, J.L., Putter, I., and Jardetzky, O. (1968). *Science* **61**, 1249.

Meadows, D. H., Markley, J. L. Cohen, J. S. and Jardetzky, O. (1967). *Proc. Natl. Acad. Sci. U.S.A.* **58**, 1307.

Meadows, D.H., Jardetzky, O., Epand, R.M., Ruterjans, H.H., and Scheraga, H.A. (1968). *Proc. Natl. Acad. Sci. U.S.A.* **60**, 766.

Nakamura, A., and Jardetzky, O. (1967). *Proc. Natl. Acad. Sci. U.S.A.* **58**, 2212.

Nakamura, A., and Jardetzky, O. (1968). *Biochemistry* **7**, 1226.

Roberts, G.C.K., and Jardetzky, O. (1970). *Adv. Prot. Chem.* **24**, 447.

Saunders, M., Wishnia, A., and Kirkwood, J.G. (1957). *J. Am. Chem. Soc.* **79**, 3289.

Takeda, M., and Jardetzky, O. (1957). *J. Chem. Phys.* **26**, 1346.

Wertz, J.E., and Jardetzy, O. (1956). *J. Chem. Phys.* **25**, 357.

3

Early Days of Biochemical NMR

R.G. Shulman

MR Center
Yale University
New Haven, CT 06520 USA

It was my pleasure to participate in Oleg's 65th birthday celebration and to reminisce about the early days of Biochemical NMR. Oleg was always there. I remember in the summer in the early 1960s sitting on lawn chairs at a Gordon Conference and discussing the need for a meeting on biochemical NMR. This was to convene those with common interests, and out of this grew the 1964 meeting in Boston, which was the first International Conference on Magnetic Resonance in Biological Systems. In organizing the 1964 meeting Oleg was stalwart, in charge of the local arrangements at the old mansion, home of the American Academy of Arts and Sciences. The venue was much appreciated by the more than 100 attendees, and the smooth arrangements and elegant, although somewhat dowdy locale, contributed to the sense, generated by the meeting, that the field had a coherent scientific core and a meaningful future.

In the early days of the 1960s the field of magnetic resonance in biological systems, brought together biannually by the society, had a coherence that was nurtured by the society. In those days the NMR and ESR methods were much less developed than they soon became, so that any reasonably competent spectroscopist could understand all the methods employed. Additionally, because the earlier studies concentrated upon the better understood biological molecules or processes, the breadth of the applications did not baffle a slightly informed biochemist. The rapid advances in definite understanding were thrilling to practitioners in the field, and individual efforts were motivated by a sense that the field was going to grow. By that time NMR was firmly established as a quantitative method in chemistry, solid state physics, and other material sciences so that with the results in hand it was logical to extrapolate to a future in which magnetic resonance could be central to biological research.

These high hopes, however, required considerable confidence in extrapolation, because the individual findings were sometimes slight when compared to the exciting cutting edges of biological research. When Oleg and Jim Fisher, for example, made their pioneering studies of relaxation changes in the ^{1}H resonances of penicillin as it bound to BSA (and we initiated parallel changes in molecules like ATP bound to paramagnetic ions) structural information of high quality was obtained but it was not comparable with rich detailed structural information being derived at the time from the early x-ray protein structures. These early individual hopes needed and received support from the excitement of more organized interplays, of which the ICMRBS was particularly helpful.

Oleg made two particularly personal contributions to the ICMRBS. First as mentioned above for the Boston meeting he worked continually to support its activities. In addition to organizing the Boston meeting, he participated in the organization of several succeeding meetings. He often wrote grants for funding and contributed his own support to the organizational matters. He kept records and kept track in a guardian way of the worldwide activities. This sense that he was available to back up the organization was particularly important because of the loose organization of these meetings in which a new group arranged all aspects of each meeting. Oleg helped form an advisory council system in which former organizers were represented on a committee whose composition turned over with time, and he was involved in the other continuing activities.

Beyond his steady hand at the oar, Oleg contributed a more formalized historical perspective both to the Society and the field by a sense of place and moment. As one can see it in retrospect, if this field were really to grow in strength, as we believed, then it truly represented an important historical flow. Oleg, perhaps because his earlier life was in a more formalized environment, became the embodiment of the more ceremonial aspects of the society and the field. He kept and preserved the records, he organized meetings to supplement the biannual conference and he spoke of the past, present and future of the field. Gradually over the years the future happened, and it is a full and rich time for NMR as we hoped.

Oleg's scientific contributions in retrospect had the same seminal values as the more historical perspectives I have been describing. His early experiments on ^{23}Na *in vivo*, his characterization in the early days of high-resolution NMR of the nucleic acids and the amino acids, his relaxation studies, and his introduction of isotopic labels were all original findings which have been the basis, not always acknowledged, of future research. Of his research at that time two directions stand out in my mind as the cutting edge, years ahead of their time.

His study, mainly by histidine titrations, of the active site of ribonuclease, with a talented group of collaborators, remains a paragon of catalytic studies by NMR. They provide an understanding of mechanism which showed how in a few short years NMR, by the late 1960's, was providing state of the art information.

In recent years NMR studies, strengthened by multi dimensional NMR methods, have generally moved away from active sites to the complete molecular structures. Many of these have been focused on the folding problem in which the pathways between random coil and folded structure are investigated. Although considerably more detail is now available, in my opinion, the basic studies were done in the 1960's on ribonuclease and staphylococcal nuclease by Oleg and his colleagues, with equally brilliant parallel studies by W.D. Phillips and his colleagues. These studies of denaturation and renaturation took advantage of the assignments of individual resonances and their difference between the two states to characterize pathways.

In his personal research as well as in his more ceremonial presentations Oleg made seminal and lasting contributions in his early days to the development of NMR in Biology and for this reason it is a pleasure to honor him on this occasion.

4

William D. Phillips Memorial Lecture

J.J.H. Ackerman

Department of Chemistry
Washington University
St. Louis, MO 63130 USA

It is a privilege to be able to share with you a few moments of reflection on William Dale Phillips, a good friend of mine and of many in this audience (Presented at a plenary session of the XVth International Conference on Magnetic Resonance in Biological Systems, August 14-19, 1994, Veldhoven, the Netherlands). Bill Phillips was a pioneer in the use of magnetic resonance for determination of protein structure. Although a major portion of his scientific career was spent in industry, primarily at EI du Pont de Nemours and Co. in Wilmington, Delaware. Bill also spent time in service to academics and the federal government. He most recently served as Associate Director for Industrial Technology in the Bush Administration's Office of Science and Technology Policy. He was 68. The cause of his death was cancer of the prostate.

I first met Bill Phillips in 1979 when George Radda, in whose laboratory I was working, suggested that I contact his good friend regarding a position at Washington University. Phillips had recently moved from DuPont, where he had been Assistant Director of Research and Development, to Washington University in St. Louis where he was Charles Allen Thomas Professor and Chairman of the Department of Chemistry.

Bill had been given the task of rebuilding the department. I was immediately struck by his vision and sense of commitment. This was a person who got things done. I was hooked.

In many ways Bill's move to St. Louis was a return home to his beloved Midwest. He was born in Kansas, City, Missouri and grew up there graduating from high school at the age of 17 in 1943. During the war he served in the U.S. Navy V-12 program achieving the rank of Lt. (jg). After the war he returned to the Midwest and in 1948 he received a B.A. in chemistry from the University of Kansas.

Following his undergraduate education, Bill left the Midwest again, this time for a long sojourn to the east coast. First stop was MIT where he studied physical chemistry (focusing on the vibrational spectroscopy of organic molecules). He received his Ph.D. in 1951. It was at MIT that Bill met Esther Parker, a Wellesley College student, better known to her friends as "Cherry". Married in 1951, Cherry was a loving partner assisting Bill in his many adventures.

Bill is widely remembered by the magnetic resonance community for the pathbreaking work he did at the DuPont where he landed after leaving Boston. DuPont was to be his home base for almost 30 years and he rose quickly through the ranks from research chemist to Associate and Assistant directorships of R&D.

Looking over Bill's contributions to magnetic resonance during his period, I am struck by both the remarkable breadth and depth of his work. Any attempt at a reasonable review of Bill's publication record during his DuPont years would take far longer than your patience or my voice would allow. Permit me, however, to make a few selective observations, illustrative of his great body of work.

Bill's first publication out of DuPont, a single authored letter to the editor in the *Journal of Chemical Physics* in 1955, described the use of proton NMR to demonstrate unequivocally the existence of restricted rotation in amides. The lead sentence reads: "Restricted rotation around the C-N bond of amides has been postulated and is an important consideration in theories pertaining to the structure of protein molecules". Already, in 1955, Bill was thinking about NMR as a tool for probing protein structure, an area that would become his central focus in later years. But this was 1955, and his spectrometer ran at 30 MHz.

In 1957, again in the *Journal of Chemical Physics,* Bill announced in a letter to the editor, "We have observed anomalous shifts in the proton magnetic resonance spectra of alcohols complexed with paramagnetic ions." He goes on to postulate bond formation between the alcohol ligand and metal ion leading to subsequent delocalization of the unpaired electrons providing a finite density of unpaired electrons at the resonating nucleus. Chemical shifts induced by paramagnetic metal ions would, in time, become a powerful tool for structural elucidation and a major focus of Bill's research.

In the 1960s Bill's research began to center on biological problems. His beautiful 1964 *Science* article entitled "Nucleic Acids: A Nuclear Magnetic Resonance Study" elegantly lays out the use of NMR - albeit at 60 MHz - as a melting point apparatus for mapping the site specific melting of double stranded DNA. Despite the importance of this work, Bill was content to conclude in understated tone that, "The observation of the behavior of individual proton magnetic resonance lines can give detailed insight into changes in molecular structure and motion during complexation or 'melting'."

In the mid 1960's DuPont installed the first Varian HRSC-1X spectrometer with the C-1024 time-averaging computer. With a quantum leap to 220 MHz, Bill turned his attention fully toward biological macromolecules. His 1967 *Science* article entitled, "High-Resolution Nuclear Magnetic Resonance Spectroscopy," gave dramatic evidence of the tremendous improvements in spectral quality and resolution over that available at the then common 60 MHz field strength. Breaking through his East Coast reserve, Bill wrote with obvious excitement, "Indeed the improvement in resolution in protein spectra at 220 MHz is so dramatic that it compels one to contemplate the possibilities in proton magnetic resonance spectra of proteins at even higher frequencies." Bill reported on this landmark work at the Second International Conference on Magnetic Resonance in Biological Systems in Stockholm.

The next ten years saw Bill Phillips publish a remarkable series of reports dealing with the structure of bio-molecules as probed by NMR, including: studies of the tertiary structure of proteins, determination of thymine nearest neighbor base sequence ratios in DNA, elucidation of resonance assignments and structure of lysozyme, and determination of the magnetic properties and structure of ferredoxins and other iron-sulfur proteins and synthetic analogs.

For this and other work he was elected to the National Academy of Sciences and the American Academy of Arts and Sciences.

Bill's six year stint at Washington University led to the rebirth and growth of the Department of Chemistry. In 1984 he returned to private enterprise as Senior Vice President of Research and Development at Mallinckrodt, Inc., in St. Louis.

Highly respected as a scientist of the first tier, Bill Phillips was also deeply in science policy issues. Moving to Washington, DC, in 1990 he served as one of five associate directors on President George Bush's science advisory board. In this capacity he chaired the National Critical Technologies Panel whose first biennial report, presented to President Bush in 1991, became a blue print for government action. In clear, lucid, direct prose the report convincingly advocated enhancing and securing for the United States, those technologies critical to its national security and economic competitiveness.

Those of us who knew Bill Phillips remember him for his great honesty and integrity and for the encouragement and support he gave young scientists. He was widely respected for his far reaching, global assessments of issues and policies both within and outside of the technology arena.

He was a man for all seasons. He is greatly missed.

Section 2: Protein Structural Studies

5

Flexibility and Function of the *Escherichia coli trp* Repressor

M.R. Gryk and O. Jardetzky

Stanford Magnetic Resonance Laboratory
Stanford University
Stanford, CA 94305 USA

The *trp* repressor from *Escherichia coli* is a DNA binding protein, which in the presence of the amino acid tryptophan inhibits the transcription of at least five operons: *trpEDCBA, trpR, aroH, mtr,* and *aroL* (Zubay *et al.*, 1972; Rose *et al.*, 1973; Zurawski *et al.*, 1981; Heatwole and Somerville, 1991, 1992). The ligand-free form (aporepressor) shows only weak binding ($K_D \sim 10^6$ - 10^7 M) to DNA, independent of the nucleotide sequence (Carey, 1988; Hurlburt and Yanofsky, 1990). The tryptophan containing form (holorepressor) binds preferentially to specific operator sequences with a much higher binding constant ($K_D \sim 10^{10}$ - 10^{11} M) (Carey, 1988; Chou *et al.*, 1989; Hurlburt and Yanofsky, 1990). The binding of the repressor is thus regulated by tryptophan, which acts as a corepressor (Rose *et al.*, 1973). With a molecular weight of approximately 25kD, the *trp* repressor is one of the smallest regulatory systems known, which makes it attractive as a prototype for the study of the molecular mechanism of allosteric regulation. In the twelve years since it was isolated and purified (Joachimiak *et al.*, 1983), it has become one of the most extensively studied allosteric systems. Although Perutz has justly pointed out that the *trp* repressor is not allosteric in a classical sense (Perutz, 1989), in fact, the control site is too close to the DNA binding site to separate direct and indirect (allosteric) effects, the system does manifest an essential feature of all allosteric control mechanisms - a structural change induced by ligand binding.

Structures of both the apo- and the holorepressor have been determined both by x-ray diffraction (Zhang *et al.*, 1987; Schevitz *et al.*, 1985; Lawson *et al.*,

1988) and by NMR (Arrowsmith *et al.*, 1991a; Zhao *et al.*, 1993). Structures of the operator DNA have also been reported (Lefèvre *et al.*, 1987; Shakked *et al.*, 1994a,b), and several structures of operator-repressor complexes are available: two crystal structures (Otwinowski *et al.*, 1988, Lawson and Carey, 1993), and a family of NMR solution structures (Zhang *et al.*, 1994).

Repressor architecture

Native *trp* repressor is a symmetric dimer of two polypeptide chains, each containing 107 residues. The molecule is highly helical, each peptide chain contains six helices labeled A-F. An unusual feature of the dimer is that it is intertwined (Figure 1) with five of the six helices involved in interchain contacts. The L-tryptophan binding pocket is formed by residues in helices C and E of one chain, and residues in the B-C turn of the other. The structure of each individual monomer is highly extended in a configuration that would be unlikely to form in the absence of contacts to the other monomer (Schevitz *et al.*, 1985).

There are five distinct subdomains in the repressor: (1) a hydrophobic "central core" formed by helices A, B, C and F of each subunit, (2) and (3) two N-terminal segments (each about fifteen residues long), and (4) and (5) the two helix-turn-helix DNA binding domains (helices D and E). The N-terminal segments appear disordered in both the crystal and the solution structures and are therefore not shown in Figure 1. The hydrophobic core provides the scaffold on

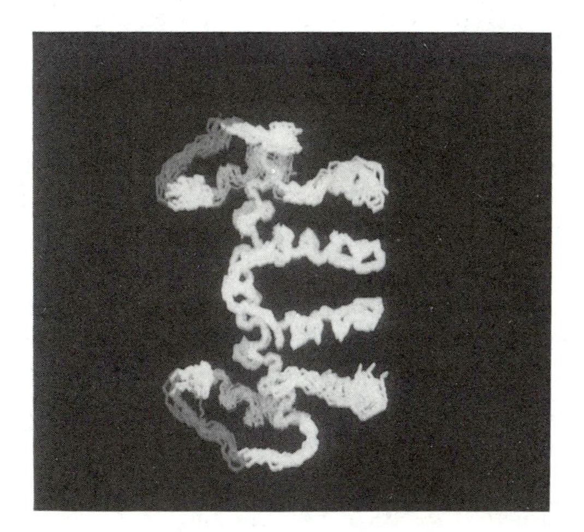

Figure 1: Refined solution structure of the *trp* holorepressor. The superposition of the backbone heavy atoms for ten structures is shown. The first 15 residues are disordered in the solution structures and are, therefore, not shown. From Zhao *et al.*, 1993.

which the DNA binding domains are mounted, and also provides a surface of positively charged residues which facilitates binding to the negatively charged phosphate backbone of the operator DNA (Guenot *et al.*, 1994). The hydrophobic core is structurally very stable and resistant to enzymatic cleavage (Carey, 1989). It remains relatively unchanged upon binding of the L-tryptophan corepressor, or of the desamino analog indole proprionate (Lawson and Sigler, 1988). Both mutation (Kelley and Yanofsky, 1985) and NMR (Tasayco and Carey, 1992) studies on truncated fragments of *trp* repressor suggest that the core residues alone (1-68) are capable of forming a stable dimer. The stability of the core seems to be imparted by the large collection of hydrophobic groups through the region, including the almost poly-leucine helix B. There are two interchain salt bridges between glutamic acid 47 and arginine 54. These salt bridges are unaffected by tryptophan binding (Zhang *et al.*, 1987).

The N-terminal segments appear to be completely unstructured in all of the free repressor structures reported. In solution these residues have very sharp resonances with no inter-residue NOEs suggesting that they are constrained to the repressor core through covalent bonds only. Similarly, in the crystal analyses, the first three residues are poorly defined, while the next twelve adopt different conformations in each of the three independent crystal structures. It has been suggested that these residues may be involved in DNA binding since mutant repressors in which these residues are missing have lower DNA binding affinity (Carey, 1989; Hurlburt and Yanofsky, 1992a). Indeed, the substitution of a negatively charged glutamic acid at position 13 by the positively charged lysine results in a repressor with greater binding affinity (Kelley and Yanofsky, 1985), presumably due to an interaction with the DNA phosphate backbone. However, as discussed later, the native N-terminal residues do not appear to be responsible for any direct DNA binding, at least not to the short operator sequences studied so far.

The structure of the DNA binding domains has several interesting features. In the crystal structures, the residues implicated in DNA binding from homology studies (Ohlendorf *et al.*, 1983) and from mutational analysis (Kelley and Yanofsky, 1985) were indeed observed to adopt a helix-turn-helix conformation as predicted, although the isotropic temperature factors for the DE helices were somewhat larger than those found for core residues (Lawson *et al.*, 1988).

The solution structures of the repressors present a similar overall picture of the DNA binding helices, when considering the effects that dynamic fluctuations have on the average structure. In a conventional NMR structure calculation based solely on NOE data, the DE region appears disordered (Arrowsmith *et al.*, 1991a). Far fewer NOEs are observed for this region than for the hydrophobic core (Arrowsmith *et al.*, 1990). In particular, many of the sequential NOEs characteristic of helical structures are missing, and a helical structure cannot be defined. The internal control of the hydrophobic core ruled out the possibility of instrumental or human error in not detecting more NOEs for the D and E helices. However, an examination of chemical shifts and ^{15}N relaxation times along the backbone indicate that the DE region is largely helical in solution (Zhao *et al.*, 1993; Zheng *et al.*, 1995). This apparent paradox can be resolved by considering the observation that the backbone protons of the DE helices exchanged with solvent much more rapidly than those in the core (Arrowsmith *et al.*, 1991b, Czaplicki *et al.*, 1991). Subsequent studies, discussed in the section on dynamics, have shown that the DE region is helical in solution as well as in the crystal, but the helices are unstable on a millisecond - second time scale (Zhao *et al.*, 1993, Gryk *et al.*, 1995).

The finding of this instability, along with the findings of larger B-factors in repressor crystal structures (Lawson *et al.*, 1988), and of large amplitudes in this region in molecular dynamics simulations (Howard and Kollman, 1992) has led to the general notion that the DNA binding domain of the repressor is "flexible." However, it must be borne in mind that in the context of NMR, the concept of flexibility encompasses a much wider range of possibilities than the simple ensemble of different conformations of rigid structures deduced from crystallographic work. Flexibility may refer to conformational mobility on a variety of timescales, with different portions of the protein being mobile on different time scales. This means that the structure may be rigid on a short time scale, but appear flexible on a longer time scale, and it may be useful to distinguish between "true" and "apparent" flexibility. The different concepts described by "flexibility" will continue to be addressed throughout this review.

Modeling studies showed that the D and E helices were in a convenient orientation to bind successive openings of the major groove of B-form DNA (Schevitz *et al.*, 1985). The first comparison of the crystal structures of the holorepressor and of the aporepressor presented a simple picture of the ligand induced conformational change - "the allosteric mechanism." The two helices in the holorepressor were found to be shifted in orientation with respect to the apo-form. This bihelical shift was described as comprising two types of movement: an en-bloc motion of the helices, and a smaller rearrangement of the helices with respect to each other. In the apo structure, the helices are collapsed onto the hydrophobic core, while in the holo structure, the ligand acts as a wedge which pushes the two helices away from the core and away from each other (Figure 2). Interestingly, although the entire D helix shifts location, only the N-terminal half of helix E was observed to change orientation between the two structures. In fact, there is actually a slight bend at position 85. This correlates well with mutation (Kelley and Yanofsky, 1985) and homology (Ohlendorf *et al.*, 1983) studies which suggest that only residues 66-86 are crucial for specific DNA binding. The mechanism of *trp* repressor action thus presented was that the L-tryptophan wedges apart the two DNA binding domains, facilitating their penetration into the major groove of operator DNA sequences (Figure 2).

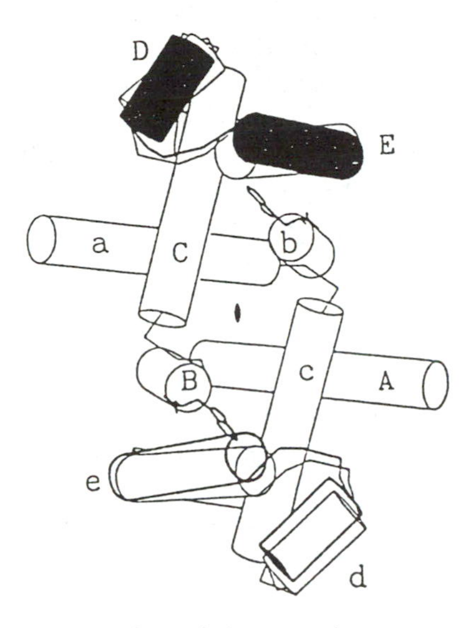

Figure 2: Cylinder representation of the crystal structures of *trp* repressor. Only helices A through E are visible in this orientation; helix F is hidden behind helix E. The upper half shows the comparison of the orientation of the DNA binding helices between the two crystal forms of the holorepressor (trigonal in black, orthogonal in white), while the lower half illustrates the shift in orientation upon ligand binding (dark outline represents the aporepressor). From Lawson *et al.*, 1988.

However, this model of the activation of *trp* repressor is oversimplified, as was demonstrated by the x-ray structure of another crystal form of the holorepressor (Lawson *et al.*, 1988). It was found that the differences in the orientation of the DNA binding helices in the two crystal forms of the *trp* holorepressor were as large as those observed between the apo and holorepressor structures (Figure 2). This observation, together with the discovery of large B-factors for the D and E helices of all three forms of the repressor, prompted the conclusion that these helices were more flexible than was previously thought (Lawson *et al.*, 1988). The conclusion regarding the action of the L-tryptophan ligand remained, as the ligand acted as a wedge in both crystal forms, albeit by wedging the DNA binding helices into different locations. This is similar to the conclusions from the solution studies, in that the term "flexibility" is used in reference to the "apparent flexibility" of rigid helices existing in a variety of possible orientations with respect to the hydrophobic core (and presumably with the DNA).

A much more complex picture emerges from NMR and crystal studies of the complex. If one compares the average of the ensemble of aporepressor to that of the ensemble of holorepressor structures, a net shift in the average position of helices D and E is seen (Zhao *et al.*, 1993). However, a comparison of the structures of the components to those of the DNA-repressor complex (Otwinowski *et al.*, 1988; Zhang *et al.*, 1994) clearly indicates that in addition to a net translocation of the helices upon complex formation, there is also a mutual adjustment in the internal structure of both the DNA and the helix-turn helix itself: an induced fit of the protein and DNA with each other. Here the internal flexibility of the helices, in addition to the "flexibility" meaning the en-bloc movement of the rigid helices mentioned above, comes into play.

Operator DNA

Although it has been debated which nucleotide sequence is the precise operator sequence recognized by *trp* repressor (Staacke *et al.*, 1990), it has been demonstrated (Carey *et al.*, 1991; Haran *et al.*, 1992; Zhang, H., unpublished results) that the principal mode of binding is observed to the palindromic consensus sequence of Lefèvre *et al.* (1985a) (Figure 3). The mutation studies of Bass *et al.* (1987) have demonstrated that the boldface CTAG sequences are the most sensitive to mutation as well as the flanking adenine and thymine bases. It should be noted that, besides the central rotation dyad, which makes the sequence palindromic (referred to as the α-symmetry element), each half of the sequence is pseudo-symmetric with a C2 dyad centered in the mutational sensitive CTAG sequence (β-symmetry).

Both solution (Lefèvre *et al.*, 1985a,b, 1987) and crystal (Shakked *et al.*, 1994a, b) studies have been reported regarding this sequence. The solution studies involved the full 20 base-pair sequence shown above and concluded that the operator was essentially B-form in solution. Dynamic studies (Lefèvre *et al.*, 1985a,b) suggested that the DNA is more flexible in the center of α-symmetry at the TA dinucleotide step. The crystal structures used a truncated 10 base-pair fragment of DNA containing only one central copy of the ACTAGT sequence. The crystal structure of this truncated sequence is mostly B-form with a bend at the central TA dinucleotide step. Thus, the important conclusions regarding the DNA structure are that it is mostly B-form (Lefèvre *et al.*, 1987; Carey *et al.*, 1991) and that it is not a rigid structure but is capable of conformational adjustments.

β α β

C G T A **C T A G** T T A A **C T A G** T A C G
G C A T **G A T C** A A T T **G A T C** A T G C
-10 -9 -8 -7 -6 -5 -4 -3 -2 -1 1 2 3 4 5 6 7 8 9 10

Figure 3: Consensus operator DNA sequence proposed by Lefèvre *et al.* (1985a) with the numbering sequence proposed by Otwinowski *et al.* (1988).

Structure of the complex

There are three reported structures of the L-tryptophan/*trp* repressor/operator DNA ternary complex, two determined crystallographically (Otwinowski *et al.*, 1988; Lawson and Carey, 1993) and one by NMR spectroscopy (Zhang *et al.*, 1994). They differ not only in experimental conditions and methods, but also in the precise operator sequence used in studying the repressor-DNA interaction. The first crystal study and the solution study used similar operator sequences, differing only in the distal base pairs. The repressor is not observed to interact with these flanking base pairs in either structure, nor is there any biochemical evidence that they play any role in the specificity of repressor binding (Bass *et al.*, 1987). The crystal structure by Lawson and Carey, however, uses a substantially different operator sequence. Their sequence is a 16 base-pair β-symmetric sequence in which only one intact CTAG sequence is contained at the center of the DNA sequence. The relevance of using such a different sequence is discussed below.

The structure of the repressor in the complex is similar to that found in the isolated repressor structures (Figure 4). The structure of the hydrophobic core is essentially identical between the different structural forms of the repressor, while the N-terminal segment was judged to be highly disordered in both the Otwinowski and Zhang structures. Helices D and E were found in a different conformation in the complex than in the isolated apo- or holo repressor, with helix E making contacts within the major groove of the DNA.

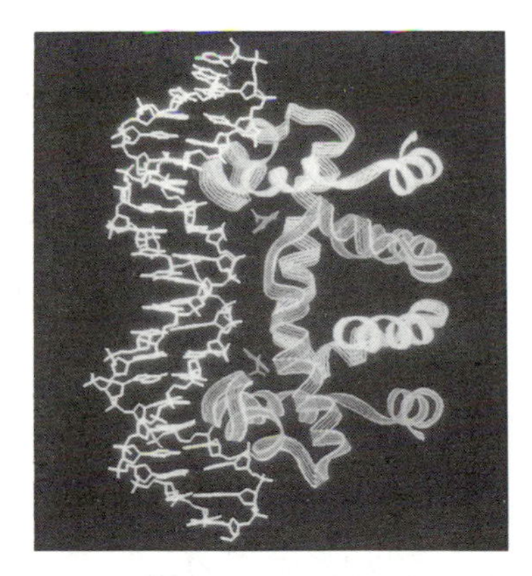

Figure 4: Solution structure of the L-tryptophan/repressor/operator ternary complex. *trp* repressor is shown in a ribbon representation. The N-terminal arms are disordered in this structure and are not shown. Adapted from Zhang *et al.*, 1994. Although the structures of the operator DNA and of the DNA binding region of the repressor change considerably upon complex formation, the overall topology of the repressor is similar to the free form (compare with Figure 1).

Otwinowski *et al.*, have attributed the stabilization of the repressor/DNA complex to three types of interactions observed in their structure: direct hydrogen bonds, water-mediated hydrogen bonds, and non-polar contacts. It should be pointed out that in all structural studies discussed here - crystallographic or NMR - the presence of a hydrogen bond has to be inferred from the proximity of neighboring donor and acceptor groups. Neither of the methods permits a direct observation of a hydrogen bond in proteins. There are twenty-eight proposed direct hydrogen bonds between the repressor dimer and the DNA duplex in the crystal structure. Twenty-four of these are to the oxygens of backbone phosphate groups while two are to the ring nitrogens of the two G_{-9}'s and two are to the G_{-9}'s carboxyl groups. The relative importance of the interactions to G_{-9} is unclear in the light of the findings that (1) a G to T mutation reduces DNA binding affinity by 700-fold *in vitro* (Marmorstein *et al.*, 1991), while (2) the same mutation results in a negligible reduction in DNA binding affinity as evaluated *in vivo* (Bass *et al.*, 1987, 1988). In addition, twelve so-called water-mediated hydrogen bonds were postulated to exist in the complex. These involve water molecules trapped in the repressor/DNA interface. An interaction is considered a water-mediated hydrogen bond, if a single water molecule is observed to have one hydrogen bond with a group on the DNA and another hydrogen bond with a group on the repressor. Six of the observed water-mediated hydrogen bonds are to oxygens of backbone phosphates, while the other six are to functional groups of mutationally sensitive DNA base pairs. As a mutation to these base pairs decreases the affinity of repressor for DNA, these six water mediated hydrogen bonds are presumed to impart some DNA binding specificity.

Finally, there are at least three non-polar contacts observed between the repressor and DNA. However, only one of these contacts (T83 to A_{-7}) can be responsible for any specificity of DNA binding (Otwinowski *et al.*, 1988).

The lack of direct hydrogen bonds to the base pairs known to be important for specificity led Otwinowski *et al.*, to propose that the mechanism of *trp* repressor/DNA recognition is a combination of water-mediated hydrogen bonds and "indirect" readout. Water-mediated hydrogen bonds can provide the necessary interaction energy (Luisi and Sigler, 1990), but do not contact all of the base pairs known from mutation and binding studies to be responsible for the specificity of interaction. The indirect readout model asserts that phosphate interactions can be used for specificity, if the precise geometry of the phosphate backbone is determined by the DNA sequence. Thus, the operator nucleotide sequence may either adopt a bent conformation, or give the DNA the conformational freedom to be bent by the repressor, bringing its phosphate groups into the proper orientation for repressor/phosphate interactions.

The proposal that water mediated hydrogen-bonds were involved in the repressor-DNA complex has been criticized as an artifact on two grounds. The crystals were grown in a highly peculiar environment containing 35% dimethyl pentanediol and less than 100 mM salt (Otwinowski *et al.*, 1988). Brennan and Matthews (1989) asserted that such a high alcohol, low salt environment would favor non-specific DNA binding. Secondly, Staacke *et al.*, (1990) argued that the α-symmetric consensus sequence used by Lefèvre *et al.*, (1985) and by Otwinowski *et al.*, was not the operator sequence recognized by *trp* repressor and therefore, by definition, the crystal structure was non-specific.

The solution structure of the *trp* repressor complex by Zhang *et al.*, (Zhang, 1993; Zhang *et al.*, 1994) provides an independent check of the validity of the original crystal work as well as the preferred DNA operator sequence. It was confirmed (Zhang, H., unpublished results) that indeed the preferred operator sequence for repressor binding was the α-symmetric sequence used by Lefèvre *et al.*, and not the β-symmetric sequence proposed by Staacke *et al.*

The overall topology observed for the solution structure of the complex agrees with that of the structures of Otwinowski *et al.*. The hydrophobic core is seen to remain relatively unchanged between the apo, holo and complex forms and between the solution and crystal studies.

As in the crystal structure, helices D and E of each repressor subunit are observed to penetrate the major groove of the operator DNA sequence and to provide the necessary contacts for DNA binding. The structures of the D and E helices are better defined in the solution structure of the complex than in the free repressor (Zhang *et al.*, 1994), mostly due to additional restraints imposed by NOE contacts between repressor and DNA; but perhaps reflecting a real reduction in the mobility of this motif as suggested by the observation of slower amide proton exchange rates (Zhang *et al.*, 1994) and a decrease in the heat capacity (Ladbury *et al.*, 1994).

As with the crystal model, the solution structure of the DNA binding region is different from that observed in the free repressor indicating that the protein undergoes a conformational rearrangement upon DNA binding. The DNA is similarly observed to distort upon DNA binding (Zhang *et al.*, 1994; Otwinowski *et al.*, 1988) suggesting that there is an "induced fit" for both the protein and DNA upon the formation of their complex.

In the solution structure, the decrease in free energy of the complex relative to the isolated species also stems from three factors: direct hydrogen bonds with the nucleotide bases, electrostatic interactions with the backbone phosphate groups, and non-polar interactions with the bases. As was seen with the one-to-one crystal complex, the observed interactions cannot account for the entire specificity of the interaction. The solution model is more successful, though, in that it explains the observed sensitivity to mutation at positions C_{+3}, A_{+5}, G_{+6} and A_{-7}. (In the crystal structure, no contacts are seen between the repressor and base C_{+3}.) However, the solution structure is unable to account for the decrease in binding affinity with mutations to bases A_{+2} or T_{+4}.

The most profound difference between the solution and crystal structures is that in the solution structure no water molecules need to be postulated to bridge the protein/DNA interaction. All of the data can be accounted for in terms of direct protein/DNA hydrogen bonds. The inability of Zhang *et al.*, to observe any water molecules trapped in the protein/DNA interface does not rule out their existence; in fact, these authors showed that a structure with the Sigler water molecules was compatible with the solution data (Zhang *et al.*, 1994). Rather, Zhang's findings put an upper limit on the time that such water molecules can spend in the protein/DNA interface. Using chemical shift and spectroscopic arguments (Otting *et al.*, 1991), Zhang *et al.*, concluded that the lifetime of any trapped water molecules can be at most 20 ms. This is consistent with NMR findings of other protein DNA complexes (Gehring *et al.*, 1994), but is an order of magnitude shorter than the lifetime of the tryptophan ligand in the complex (Zhang *et al.*, 1994) and 1-4 orders of magnitude shorter than the lifetime of the complex itself (Carey, 1988; Hurlburt and Yanofsky, 1992b; Zhang *et al.*, 1994).

The crystal structure of Lawson and Carey (1993) provided additional insights into the repressor/operator interaction. It provides additional example for the existence of water mediated contacts, while refuting the argument of Brennan and Matthews (1989) that the high alcohol content of the crystal artificially hydrated the crystal. The conditions for crystal growth in the study of Lawson and Carey were far less severe than in the work of Otwinowski *et al.* The crystals were grown in aqueous solution containing moderate amounts of sodium chloride and other salts in addition to the repressor, DNA and L-tryptophan (Carey *et al.*, 1993). While Otwinowski's crystals took a year to grow, Carey's required only a few weeks to form. In addition, the operator sequence used by Lawson and Carey was a β-symmetric operator similar to that suggested by Staacke *et al.*, confirming that the original binding mode proposed by Otwinowski *et al.*, was correct and that a specific complex can be formed with both types of DNA sequences. Even under such different conditions with different operator DNA sequences, the same water-mediated hydrogen bonds are seen in the two crystal structures. However, the functional relevance of the water molecules in solution is yet to be shown.

In the structure of Lawson and Carey, two repressors are found to bind the operator in a tandem sequence (see Figure 5). That is, the consensus operator sequence is constructed by a linear repeat of the sequence ACTAGT. This sequence is palindromic which gives the 18 base-pair sequence its α C2 symmetry at the central Pribnow box (see Figure 3). However, as each of the half-sites is symmetric, and since the repressor is observed to only bind the second half of the sequence 'AGT', it is possible for two repressors to share this 6 base pair sequence in tandem. Thus, although a single repressor does not recognize the ACT portion of the β-site, two repressors in tandem will recognize the two AGT sequences of different strands. The effect of this is that repressor binding affinity is sensitive to all six base pairs. Thus, the argument that DNA binding specificity is maintained through a mechanism of indirect readout is not needed, since all of the mutationally important base pairs are seen to contact the repressor in the tandem complex (Lawson and Carey, 1993).

Although this explanation of the sensitivity to mutation at all six base pairs is appealing, it is clearly over-simplified. If the specificity of the ACT sequence was actually due solely to repressor specificity to the AGT sequence on the complementary DNA strand, one would expect similar losses of binding affinity when complementary base mutations are made. For instance, as an A_{+5} to G_{+5} mutation results in a 1000-fold reduction in repressor binding (Bass *et al.*, 1987), one would conclude from Carey's argument that the symmetric T_{+4} to C_{+4} mutation should result in a comparable loss of binding affinity by mutating the A_{-4} to G_{-4}. (In the β-symmetry reference, A_{+5} is A_{+1}, T_{+4} is T_{-1}, such that mutating T_{-1} also changes A_{+1}.) However, this is not observed; rather, the T_{+4} to C_{+4} mutation produces a negligible (less than 5-fold) effect on repressor binding (Bass *et al.*, 1987).

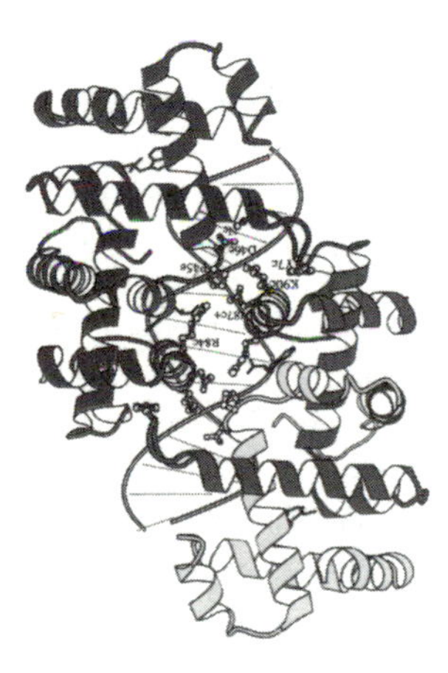

Figure 5: Crystal structure of the 2:1 repressor/operator tandem complex. Ladder representation of the operator DNA and the heavy atoms of the tryptophan corepressor. The two repressor dimers are shown in a ribbon represenation. Selected residues are labeled and shown in a ball and stick representation. From Lawson and Carey, 1993.

Finally, the tandem structure provides insight into the observed importance of the N-terminal arms (Carey, 1989; Hurlburt and Yanofsky, 1992a). In the two-to-one repressor/DNA complex, the repressors are seen to interact with one another. A large part of this interaction is between the N-terminal segment of one dimer and a hydrophobic pocket of the neighboring dimer. Thus, although the N-terminal arms do not impart any specificity to the interaction, they do stabilize the two-to-one tandem complex through protein/protein interactions (Lawson and Carey, 1993).

As the 20 base-pair α-symmetric DNA sequence of Lefèvre is too short to allow tandem binding (Zhang *et al.*, 1994), a 1:1 stoichiometry was observed for the solution complex. Thus, the N-terminal arms were not seen in this study, and are presumed to undergo rapid dynamic averaging. A study of the 2:1 repressor/operator complex using NMR techniques is presently underway which may be able to confirm the role of the N-terminal arms in tandem binding (Arrowsmith, C., personal communication).

Dynamics and function

There are several time regimes which have been studied with regard to the dynamics of the repressor. However, the starting point in analysis must be a static view of the average structure. In the case of the DNA binding domain, it has been shown in both NMR and crystal studies that this domain is comprised of two helices. The objective of examining the dynamics of this region is to determine the amplitude and frequency of fluctuations of these helices, and to determine if and how these fluctuations effect function (as the "en-bloc" shift of Zhang *et al.*, is proposed to do).

Motions occurring across three different timescales have been studied so far (Figure 6): 1) the subnanosecond timescale which reflects the overall tumbling of the protein and rapid internal motions characteristic of disordered, coil-like peptide segments (Zheng *et al.*, 1995), 2) the millisecond to second timescale during which exchange of backbone amide protons of the DNA binding domain (Gryk *et al.*, 1995; Finucane and Jardetzky, 1995a) as well as the tryptophan corepressor occur (Chou *et al.*, 1989; Schmitt *et al.*, 1995), and 3) the kilosecond to megasecond timescale which encompasses backbone amide proton exchange of residues in the hydrophobic central core (Czaplicki *et al.*, 1991; Finucane and Jardetzky, 1995b) and dimer dissociation (Hurlburt and Yanofsky, 1993).

^{15}N relaxation measurements have been used to study motions of the order of the correlation time of the protein and faster (King *et al.*, 1978; Lipari and Szabo, 1982; Kay *et al.*, 1989). The correlation time of *trp* repressor is about 12 ns, and thus frequencies of motion greater than a few megahertz can be observed. Zheng *et al.*, (1995) have applied the analysis of ^{15}N T_1, T_2 relaxation and ^{1}H-

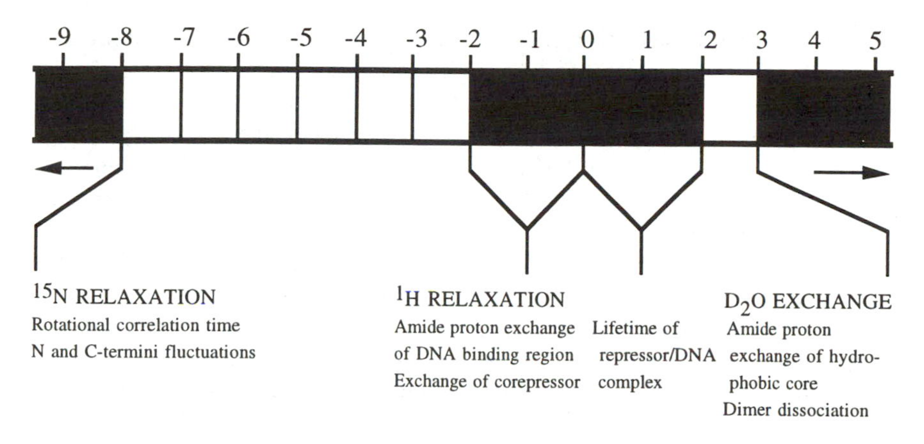

Figure 6: Timescale of motion relevent to the *trp* repressor system. Regions are denoted in log seconds such that fast processes occur on the negative side of the scale and slow processes occur on the positive side. The NMR observables are denoted in capital letters below the timeline, with the repressor motions occuring during the corresonding time frame listed below.

^{15}N heteronuclear NOE to study these ultra-fast motions of *trp* repressor. These authors have established three crucial results. First, the N- and C-termini undergo high frequency motions typical of unfolded, disordered peptides. They observe statistically large values of T_1 and T_2 with a correspondingly low value for NOE, which result in the determination of significantly large librational angles (Zheng *et al.*, 1995). Second, the motion of the rest of the molecule, helices A through F, can be described as a fairly rigid anisotropic rotor. The degree of anisotropy of tumbling correlates well with the known structural observation that the repressor resembles an ellipsoid of axial ratio 1.6:1. More importantly, there appears to be no difference in the mobility of the D and E helices relative to the core on the nanosecond timescale. This is in contrast the chain terminals and a clear indication that the DNA binding region is structured and ordered (Zheng *et al.*, 1995). Third, Zheng detects no difference in mobility between the apo and holorepressor (Zheng, Z., personal communication) suggesting that there is no "folding" process which is associated with ligand binding.

The amplitude and frequency of helical motions on a millisecond to second timescale have been inferred from the fast amide proton exchange rates in the DNA binding domain (Gryk *et al.*, 1995). It has long been appreciated that many backbone amide protons of a protein exchange much slower in a folded peptide than would be expected in an extended, unfolded structure (Lenormant and Blout, 1953; Hvidt and Linderstrøm-Lang, 1954). The ratio of the observed exchange rate from the folded protein to the exchange rate expected from an unfolded peptide, termed the intrinsic exchange rate, is commonly reported as a protection factor. Protection factors for many different amide protons in proteins have been measured, and range from unity (no protection) to 10^{10} (Englander and Kallenbach, 1984). This relative protection from exchange has been used to infer the presence of internal motions on the timescale of the observed exchange or faster (Hvidt and Nielsen, 1966; Englander and Kallenbach, 1984), although the amplitude of motions and mechanisms of exchange are still open to debate.

The method for measuring amide proton exchange in the millisecond to second timescale employed by Gryk *et al.*, (1995) relies on the effects that exchange with solvent have on a proton's magnetization and T_1 relaxation (Waelder *et al.*, 1975; Krishna *et al.*, 1979). By selectively perturbing the magnetization of both the solvent (saturation transfer) and the protein (relaxation), the rate of exchange can be estimated by monitoring the transfer of magnetization between the two populations, provided that the exchange is faster than the intrinsic longitudinal relaxation rate of the amide protons (Krishna *et al.*, 1979; Spera *et al.*, 1991; Gryk *et al.*, 1995). These techniques are accurate in monitoring amide proton exchange on the timescale of proton longitudinal relaxation, which for *trp* repressor is about one S^{-1} at 45°C. Correspondingly, the measured exchange rates for amide protons in the DNA binding region of *trp* holorepressor are between 1 and 100 S^{-1} (see Figure 7) (Gryk *et al.*, 1995).

Interestingly, these authors found that the observed hydrogen exchange in the DNA binding region did not follow first order kinetics. This was not unexpected as the commonly accepted mechanism of the protection of amide protons from

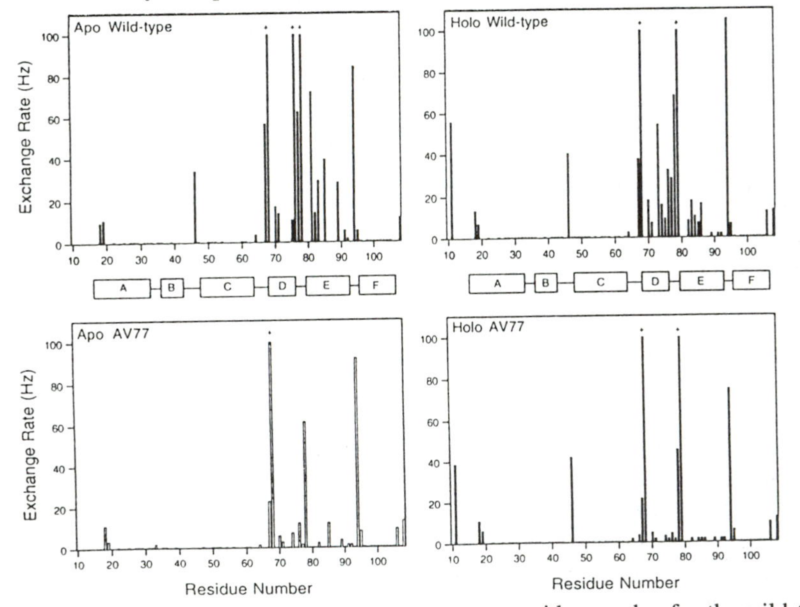

Figure 7: Measured amide proton exchange rates vs. residue number for the wild-type and AV77 apo and holorepressors. Exchange rates were measure at 45°C at pH 7.6.

$$NH_A \underset{k_{close}}{\overset{k_{open}}{\rightleftharpoons}} NH_B \underset{k'_{intrinsic}}{\overset{k_{intrinsic}}{\rightleftharpoons}} H_2O$$

Figure 8: Linderstrøm-Lang model for amide proton exchange (Berger and Linderstrøm-Lang, 1957). Amide protons are presumed to exchange between a closed state (NHA) and an open state (NHB). Exchange from the closed state is prohibited, while in the open state, amide protons exchange at the rate ($k_{intrinsic}$) observed for small peptide analogs.

exchange in proteins (Berger and Linderstrøm-Lang, 1957) postulates an interconversion between a protected state in which exchange cannot occur and an open state in which protons exchange at their intrinsic rate (see Figure 8). However, the observed *intrinsic* exchange rates for the DE helices were also slower than expected implying that both interconverting states are protected from exchange. As chemical shift data had suggested that the region is highly helical on the average (Zhao *et al.*, 1993), it was concluded that both of these protected states are helical, and only one is allowed to open for exchange. The estimated helicity for the DNA binding helices was confirmed to be greater than 90% (Gryk *et al.*, 1995).

A simple method of monitoring amide proton exchange in the suprakilosecond timescale (hours to weeks) is the observation of isotope replacement by NMR (Saunders and Wishnia, 1958; Wagner and Wüthrich, 1982). In these experiments a protonated sample is dissolved into D_2O buffer. (The converse experiment dissolving a deuterated sample in H_2O is rarely done as the faster exchanging protons complicate the later spectra by appearing first. When exchanging into D_2O, the faster exchanging deuterons are unobservable, simplifying the analysis of the remaining spectra.) Hydrogen exchange at each amide proton site is monitored as the decrease in intensity of the amide proton resonance line. Czaplicki *et al.*, (1991) have used this technique for studying exchange of amide protons of *trp* repressor with lifetimes greater than ten minutes (corresponding to exchange rates slower than 1×10^{-3} Hz.) It was found that most of the amide protons in the hydrophobic core had exchange rates small enough to be determined by this method. However, the exchange rates for residues in the DNA binding domains were too fast to be measured by isotope replacement (Arrowsmith, 1991b).

Finucane and Jardetzky (1995b) have extended the analysis of the hydrogen exchange for residues in the hydrophobic core by measuring the pH-dependence of the observed exchange rates. They find that generally the logarithm of the exchange rates increases monotonically with pH in the range of 6.3 to 7.2. This is consistent with the postulated mechanism of base-catalyzed exchange (Berger *et al.*, 1959) and with studies on small peptide analogs (Molday *et al.*, 1972; Bai *et al.*, 1993). However, at low pH, residues in helices B and C deviate from linearity. The authors ascribe this deviation to a destabilization of the protein at low pH such that the unfolding of the core becomes the rate limiting step in exchange (Kim and Woodward, 1993). As residues in the B and C helices have the slowest exchange rates, the unfolding process is observed only in this region, with rates in regions that exchange rapidly anyway remaining unaffected.

Effects of corepressor binding and mutation

The nature of the dynamic alteration occurring upon ligand binding is clarified by the hydrogen exchange studies of *trp* repressor (Czaplicki *et al.*, 1991; Gryk *et al.*, 1995; Finucane and Jardetzky, 1995a). Finucane and Jardetzky (1995a) have carried out a study of amide proton exchange rates on the DNA binding domain of the aporepressor. Upon comparing their results from the aporepressor with those from the holorepressor, the authors find that the binding of L-tryptophan slows down the rate of amide proton exchange by a factor of 2 to 9 for residues in the DNA binding domain. This is comparable to the 3 to 10-fold reduction in exchange for residues in the hydrophobic core (Czaplicki *et al.*, 1991). This global stabilization led Finucane and Jardetzky (1995a) to the conclusion that amide proton exchange is due to concerted backbone motions which periodically expose the amide protons of the D and E helices to solvent. The action of L-tryptophan is to damp these concerted motions throughout the molecule including the DNA binding helices.

Similar results were obtained for the mutant repressor AV77, in which an alanine residue in the turn of the helix-turn-helix motif is substituted with valine (Gryk and Jardetzky, 1996). This mutant is observed to be a superrepressor *in vivo* (Kelly and Yanofsky, 1985; Arvidson *et al.*, 1993). That is, the AV77 mutant requires less tryptophan for specific binding to the operator sequence. Interestingly, no difference is found between wild-type and AV77 apo or holorepressor operator DNA binding *in vitro* (Hurlburt and Yanofsky, 1990; Liu and Matthews, 1994). This has led Arvidson *et al.* to conclude that binding specificity is a competition between specific and non-specific DNA sites, and the superrepressor phenotype of AV77 is due to a decrease in the non-specific binding affinity of the mutant aporepressor (Arvidson *et al.*, 1993).

Gryk and Jardetzky (1996) have measured the amide proton exchange rates for the DNA binding domains of the AV77 mutant apo and holo repressors. They find that the hydrogen bonded protons in the DNA binding domains of both forms of the mutant repressors are more protected from amide proton exchange than the corresponding wild-type repressors by at least an order of magnitude (see Figure 7). Thus, similar to the effect of corepressor binding, the substitution of a valine for an alanine residue protects not only neighboring protons, but the entire helices that flank the mutation. Also similar to L-tryptophan binding, the mutation enhances the binding specificity of the repressor (Arvidson *et al.*, 1993). However, unlike tryptophan binding, the mutation seems only to affect the DNA binding domain of the mutant holorepressor, as the amide proton exchange rates of residues in the hydrophobic core (unpublished results) and the exchange rate of the ligand (Schmitt *et al.*, 1995) are unaffected.

Thus, it appears that the intrinsic flexibility of the DNA binding regions of the wild-type *trp* repressor is required for regulation. Flexibility decreases the relative specificity of the repressor for operator, shifting the equilibria of free and bound repressors, as well as of free and bound operators, into the required physiological regime. The AV77 repressor is less flexible, giving it the ability to bind specifically to its operator sequence in the presence of a lower concentration of L-tryptophan. However, this enhanced binding is not desired by the cell as tryptophan is no longer effective as a regulator. (The conclusion that *E. coli* has chosen the wild-type repressor through selection is supported by the ease in which a single nucleotide mutation changes the alanine to valine (Kelley and Yanofsky, 1985) and the contrasting retention of alanine in the *trp* repressors of other bacteria (Arvidson *et al.*, 1994)).

It is obvious from the preceding discussion that the *trp* system presents a rather complicated set of nested equilibria, the interplay of which must be examined with regard to the stability and function of the complex. It is clear from the structural changes observed in both the repressor and DNA upon complexation that both species are capable of, and in need of, quite significant degrees of conformation freedom. The most apparently flexible parts of the repressor molecule are the N-terminal arms and the DNA binding domains. The operator DNA duplex seems to be flexible throughout the sequence, but perhaps more so at the TA dinucleotide steps in the middle of the operator sequence (Lefèvre *et al.*, 1985b) and in the middle of the DNA binding half-sites (Lefèvre *et al.*, 1985a; Shakked *et al.*, 1994a, b).

It has been suggested that ligand binding might provide the necessary thermodynamic energy to drive the folding of different regions of the protein structure upon ligand binding (Jin *et al.*, 1993; Spolar and Record, 1994). In this proposal, the binding of L-tryptophan to the aporepressor is associated with the folding of helix E, while the binding of DNA to the holorepressor is associated with the folding of helix D (Jin *et al.*, 1993). To extend this model, the binding of two repressor molecules in tandem is associated with the folding of the N-terminal regions.

However simple and suggestive this model of the coupling of binding to folding may be, it is clear from the NMR studies of the structure and dynamics, that for *trp* repressor it is not correct. The DNA binding domain is helical in all forms of the repressor: unliganded, complexed with L-tryptophan and complexed with DNA (Zhao *et al.*, 1993, Zhang *et al.*, 1994, Gryk *et al.*, 1995, Zheng *et al.*, 1995). Rather than ligand binding being responsible for the folding of helices in the DNA binding regions of *trp* repressor, ligand binding appears to be responsible for the *stabilization* of these domains and of the protein as a whole (Finucane and Jardetzky, 1995a).

Stabilization of the molecular structure appears to be essential to optimal function in this system. It is the stabilized tryptophan-containing form of the repressor that optimally binds to DNA. The even further stabilization by mutation leads to the mutants functioning as super-repressors, controlling transcription at much lower tryptophan concentrations. It seems that the basic design of the repressor molecule allows too much flexibility for the molecule to function effectively and it is the task of the ligand to filter out some of the excess flexibility, leaving just enough for optimal control.

An optimum of flexibility may be quite generally required for the function of allosteric proteins. Protein function as a rule depends on the successive and reversible formation and braking of specific atomic contacts. Rigidity is required to make good contacts, but flexibility is required to break them. Flexibility is also required to correct for imperfections of geometric fit and to allow versatility in making contacts with more than one partner - for example specifically binding to more than one DNA sequence, which in the case of the *trp*-repressor is known to occur. In this context it is of interest to note that stabilization of the repressor molecule by mutation decreases the selectivity of the AV77 mutant for different operator sequences (C. Yanofsky, personal communication). Stabilization thus optimizes the binding to a specific DNA sequence at the expense of the others and thus decreases versatility.

Finally, we must return to the water molecules presumed to mediate specific DNA binding. It seems that these water molecules are not simply an artifact of crystallization, but can reside in the protein/DNA interface. This is not without precedent, as several similar water molecules are observed in the solution studies of the Antennapedia homeodomain DNA complex (Gehring *et al.*, 1994), although not providing the majority of base specific contacts as in *trp* repressor. However, similar to that observed for the homeodomain complex (Gehring *et al.*, 1994), the rate of exchange of any interfacial water molecules in the *trp* repressor complex is larger than fifty S^{-1} (Zhang *et al.*, 1994). Such exchange is much faster than the rate of complex dissociation (Hurlburt and Yanofsky, 1992b; Zhang *et al.*, 1994), ligand dissociation (Chou *et al.*, 1989; Schmitt *et al.*, 1995) or helical fluctuations (Arrowsmith, 1990; Gryk *et al.*, 1995) in the isolated repressors.

Thus, the precise role of these bridging water molecules is still unclear. In principle it is not the exchange rate that is important for binding energy but rather the occupancy. However, for a system dynamic enough to expel such water molecules at a rate one hundred times faster than the corepressor, and at least a thousand times faster than the lifetime of the complex, it seems unavoidable that there is a large number of similar binding modes, with similar specificities, through which the repressor/operator complex rapidly interconverts. It, therefore, may be that, in the case of *trp* repressor, our precision in defining structures has surpassed the precision of nature.

References

Arrowsmith, C.H., Pachter, R., Altman, R., Iyer, S., and Jardetzky, O. (1990). *Biochemistry* **29**, 6332.

Arrowsmith, C., Pachter, R., Altman, R., and Jardetzky, O. (1991a). *Eur. J. Biochem.* **202**, 53.

Arrowsmith, C.H., Czaplicki, J., Iyer, S.B., and Jardetzky, O. (1991b). *J. Am. Chem. Soc.* **113**, 4020.

Arvidson, D.N., Pfau, J., Hatt, J.K., Shapiro, M., Pecoraro, F.S., and Youderian, P. (1993). *J. Biol. Chem.* **268**, 4362.

Arvidson, D.N., Arvidson, C.G., Lawson, C.L., Miner, J., Adams, C., and Youderian, P. (1994). *Nuc. Acids Res.* **22**, 1821.

Bai, Y., Milne, J.S., Mayne, L., and Englander, S.W. (1993). *Proteins: Struct. Funct. and Genet.* **17**, 75.

Bass, S., Sugiono, P., Arvidson, D.N., Gunsulas, R.P., and Youderian, P. (1987). *Genes Dev.* **1**, 565.

Bass, S., Sorrells, V., and Youderian, P. (1988). *Science* **242**, 240.

Berger, A., and Linderstrøm-Lang, K. (1957). *Arch. Biochem. Biophys.* **69**, 106.

Berger, A., Loewenstein, A., and Meiboom, S. (1959). *J. Am. Chem. Soc.* **81**, 62.

Brennan, R.G., and Matthews, B.W. (1989). *J. Biol. Chem.* **264**, 1903.

Carey, J. (1988). *Proc. Natl. Acad. Sci. U.S.A.* **85**, 975.

Carey, J. (1989). *J. Biol. Chem.* **264**, 1941.

Carey, J., Lewis, D.E.A., Lavoie, T.A., and Yang, J. (1991). *J. Biol. Chem.* **266**, 24509.

Carey, J., Combatti, N., Lewis, D.E.A., and Lawson, C.L. (1993). *J. Mol. Biol.* **234**, 496.

Chou, W.-Y., Bieber, C., and Matthews, K.S. (1989). *J. Biol. Chem.* **264**, 18309.

Czaplicki, J., Arrowsmith, C.H., and Jardezky, O. (1991). *J. Biomol. NMR* **1**, 349.

Englander, S.W., and Kallenbach, N.R. (1984). *Quart. Rev. Biophys.* **16**, 521.

Finucane, M.D., and Jardetzky, O. (1995a). *J. Mol. Biol.*, **253**, 576.

Fincuane, M.D., and Jardetzky, O. (1995b). *in preparation.*

Gehring, W.J., Qian, Y.Q., Billeter, M., Furukubo-Tokunaga, K., Schier, A.F., Resendez-Perez, D., Affolter, M., Otting, G., and Wüthrich, K. (1994). *Cell* **78**, 211.

Gryk, M.R., and Jardetzky, O. (1996). *J. Mol. Biol.* **255**, 204.

Gryk, M.R., Finucane, M.D., Zheng, Z., and Jardetzky, O. (1995). *J. Mol. Biol.* **246**, 618.

Guenot, J., Fletterick, R.J., and Kollman, P.A. (1994). *Protein Sci.* **3**, 1276.

Haran, T.E., Joachimiak, A., and Sigler, P.B. (1992). *EMBO J.* **11**, 3021.

Heatwole, V.M., and Somerville, R.L. (1991). *J. Bacteriol.* **173**, 3601.

Heatwole, V.M., and Somerville, R.L. (1992). *J. Bacteriol.* **174**, 331.

Howard, A.E., and Kollman, P.A. (1992). *Protein Sci.* **1**, 1173.

Hurlburt, B.K., and Yanofsky, C. (1990). *J. Biol. Chem.* **265**, 7853.

Hurlburt, B.K., and Yanofsky, C. (1992a). *Nuc. Acids. Res.* **20**, 337.

Hurlburt, B.K., and Yanofsky, C. (1992b). *J. Biol. Chem.* **267**, 16783.

Hurlburt, B.K., and Yanofsky, C. (1993). *J. Biol. Chem.* **268**, 14794.

Hvidt, A., and Linderstrøm-Lang, K. (1954). *Compt.-rend. Lab. Carlsbe. Sér. Chim.* **29**, 385.

Hvidt, A., and Nilsen, S.O. (1966). *Adv. Prot. Chem.* **21**, 287.

Jin, L., Yang, J., and Carey. J. (1993). *Biochemistry* **32**, 7302.

Joachimiak, A., Kelley, R.L., Gunsalus, R.P., Yanofsky, C., and Sigler, P.B. (1983). *Proc. Natl. Acad. Sci. USA* **80**, 668.

Kay, L.E., Torchia, D.A., and Bax, A. (1989). *Biochemistry* **28**, 8972.

Kelley, R.L., and Yanofsky, C. (1985). *Proc. Natl. Acad. Sci. U.S.A.* **82**, 483.

Kim, K.-S., and Woodward, C. (1993). *Biochemistry* **32**, 9009.

King, R., Maas, R., Gassner, M., Nanda, P.K., Conover, W.W., and Jardetzky, O. (1978). *Biophys. J.* **24**, 103.

Krishna, N.R., Huang, D.H., Glickson, J.D., Rowan III, R., and Walter R. (1979). *Biophys. J.*, **26**, 334.

Ladbury, J.E., Wright, J.G., Sturtevant, J.M., and Sigler, P.B. (1994). *J. Mol. Biol.* **238**, 669.

Lawson, C.L., Zhang, R.-G., Schevitz, R.W., Otwinowski, Z., Joachimiak, A., and Sigler, P.B. (1988). *Proteins: Struct.Funct. and Genet.* **3**, 18.

Lawson, C.L., and Sigler, P.B. (1988). *Nature* **333**, 869.

Lawson, C.L., and Carey, J. (1993). *Nature* **366**, 178.

Lefèvre, J.-F., Lane, A.N., and Jardetzky, O. (1985a). *J. Mol. Biol.* **185**, 689.

Lefèvre, J.-F., Lane, A.N., and Jardetzky, O. (1985b). *FEBS Lett.* **190**, 37.

Lefèvre, J.-F., Lane, A.N., and Jardetzky, O. (1987). *Biochemistry* **26**, 5076.

Lenormant, H., and Blout, E.R. (1953). *Nature* **172**, 770.

Lipari, G., and Szabo, A. (1982). *J. Am. Chem. Soc.* **104**, 4546.

Liu, Y.-C., and Matthews, K.S. (1994). *J. Biol. Chem.* **269**, 1692.

Luisi, B.F., and Sigler, P.B. (1990). *Biochem. Biophys. Acta* **1048**, 113.

Marmorstein, R.Q., Sprinzl, M., and Sigler, P.B. (1991). *Biochemistry* **30**, 1141.

Molday, R.S., Englander, S.W., and Kallen, R.G. (1972). *Biochemistry* **11**, 150.

Ohlendorf, D.H. Anderson, W.F., and Matthews, B.W. (1983). *J. Mol. Evol.* **169**, 109.

Otting, G., Liepinsh, E., and Wüthrich, K. (1991). *J. Am. Chem. Soc.* **113**, 4363.

Otwinowski, Z., Schevitz, R.W., Zhang, R.-G., Lawson, C.L., Joachimiak, A., Marmorstein, R.Q., Luisi, B.F., and Sigler, P.B. (1988). *Nature* **335**, 321.

Perutz, M. (1989). *Quar. Rev. Biophys.* **22**, 139.

Rose, J.K., Squires, C.L., Yanofsky, C., Yang, H.-L., and Zubay, G. (1973). *Nature New Biology* **245**, 133.

Saunders, M., and Wishnia, A. (1958). *Ann. N.Y. Acad. Sci.* **70**, 870.

Schevitz, R.W., Otwinowski, Z., Joachimiak, A., Lawson, C.L., and Sigler, P.B. (1985). *Nature* **317**, 782.

Schmitt, T.H., Zheng, Z., and Jardetzky, O. (1995). *Biophys. J. Abstr.* **68**, A297.

Shakked, Z., Guzikevich-Guerstein, G., Frolow, F., Rabinovich, D., Joachimiak , A., and Sigler, P.B. (1994a). *Nature* **368**, 469.

Shakked, Z., Guzikevich-Guerstein, G., Frolow, F., Rabinovich, D., Joachimiak , A., and Sigler, P.B. (1994b). in Structural Biology: The State of the Art, Academic Press, pp 199-215.

Spera, S., Ikura, M., and Bax, A. (1991). *J. Biomol. NMR* **1**, 155.

Spolar R.S., and Record, Jr., M.T. (1994). *Science* **263**, 777.

Staacke, D., Walter, B., Kisters-Woike, B., Wilcken-Bergmann, B.V., and Müller-Hill, B. (1990). *EMBO J.* **9**, 1963.

Tasayco, M.L., and Carey, J. (1992). *Science* **255**, 594.

Waelder, S., Lee, L., and Redfield, A.G. (1975). *J. Am. Chem. Soc.* **97**, 2927.

Wagner, G., and Wüthrich, K. (1982). *J. Mol. Biol.* **160**, 343.

Zhang, H. (1993). Structural Investigation of trp Repressor-DNA Complex by NMR. Ph.D. thesis, Stanford University, Stanford, CA.

Zhang, H., Zhao, D., Revington, M., Lee, W., Jia, X., Arrowsmith, C., and Jardetzky, O. (1994). *J. Mol. Biol.* **238**, 592.

Zhang, R.-G., Joachimiak, A., Lawson, C.L., Schevitz, R.W., Otwinowski, Z., and Sigler, P.B. (1987). *Nature* **327**, 591.

Zhao, D., Arrowsmith, C.H., Jia, X., and Jardetzky, O. (1993). *J. Mol. Biol.* **229**, 735.

Zheng, Z., Czaplicki, J., and Jardetzky, O. (1995). *Biochemistry* **34**, 5121.

Zubay, G., Morse, D.E., Schrenk, W.J., and Miller, J.H.M. (1972). *Proc. Natl. Acad. Sci. U.S.A.* **69**, 1100.

Zurawski, G., Gunsalus, R.P., Brown, K.D., and Yanofsky, C. (1981). *J. Mol. Biol.* **145**, 47.

6

Heteronuclear Strategies for the Assignment of Larger Protein/DNA Complexes: Application to the 37 kDa *trp* Repressor-Operator Complex

M.J. Revington, W. Lee, and C.H. Arrowsmith

Division of Molecular and Structural Biology
Ontario Cancer Institute

Department of Medical Biophysics
University of Toronto
Toronto, Ontario, Canada

The sequence-specific DNA binding function of many proteins is recognized as one of the central mechanisms of regulating transcription and DNA replication and repair. The ability of these proteins to select a short (usually 10 to 20 basepair) sequence out of the entire genome with which to form a stable complex is a prime example of molecular recognition. Atomic resolution structural studies using NMR and X-ray crystallography have emerged as essential techniques in understanding the basis of specificity and stability in these systems.

While NMR studies of small DNA-binding domains of proteins have become almost routine (see Kaptein, 1993 for a review) relatively few NMR studies of protein-DNA complexes have been reported. These include the *lac* repressor headpiece complex (Chuprina *et al.*, 1993). the *Antennapedia* homeodomain complex (Billeter *et al.*, 1993), the GATA-1 complex (Omichinski *et al.*, 1993). and the Myb DNA binding domain complex (Ogata *et al.*, 1993); all of these complexes are smaller than 20 kDa. In most cases, size limitations have meant that only the DNA binding domain of the protein in complex with a single binding element have been studied. *In vivo*, however, most DNA binding proteins are much larger than these domains and often function as oligomers.

The decrease in quality and increase in complexity of spectra as the molecular weight of the sample increases, limits the number of systems amenable to study using NMR and influences the decision to focus on single domains of multidomain proteins. However, since many DNA-binding proteins are regulated by the binding of ligands, other proteins or phosphorylation, often at sites distal from the DNA-binding domain, it is preferable to study as much of the intact protein as possible in order to characterize allosteric and regulatory mechanisms (Pabo and Sauer, 1992).

E. coli trp repressor is a 25 kDa homodimer that regulates operons involved in tryptophan biosynthesis. The dimer is one of the smallest intact proteins that binds sequence specifically to DNA and whose affinity is modulated by an effector (L-tryptophan). The complex between this protein and the shortest relevant oligomer of DNA containing the consensus binding sequence has a total molecular weight of 37,000 and presents a considerable challenge for an NMR study. Recently, the solution structure of this complex was solved by a combination of selective deuteration and heteronuclear (^{13}C and ^{15}N) techniques (Zhang *et al.,* 1994). Here we report further details of the heteronuclear techniques and outline strategies for the assignment of large protein-DNA complexes.

Methods

Sample preparation

Trp repressor (trpR) was purified from *E. coli* strain CY15070 carrying plasmid pJPR2 (Paluh and Yanofsky, 1986). Isotopically labeled protein was purified from cells grown in M9 media (Maniatis *et al.,* 1982) with ^{15}N labeled NH_4Cl and/or D-glucose-$^{13}C_6$ as the sole nitrogen and carbon sources, respectively. In order to isolate a ~70% deuterated protein, the cells were first adapted to growth in D_2O by successive growths and plating of the strain on M9 media with 33%, 56%, and 70% D_2O as the solvent. NMR samples of the holorepressor were prepared at concentrations between 2 and 4 mM trpR (subunit concentration) in 500 mM NaCl, 50 mM Na_2PO_4 adjusted to pH 6. Protein concentrations were determined by absorption of the sample at 280 nm with 1 a.u. being equivalent to a 1.2 mg/ml of protein (Joachimiak *et al.,* 1983). The co-repressor concentration was 1.5 to 2 times that of the trpR subunit concentration to ensure saturation of the L-trp binding sites. Since 5-methyl-tryptophan was only available in a racemic mixture, the samples with this alternate co-repressor were made in a 2.4:1 ratio of 5-methyl-tryptophan : trpR subunit. The D-isomer has a low affinity for the protein (Marmorstein *et al.,* 1987), and it was assumed that only the L-isomer bound to the protein. The ^{15}N and ^{13}C,^{15}N labeled protein samples were studied in 90% H_2O/10% D_2O solvent, while the ^{13}C labeled sample was dissolved in 99.996% D_2O solvent.

The 20 basepair palindromic operator DNA

(5'CGTACTAGTT!AACTAGTACG 3')

was prepared synthetically using standard phosphoramidite chemistry (Beacage *et al.,* 1981) in 10 uM quantities. Purification for NMR samples was achieved by reverse phase HPLC, acetic acid detrytilation, ether separation, and ethanol precipitation (ABS Bulletin, 1988). The concentration of DNA was measured by absorbance at 260 nm with 1 a.u. being equivalent to 39 ug/ml (Zhang, H. and Arrowsmith, C., unpublished). Final purity was checked by using 1D ^1H and 2D NOESY NMR spectra. Ternary complexes were prepared by addition of equimolar amounts of ds-DNA to an NMR sample of trp holorepressor in a high salt buffer to prevent protein aggregation. The complex was then dialysed into a 50 mM Na_2PO_4, pH 6 solution by repeated dilution and concentration in a microconcentrator cell (MW cutoff 3 kDa). Additional corepressor was added in the final round of concentrating to replace dialysis losses.

NMR experiments

All NMR experiments were performed on either Varian Unity 600 or Varian Unity$^+$ 500 spectrometers. The 600 MHz instrument was equipped with a triple resonance probe with an additional PTS synthesizer as a pseudo fourth channel. The 500 MHz spectrometer was a four channel instrument with a triple resonance probe with an actively shielded pulsed field gradient coil. All experiments were performed at 37°C. Data processing was performed using NMRZ software (New Methods Research, Syracuse, N.Y.) with routines described by Marion *et al.* (1989a) for processing of 3D data, or using NMRPipe software (Delaglio, 1993).

All experiments on the holorepressor were performed at 600 MHz with presaturation of the solvent resonance. HSQC (Bodenhausen and Reuben, 1980) spectra employed a ^1H spectral width of 13.3 ppm centered about the water resonance using 512 or 1024 complex points. The ^{15}N dimension had a spectral width of 22 ppm centered at 119 ppm collected as 128 complex points. The ^{15}N NOESY-HMQC (Zuiderweg and Fesik, 1989; Marion *et al.,* 1989b) spectrum was collected as a 128 x 32 x 512 complex matrix with $t_1max(^1H)$=19.7 ms, $t_2max(^{15}N)$=24 ms, and $t_3max(^1H)$=128 ms. The mixing time was 110 ms and a relaxation delay of 0.9-1.1 s was used between transients. The HNCA (Kay *et al.*, 1990) experiment was acquired as a 32 x 32 x 512 complex matrix with $t_1max(^{13}C)$=4 ms, $t_2max(^{15}N)$=24 ms, and $t_3max(^1H)$=128 ms and a delay of 1 s between transients.

The ^{15}N-^1H HSQC and ^{15}N-edited NOESY-HMQC spectra of the protein/DNA complex were performed using the same spectral windows as the holorepressor spectra. However, for the ^{15}N-edited NOESY-HMQC of the complex, 32 transients were recorded, and the number of increments collected was reduced to 84 complex points in t_1 and to 56 complex points in t_2. The t_1 dimension was zero-filled to 128 increments before processing, and t_2 was extended to 112 planes using linear prediction (Marion and Bax, 1989) and then zero-filled to 128 points. Presaturation was used to suppress the solvent signal. The ^{13}C HCCH-TOCSY (Bax *et al.*, 1990) was collected on a sample in 99.996% D_2O with a data size of 96 x 26 x 512 with $t_{1max}(^1H)$ = 16 ms, $t_{2max}(^{13}C)$ = 4.3 ms, and $t_{3max}(^1H)$ = 128 ms. Two spectra with identical parameters were recorded with differing spin lock mixing times (10 ms, 15 ms) in order to optimize transfer for different types of sidechains. The ^{13}C-edited NOESY-HSQC (Muhandiram *et al.*, 1993), using pulsed field gradients for solvent suppression and coherence selection, was performed on a ^{13}C sample in 90% H_2O/10% D_2O solvent with 16 transients per increment recorded. The F3(^1H) spectral width was 16 ppm using 512 complex points, and the spectral width in F1(^1H) was 11 ppm using 128 complex increments. In F2(^{13}C) 32 complex points were collected over 24 ppm centred at 43 ppm with subsequent linear prediction to 64 complex points. The acquisition parameters for ^{13}C NOESY-HSQC were $t_{1max}(^1H)$ = 23 ms, $t_{2max}(^{13}C)$ = 10.6 ms, $t_{3max}(^1H)$ = 64 ms, with a relaxation delay, of 1 s. HBCBCACOHA (Kay, 1993) data were collected as a 52 ($^{13}C\alpha,\beta$) x 32 ($^{13}C'$) x 512 (^1H) complex matrix with a relaxation delay of 0.9s. A HNCO (Kay *et al.*, 1990) spectrum was acquired with the following acquisition parameters: $t_{1max}(^{13}C')$ = 42.7 ms, $t_{2max}(^{15}N)$ms = 24 ms, $t_{3max}(^1H)$ = 64 ms, and a relaxation delay of 1.0s.

Isotope-double-filtered NOESY ([F1-C/N,F2-C/N]-NOESY and [F1-C/F2-C]-NOESY) experiments were performed as described by Ikura *et al.* (1991) on complexes containing $^{13}C/^{15}N$-labeled trpR in 90% H_2O/10% D_2O and ^{13}C-labeled trpR in D_2O, respectively. A 3D ^{13}C-F3-filtered HMQC-NOESY experiment on the ^{13}C labeled trpR-DNA complex provided solely protein-DNA or protein-ligand NOEs separated by ^{13}C frequencies of the attached carbon in the protein (Lee *et al.*, 1994). The carriers were centered at 3.0 ppm (^1H) and 43 ppm (^{13}C) with SEDUCE-1 decoupling for carbonyl decoupling during the $t_2(^{13}C)$ evolution period. Data were collected as a 26 x 84 x 512 complex matrix. The acquisition parameters for F3-isotope-filtered NOESY were: $t_{1max}(^1H)$ = 20 ms, $t_{2max}(^{13}C)$ = 8.7 ms, $t_{3max}(^1H)$ = 64 ms. A relaxation delay of 1.1 sec and a NOE mixing time of 110 ms were used.

Triple resonance data of the ^{13}C, ^{15}N, ~70% ^2H labeled protein complex were collected as described by Yamazaki *et al.* (1994a,b). These experiments were performed on a three channel Varian Unity500 spectrometer modified to perform the ^{15}N, pulses, ^{15}N decoupling and ^2H decoupling on a single channel. Alternatively, these could be performed on a 4-channel instrument without modification. Repetition delays of 1.8-2 s were necessary due to the longer T_1 relaxation times of amide protons in the deuterated protein ($T_1 \approx 1.4$s) compared to the nondeuterated protein (T1 ≈ 0.9 s).

Results

The initial step in the assignment of the complex was the assignment of the 25 kDa protein separately. This involved no extra expense, as the samples used for this step were later used to make the DNA complex. The assignments from spectra of the holorepressor were very useful in interpreting the spectra of the protein complexed with DNA. Nuclei that are not located at the protein-DNA interface experienced only small chemical shift changes. Comparison of the ^{15}N HSQC and ^{15}N NOESY-HMQC spectra of the holorepressor with those of the protein in the complex allowed the assignment of the majority of signals from residues not in direct contact with the DNA, making the assignment of the remainder of the signals in the complex a simpler task.

Holorepressor assignments

The ^{15}N HSQC spectrum of the holorepressor was the initial experiment recorded in this study. The HSQC is a fast, sensitive experiment which serves as a reference for 3D spectra which correlate amide ^{15}N and ^{1}H resonances. The ^{15}N NOESY-HMQC was the first 3D experiment acquired. This spectrum showed both intraresidue NH^i-NH^{i+1} and NH^i-$H\alpha^{i-3}$ connectivities of this mostly α helical protein, as well as NH-sidechain connectivities which help identify spin systems. Since almost complete sequence specific ^{1}H assignments existed previously (Arrowsmith *et al.*, 1990), the chemical shifts observed in this spectra were simply matched to the previous assignments, allowing quick sequence specific identification of 60% of the backbone ^{15}N resonances. The HNCA experiment gave both the intraresidue Cα correlation to the backbone amide proton and a Cα correlation from the preceding residue. Figure 1 shows sample "strips" from the HNCA of Helix F. Building upon the assignments from the NOESY-HMQC spectra, these sequential correlations clarified many of the assignments and 93 of the 107 backbone N and Cα resonances were assigned from these two spectra. Total collection time for the NOESY-HMQC and HNCA was 5 1/2 days. Of the 14 unassigned residues, four were prolines which lack the amide protons and do not give signals in these experiments, and most of the remainder were in the N terminal "arm" which is unstructured in solution (Arrowsmith *et al.*, 1989). It should be noted that this high level of assignment was possible from only two experiments because of the existence of the previous assignments. In a study of an unassigned protein, a more thorough analysis would be necessary for backbone and sidechain assignments (see Grzesiek and Bax, 1993 for a summary of these techniques). Our results for the HNCA experiment as well as other recent reports (Remerowski *et al.*, 1994;

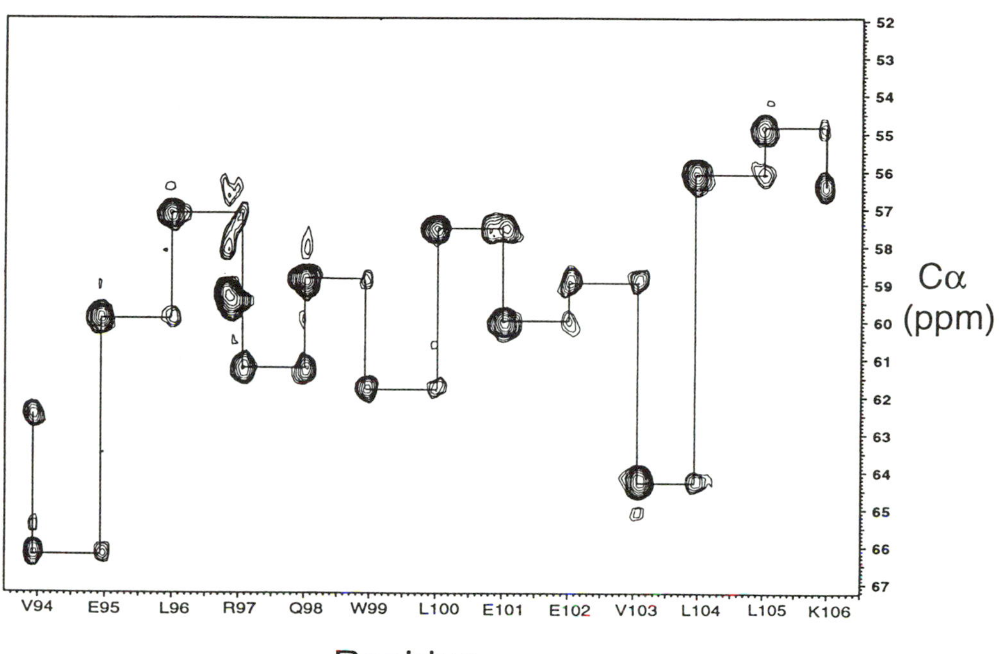

Figure 1: Strips from different ^{15}N slices of the HNCA spectrum of trpR. Residue number is indicated across the x axis and ^{13}Cα frequency on the y axis. Each strip contains the ^{13}Cα of the indicated residue and the preceding residue.

Fogh *et al.*, 1994) demonstrate the applicability of triple resonance techniques for the assignment of proteins of molecular weight in the 25,000 - 30,000 range.

Characterization of the complex

Operator DNA was titrated into the ^{15}N labeled trpR sample and monitored using HSQC spectra. These spectra showed that most (70%) of the NH groups did not change resonant frequencies significantly upon complex formation (shifts of <0.5 ppm (^{15}N) and/or <0.2 ppm (^{1}H)). HSQC spectra were used to determine the optimum solution conditions for the study of the complex. Whereas free trpR requires high salt (500 mM NaCl) to prevent aggregation, the complex is more stable at low salt concentrations (<50 mM) (Record *et al.*, 1994). Figure 2 compares the HSQC spectra of trpR free and complexed with DNA. It can be seen from this spectrum that most of the resonances that have shifted significantly are those in the helix-turn-helix DNA-binding motif, residues 68-90.

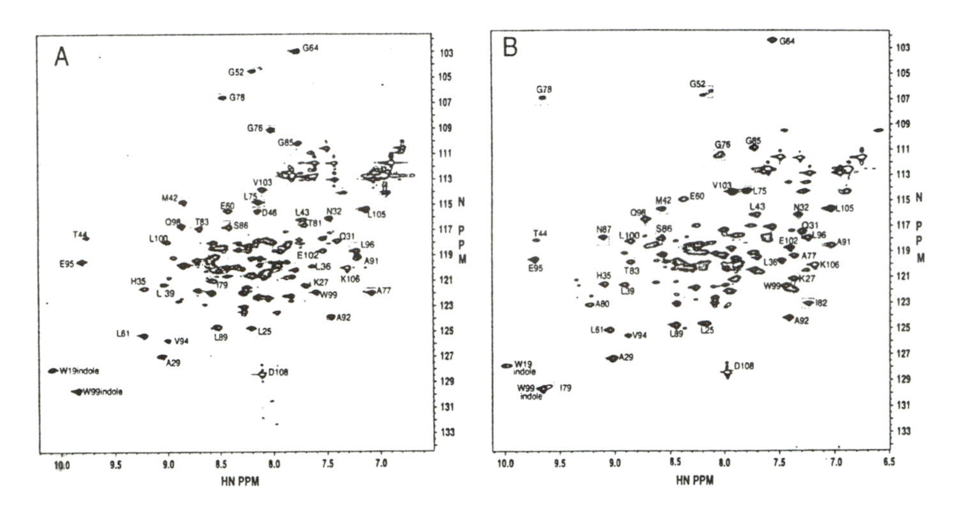

Figure 2: HSQC spectra of (A) holotrpR and (B) trpR-operator complex at 37% C. Selected resolved peaks are labeled with residue number. Peaks that shift significantly upon binding to DNA are boxed.

Operator DNA

The operator DNA used in this study had been assigned and the solution structure determined previously (Lefevre *et al.*, 1986). In order to assign the protein-bound operator the ^{1}H signals of the DNA were observed selectively using [F1,F2]-filtered NOESY experiments (Ikura and Bax, 1991). Most of the DNA resonances change significantly in the complex, and a complete reassignment of the oligomer was made using conventional NOE-based analysis (Wüthrich, 1986) of the 2D isotope filtered NOESY spectra. The results were confirmed by comparison to the assignments derived from homonuclear NOESY spectra of a completely deuterated protein bound to the operator (Zhang *et al.*, 1994). The large number of chemical shift changes reflects the significant deformation of the DNA in the complex along with a large contact surface with the protein (Zhang *et al.*, 1994).

TrpR complex assignments

In all the 3D spectra collected on the complex, 32 transients per increment were necessary to achieve a signal-to-noise ratio similar to that for the holorepressor. This forced us to compromise on the number of increments collected in the indirectly-detected dimensions. To regain resolution lost in these compromises and to extend the resolution of the data sets, the use of linear prediction in F2 was critical. Many of the triple resonance experiments, especially the HNCA, when performed on the complex, did not have sufficient signal to noise to be of use. This is due to rapid T_2 relaxation during the pulse sequence, especially when trying to correlate the small $J_{C\text{-}N}$ couplings which require ^{15}N and ^{13}C magnetization to remain in the transverse plane for longer periods of time than for other triple resonance experiments. Therefore, most of the assignment process relied upon double resonance experiments (^{15}N-H or ^{13}C-H correlations), particularly NOE correlations.

^{15}N edited NOESY-HMQC spectra were collected on the complex using identical spectral ranges to those used for the holorepressor to allow a direct comparison of the data sets. Comparison of these spectra led to direct assignment of nearly 70% of the backbone amide signals from similarity. A further 15% of the NH signals that had shifted were assigned from a combination of sequential and intraresidue NOE connectivities. Consequently, a large percentage of the protein assignments could be derived from this spectrum alone. The ^{13}C NOESY-HSQC experiment in H_2O provided fewer backbone sequential NOE contacts than the ^{15}N NOESY-HMQC and hence would have been difficult to use without the previous assignments. However, ~50% of the Cα assignments for the holorepressor could be carried over to the complex, and these assignments formed a starting point for the analysis of the 3D ^{13}C spectra. Additional Cα and sidechain assignments were derived from this spectrum on the basis of the many ^1H assignments from the ^{15}N-edited NOESY-HMQC combined with the intra-residue NOE connectivities (including those to backbone amide protons) in the ^{13}C spectrum. The ^{13}C-NOESY-HSQC displays more clearly the intermolecular and many of the long range protein interactions which are sidechain mediated. The HCCH-TOCSY experiment was acquired to help differentiate intra-residue NOE interactions from inter-residue NOEs. However, the connectivities in this experiment were incomplete so that assignment of sidechains based solely on scalar connectivities was not possible. Thus, using essentially the ^{13}C NOESY-HSQC and ^{15}N NOESY-HMQC experiments, more than 80% of the Cα, backbone ^{15}N and sidechain aliphatic ^{13}C resonances and their attached protons were assigned, (excluding the N-terminal arm, see discussion) and a large number of sequential and long-range NOEs identified.

Intermolecular NOEs

The ^{13}C edited NOESYs of the complex contain many intermolecular NOEs between labeled protein and unlabeled DNA and co-repressor. These intermolecular NOEs were identified in the 3D ^{13}C-F3-filtered-HMQC-NOESY which shows only NOEs between ^{13}C labeled and nonlabeled species. This experiment is very important for characterizing the protein-DNA interaction surface and corepressor binding pocket. Our first attempts at running this experiment on the complex were only successful when it was performed in the 2D mode (*i.e.,* the carbon evolution time was set to zero), owing to rapid T_2 losses during the evolution and "purge" periods of the pulse sequence. A modification of this experiment as described by Lee *et al.* (1994) designed to minimize the time spins spend in the transverse plane, allowed us to obtain a 3D spectrum with good sensitivity. By comparing carbon planes of this spectrum to those of the ^{13}C-edited NOESY, one could readily identify and assign intermolecular NOEs.

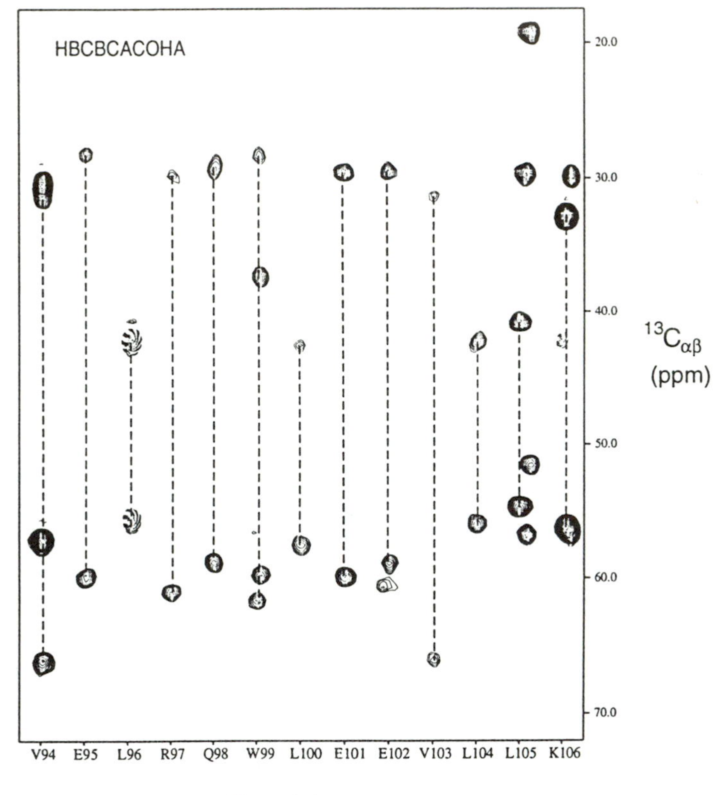

Figure 3: "Strips" from the (HB)CBCACOHA spectrum of the $^{13}C/^{15}N$ labeled, 5-methyl-tryptophan complex. Each strip is taken at the carbonyl frequency of the indicated residue showing the correlations to the intraresidue Cα and Cβ chemical shifts.

5-Methyl-Tryptophan complex

After the structure of the complex had been calculated (Zhang *et al.*, 1994) further studies were carried out with 5-methyl-tryptophan as the corepressor. 5-methyl-tryptophan has been shown to bind trpR an order of magnitude more tightly than L-trp (Marmorstein and Sigler, 1989), and comparison of the HSQC spectra of the L-trp and 5-methyl-tryptophan complexes indicated that the protein conformation is the same in both complexes. However, spectra of the 5-methyl-tryptophan complex were of significantly better quality, likely due to the slower dynamics associated with ligand exchange (see below). Using this complex we were able to collect an HNCO spectrum with reasonable sensitivity. This spectrum allowed the identification of all of the carbonyl resonances of residues for which the succeeding amide was assigned. These assignments were then used to interpret an (HB)CBCACO(CA)HA spectrum (Kay, 1993) as shown in Figure 3. By matching the carbonyl resonances of these last two experiments it was possible to confirm the vast majority of the previously assigned backbone and Cß assignments.

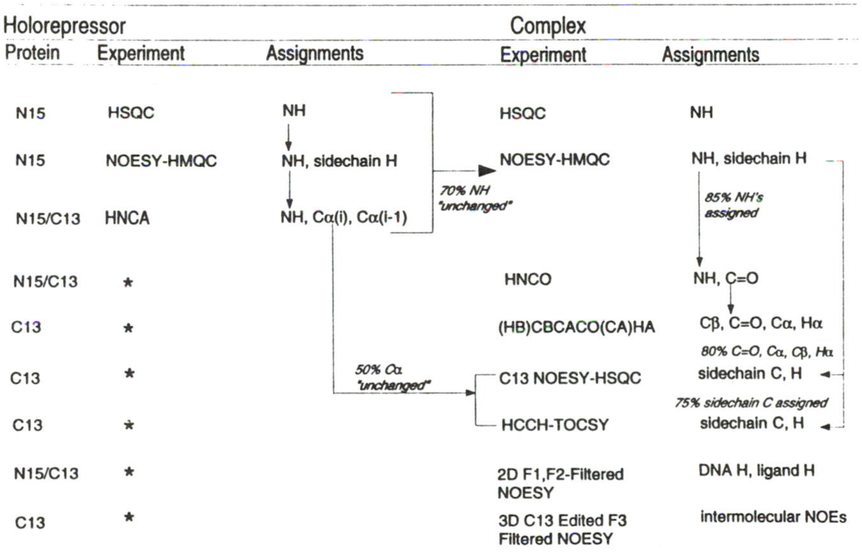

Holorepressor			Complex	
Protein	Experiment	Assignments	Experiment	Assignments
N15	HSQC	NH	HSQC	NH
N15	NOESY-HMQC	NH, sidechain H	NOESY-HMQC	NH, sidechain H
N15/C13	HNCA	NH, Cα(i), Cα(i-1)	*70% NH "unchanged"*	*85% NH's assigned*
N15/C13	*		HNCO	NH, C=O
C13	*		(HB)CBCACO(CA)HA	Cβ, C=O, Cα, Hα
				80% C=O, Cα, Cβ, Hα
C13	*	*50% Cα "unchanged"*	C13 NOESY-HSQC	sidechain C, H
				75% sidechain C assigned
C13	*		HCCH-TOCSY	sidechain C, H
N15/C13	*		2D F1,F2-Filtered NOESY	DNA H, ligand H
C13	*		3D C13 Edited F3 Filtered NOESY	intermolecular NOEs

Figure 4: Schematic representation of the assignment strategy used in this study. In total, 87% of backbone assignments were obtained for the holorepressor from the three experiments shown, combined with previous proton and ^{15}N assignments (Arrowsmith *et al.*, 1990; Czaplicki *et al.*, 1991). The star indicates further triple resonance and isotope edited experiments that would be desirable for the assignment of the uncomplexed species, if necessary. The progression of assignments made in this study is indicated in italics.

On the basis of our experience with this complex, we propose the scheme shown in Figure 4 as a method for the assignment and gathering of NOE data for large macromolecular complexes with solely $^{15}N/^{13}C$ labeled protein. These experiments are applicable to complexes in the 30-40 kDa range in which the protein portion of the complex can be uniformly isotopically labeled with ^{13}C and ^{15}N and only one protein sample need be prepared. The HNCO (Kay *et al.*, 1990) and (HB)CBCACO(CA)HA (Kay, 1993) experiments are a particularly useful combination. The former is the most sensitive triple resonance experiment and gives sequential connectivities from the carbonyl carbon through the peptide bond to the amide proton of the next residue. The latter relies on the larger carbon-carbon and carbon-proton couplings and gives intra-residue connectivities from the carbonyl carbon. Thus, by matching the carbonyl resonances in each spectrum it is possible, in principle, to assign all backbone and Cß resonances. The carbon resonances then serve as a good starting point for analysis of the ^{13}C-edited NOESY and HCCH-TOCSY.

$^{2}H/^{13}C/^{15}N$ labeled protein complex

The two major difficulties with assigning spectra of larger complexes are spectral overlap and poor sensitivity of many experiments due to rapid transverse relaxation times. In an attempt to address the latter problem we prepared $^{15}N/^{13}C$ labeled protein with ~70% random deuteration throughout the protein. Thus, ~70% of the Cα nuclei were bound to deuterium rather than hydrogen. This results in a dramatic improvement of the Cα T_2 relaxation times from ~17ms in the complex of the $^{13}C/^{15}N$ labeled protein to ~130 ms in the $^{2}H/^{13}C/^{15}N$ complex (Yamazaki *et al.,* 1995). This complex was used to perform a series of triple resonance experiments which allowed the relatively straight-forward assignment of the backbone HN, ^{15}N, $^{13}C\alpha$, and $^{13}C\beta$ resonances of trpR bound to DNA and the corepressor, 5-methyl tryptophan (Yamazaki *et al.,* 1994; 1995). These experiments, 3D constant-time (CT)-HNCA, CT-HN(CO)CA, HN(CA)CB, HN(CACO)CB and 4D HNCACB, are similar to previously published versions (Ikura *et al.,* 1990; Kay *et al.,* 1990; Grzesiek and Bax, 1992) except that they are designed to exploit the line-narrowing effect of deuterium on ^{13}C resonances and also require deuterium decoupling. The first two experiments provide correlations between the ^{1}HN, ^{15}N and $^{13}C\alpha$ of the same and previous residues. The third and fourth experiments provide correlations between HN, ^{15}N, and $^{13}C\beta$ resonances. The 4D experiment correlates HN and ^{15}N chemical shifts with both inter and intra residue $^{13}C\alpha$, $^{13}C\beta$ pairs. The high sensitivity of all five experiments is due to the fact that magnetization originates and is detected on the ^{1}HN spin (90% populated in a 10% D_2O solution) and that none of the scalar transfer steps or evolution times involves ^{1}H-C magnetization. Thus, the high sensitivity of protons (^{1}HN) combined with long $^{13}C\alpha$ T_2 relaxation times results in very

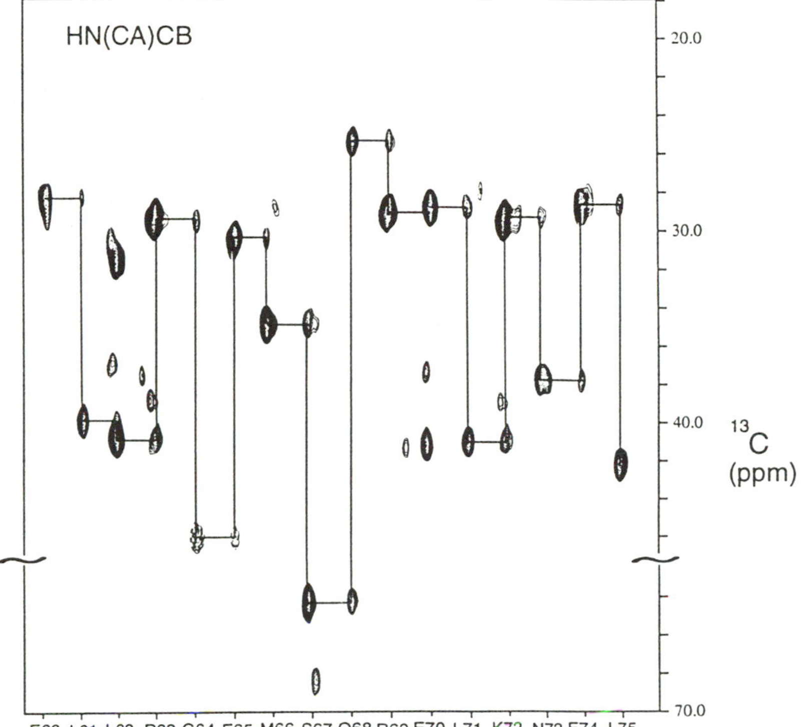

Figure 5: "Strips" from the HN(CA)CB spectrum of the $^{2}H(\sim70\%)/^{13}C/^{15}N$-trpR/operator/5-methyl-trptophan complex taken from ^{15}N planes at the amide proton frequencies of the indicated residues. The vertical axis displays the $^{13}C\beta$ chemical shift.

good quality spectra. Portions of the 3D HN(CA)CB spectrum are shown in Figure 5. The improved $^{13}C\alpha$ T_2 relaxation times of the deuterated protein also allow one to use constant-time evolution of the $^{13}C\alpha$ dimension, resulting in much better resolution compared to previous versions of the HNCA and HN(CO)CA experiments. This series of experiments provided >99% of the HN, ^{15}N, $^{13}C\alpha$, and $^{13}C\beta$ assignments of the protein in the 5-methyl-tryptophan/trpR /operator complex based solely on scalar connectivities. The vast majority of these were in agreement with the previous assignments of both the wild type and 5-methyl tryptophan complexes, with discrepancies occurring almost exclusively for residues either in the ligand binding pocket which likely reflect differences in protein-ligand binding or for residues whose previous assignments were tentative.

Discussion

The initial assignment of ^{1}H, ^{13}C and ^{15}N resonances in the trpR/DNA complex was a collaborative effort using both homonuclear NOESY analysis of selectively deuterated proteins (Zhang *et al.,* 1994) and heteronuclear experiments as described here and elsewhere (Zhang *et al.,* 1994, Lee *et al.,* 1994). While the majority of assignments could be obtained by either method alone, a "consensus assignment" based on both methods proved the most reliable. This is because both assignment strategies relied mostly on NOE connectivities and are therefore less rigorous than assignments based solely on scalar coupling connectivities. Furthermore, the distribution of the assignments through the sequence was not uniform. In an assignment strategy based mostly upon NOESY spectra, the extent of assignment in any part of a molecule will be affected by mechanisms that influence cross relaxation, such as motional processes. In the case of this protein, the unstructured and flexible nature of the N-terminal arm meant that very few NOE-derived assignments in those residues could be obtained. The other region of the protein with a low level of assignments was in parts of helix C (residues 45-49 and 54-58), which form the co-repressor binding site and interact nonspecifically with the phosphate backbone of the DNA. NMR studies of co-repressor exchange show that the ligand is exchanging in and out of the complex on the millisecond timescale (Lee *et al.*, submitted) while the complex itself has a lifetime on the order of several minutes (Hurlburt and Yanofsky, 1992). Residues in this region of the protein have proton signals with very poor signal to noise in all the NMR spectra. This is likely due to chemical shift averaging associated with co-repressor exchange as is supported by the fact that these signals improved somewhat with 5-methyl tryptophan which has a slower ligand off-rate (Lee *et al.*, unpublished).

The bulk of the assignments for the complex have been published along with the structure (Zhang *et al.*, 1994). However, it is instructive to compare chemical shift values for the complex with those of the free holorepressor. Figure 6 shows a comparison of amide nitrogen and $C\alpha$ chemical shifts for the DNA-free and bound forms of the protein. For both backbone positions the largest changes occur in the DNA binding domain, helices D and E, and the smallest changes are in helices A and F. The latter helices are on the side of the protein farthest from the DNA. Helix C and the turn between B and C form part of the corepressor binding site and are in contact with the phosphate backbone of the DNA. The changes in ^{1}H chemical shifts show a similar profile (Zhang *et al.*, 1994). It is interesting to note that the ^{15}N chemical shifts show some of the largest changes in absolute terms, but these changes are mostly localized to the DNA binding region. The changes in $C\alpha$ chemical shifts, on the other hand, are generally larger relative to the range within which the $C\alpha$'s resonate and are distributed more uniformly throughout the sequence. The latter may reflect small adjustments of the ϕ,ψ torsion angles throughout the sequence (de Dios *et al.*, 1993) while the ^{15}N shifts are more likely influenced by changes in χ^1 torsion associated with conformational changes in the side chains that interact with DNA (de Dios *et al.*, 1993) or changes in the strength of hydrogen bonds from these residues.

Our strategy relied heavily on the fact that the protein in the DNA-bound and free states had a very similar conformation and, therefore, that many of the assignments could be carried over from the smaller molecule. Many small DNA-binding domains undergo significant chemical shift changes upon interaction

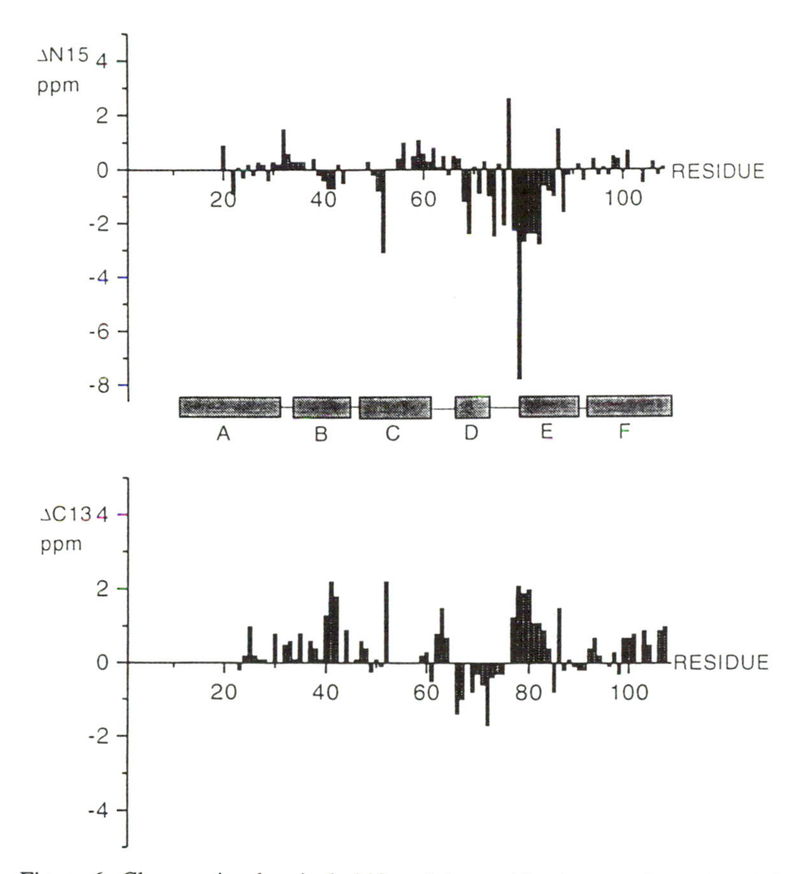

Figure 6: Changes in chemical shifts of the amide nitrogen (upper) and Cα (lower) resonances upon binding DNA (δ(holotrpR) - δ(complex)). Helices A-F are indicated. Helices D and E comprise the helix-turn-helix DNA-binding motif and are in direct contact with DNA. Helix C and the B-C turn make non-specific contacts with DNA and also comprise part of the ligand binding site.

with DNA because essentially the whole protein is in a new environment (Chuprina *et al.,* 1993, Guntert *et al.*, 1993, Bajela *et al.,* 1992). However, for larger DNA-binding proteins and domains it is likely that only those regions in contact with DNA will change as in the present case. To date there is little NMR evidence to support this; however, crystallographic studies of several medium-sized proteins bound to DNA, such as *E. coli* CAP (Schultz *et al.*, 1991), *E.coli Met* repressor (Somers *et al.*, 1992) and *E. coli trp* repressor (Otwinowski *et al.*, 1988) have shown that, in these cases, the conformation of the protein remains very much the same when bound to DNA. Therefore, one would expect major chemical shift changes only for regions in contact with DNA for these proteins.

The structure determination of the trpR/operator complex (Zhang *et al.*, 1994) demonstrates that it is possible to assign a large complex even without extensive triple resonance data. However, the use of triple resonance experiments, when possible, greatly facilitates the assignment process. Recently, Remerowski *et al.* (1994) have reported backbone assignments for a 28 kDa protein and point out the low sensitivity of some of the triple-resonance experiments in their study. It appears that conventional triple-resonance experiments will be very sensitive to size and dynamics for molecules over 25 kDa and are unlikely to be useful for molecular weights over ~35 kDa. Therefore, more general alternative strategies for assignments and structural determinations of large proteins and complexes are desirable. We and others have shown that one step in this direction is fractional deuterium labeling to improve the carbon T_2 times and, therefore, the sensitivity of multinuclear experiments (Grzesiek *et al.*, 1993; Kushlan and LeMaster, 1993; Yamazaki *et al.,* 1994a,b). Our recent results for a ~70% ^{2}H, 98% ^{13}C,^{15}N trpR/operator complex have demonstrated that triple resonance experiments combined with deuterium labeling can be performed with high sensitivity on this complex yielding rapid backbone assignments (Yamazaki *et al.,* 1994; 1995). These experiments, which allow one to acquire spectra with good digital resolution in the carbon dimension(s), are applicable to larger single-chain proteins in the range of 30-35 kDa. In this respect they are more advantageous than methods that involve 2D spectroscopy of selectively labeled proteins, because the latter will suffer a great deal from spectral overlap. However, the triple-resonance experiments only provide NH and heavy atom assignments of the backbone and, therefore, provide none of the NOE data needed to determine the structure. Therefore, a *general* method for the assignment of sidechain resonances of larger proteins or complexes is still desirable. It will be interesting to see if this can be achieved with random deuteration combined with ^{13}C and ^{15}N labeling so that only a single, triply-labeled protein sample will be needed to provide all the data necessary to assign the NMR resonances and determine the three-dimensional structure of larger proteins or complexes.

Acknowledgements

This research was supported by the National Cancer Institute of Canada, the Human Frontier Science Program and the Princess Margaret Hospital Foundation (NMR instrumentation). M.J.R. was supported by a studentship from the Natural Science and Engineering Research Council of Canada. We acknowledge our collaborations with the laboratory of O. Jardetzky on the initial assignment of the trpR complex and its structure determination and with the laboratory of L.E. Kay which developed the deuterium-decoupled triple-resonance experiments.

References

Applied Biosystems; User Bulletin No. 50, August 26 1988.

Arrowsmith, C.H., Carey, J., Treat-Clemons, L., and Jardetzky, O. (1989). *Biochemistry* **28**, 3875.

Arrowsmith, C.H., Pachter, R., Altman, R.B., Iyer, S.B., and Jardetzky, O. (1990). *Biochemistry* **29**, 6332.

Bajela, J.D., Marmorstein, R., Harrison, S.C., and Wagner, G. (1992). *Nature* **356**, 450.

Bax, A., Clore, M., and Gronenborn, A.M. (1990). *J. Magn. Reson.* **88**, 425.

Beacage, S.L., and Carruthers, M.H. (1981). *Tetrahedron Letters* **22**, 1859.

Billeter, M., Qian,Y.Q., Otting, G., Muller, M., Gehring, W., and Wüthrich, K. (1993). *J. Mol. Biol.* **234**, 1084.

Bodenhausen, G., and Reuben, D.J. (1980). *Chem. Phys. Lett.* **69**, 185.

Chuprina, VP., Rullmann, J,A., Lamerichs, R.M., van Boom, J.H., Boelens, R., and Kaptein, R. (1993). *J. Mol. Biol.* **234**, 446.

Czaplicki, J., Arrowsmith, C.H., and Jardetzky, O. (1991). *J. Biomol. NMR* **1**, 349.

de Dios, A.C., Pearson, J.G., and Oldfield, E. (1993). *Science* **260**, 1491.

Fogh, R.H., Schnipper, D., Boelens, R., and Kaptein, R. (1994). *J. Biomol. NMR* **4**, 123.

Grzesiek, S., Anglister, J., Ren, H., and Bax A. (1993). *J. Am. Chem. Soc.* **115**, 4369.

Grzesick, S., and Bax, A. (1992). *J. Magn. Reson.* **96**, 432.

Grzesick, S., and Bax, A. (1993). *J. Biomol. NMR* **3**, 185.

Guntert, P., Qian, Y.Q., Otting, G., Muller, M., Gehring, W., and Wüthrich, K. (1991). *J. Mol. Biol.* **217**, 531.

Hurlburt, B.K., and Yanofsky, C. (1992). *J. Biol. Chem.* **267**, 16783.

Ikura, M., and Bax, A. (1991). *J. Am. Chem. Soc.* **7**, 2433.

Ikura, M., Kay, L.E., and Bax, A. (1990). *Biochemistry* **20**, 4569.

Jamin, N., Grielsen, O.S., Gilles, N., Lirsac, P., and Toma, F. (1993). *Eur. J. Biochem.* **216**, 147.

Joachimiak, A., Kelly, R.L., Gunsalus, R.P., Yanofsky, C., and Sigler, P.B. (1983). *Proc. Nat. Acad. Sci. U.S.A.* **80**, 668.

Kaptein, R. (1993). *Curr. Op. Struct. Biol.* **3**, 50.

Kay, L.E. (1993). *J. Am. Chem. Soc.* **115**, 2055.

Kay, L.E., Ikura, M., Tschudin, M., and Bax, A. (1990). *J. Magn. Reson.* **89**, 496.

Kay, L.E., Marion, D., and Bax, A. (1989). *J. Magn. Reson.* **84**, 72.

Kushlan, D.M., and LeMaster, D.M. (1993). *J. Biomol. NMR.* **3**, 701.

Lee, W., Revington, M. J., Arrowsmith, C.H., and Kay, L.E. (1994). *FEBS Lett.* **350**, 87.

Lee, W., Revington, M., Farrow, N., Utsunomiya-Tae,N., Miyake, Y., Kainosho, M., and Arrowsmith, C.H. (1995). *submitted.*.

Lefevre, J.F., Lane, A.N., and Jardetzky, O. (1986). *Biochemistry* **26,** 5076.

Maniatis T., Fritsch, E.F., and Sambrook, J. (1982). *Molecular Cloning-A Laboratory Manual.*

Marion, D., and Bax, A. (1989). *J. Magn. Reson.* **83**, 205.

Marion, D., Kay, L.E., Bax, A. (1989a). *J. Magn. Reson.* **84**, 72.

Marion, D., Kay, L.E., Sparks, S.W., Torchia, D.A., and Bax, A. (1989b). *J. Am. Chem. Soc.* **111**, 1515.

Marmorstein, R.Q., Joachimiak, A., Sprinzl, M., and Sigler, P.B. (1987). *J. Biol. Chem.* **262**, 4922.

Marmorstein, R.Q., and Sigler, P.B. (1989). *J. Biol. Chem.* **264**, 9149.

Muhandiram, D.R., Farrow, N.A., Xu, G., Smallcombe, S.H., and Kay, L.E. (1993). *J. Magn. Reson., Series B* **102**, 317.

Ogata, K., Kanai, H., Inoue, T., Sekikawa, A., Sasaki, M., Nagadoi, A., Sarai, A., Ishii, S., and Nishimura, Y. (1993). *Nuc. Acids Symp. Ser.* **29**, 201.

Omichinski, J.G., Clore, G.M., Schaad, O., Felsenfeld, G., Trainor, C., Appella, E., Stahl, S.J., and Gronenborn, A.,M. (1993). *Science* **261**, 438.

Otwinowski, Z., Schevitz, R.W., Zhang, R.G., Lawson, C.L., Joachimiak, A., Marmorstein, R.Q., Luisi, B.F., and Sigler, P.B. (1988). *Nature* **335**, 321.

Pabo, C., and Sauer, R.T. (1992). *Ann. Rev. Biochem.* **61**, 1053.

Paluh, J. L., and Yanofsky, C. (1986). *Nucl. Acids Res.* **14**, 7851.

Remerowski, M.L., Domeke, T., Groenewegen, A., Pepermans, H.A.M., Hilbers, C.W., and van de Ven, F.J.M. (1994). *J. Biomol. NMR* **4**, 257.

Somers, W.S., and Phillips S.E.V. (1992). *Nature* **359**, 387.

Schultz, S.C., Sheilds, G.C., and Steitz, T.A. (1991). *Science* **253**, 1001.

Wüthrich, K. (1986). *NMR of Proteins and Nucleic Acids*, John Wiley and Sons.

Yamazaki, T., Lee W., Revington, M., Mattiello, D., Dahlquist, W., Arrowsmith, C.H., and Kay, L.E. (1994). *J. Am. Chem. Soc.* **116**, 6464.

Yamazaki, T., Lee, W., Arrowsmith, C.H., Muhandiram, D.R., and Kay, L.E. (1995). *J. Am. Chem. Soc., in press.*

Zhang, H., Zhao, D., Revington, M., Lee, W., Jia,X., Arrowsmith, C.H., and Jardetzky, O. (1994). *J. Mol. Biol.* **238**, 592.

Zhao, D., Arrowsmith, C.H., Jia, X., and Jardetzky, O. (1993). *J. Mol. Biol.* **229**, 735.

Zuiderweg, E.R.P., and Fesik, S.W. (1989). *Biochemistry* **28**, 2387.

7

Design of Novel Hemoglobins

C. Ho and H.-W. Kim

Department of Biological Sciences
Carnegie Mellon University
Pittsburgh, PA 15213 USA

Human normal adult hemoglobin (Hb) A, the oxygen carrier of blood, is a tetrameric protein consisting of two α chains of 141 amino acid residues each and two β chains of 146 amino acid residues each. Each Hb chain contains a heme group which is an iron complex of protoporphyrin IX. Under physiological conditions, the heme-iron atoms of Hb remain in the ferrous state. In the absence of oxygen, the four heme-irom atoms in Hb A are in the high-spin ferrous state [Fe(II)] with four unpaired electrons each. Each of the four heme-iron atoms in Hb A can combine with an O_2 molecule to give oxyhemoglobin (HbO_2) in which the iron atom is in a low-spin, diamagnetic ferrous state. The oxygen binding of Hb exhibits sigmoidal behavior, with an overall association constant expression giving a greater than first-power dependence on the concentration of O_2. Thus, the oxygenation of Hb is a cooperative process, such that when one O_2 is bound, succeeding O_2 molecules are bound more readily. Hb is an allosteric protein, *i.e.*, its functional properties are regulated by a number of metabolites [such as hydrogen ions, chloride, carbon dioxide, 2,3-diphosphoglycerate (2,3-DPG)] other than its ligand, O_2. It has been used as a model for allosteric proteins, and indeed, hemoglobins of vertebrates are among the most extensively studied allosteric proteins. Their allosteric properties are physiologically important in optimizing O_2 transport by erythrocytes. The large number of mutant forms of Hb available provides an array of structural alterations with which to correlate effects on function. For details, see Dickerson and Geis (1983), Bunn and Forget (1986), Ho (1992), Ho and Perussi (1994).

There are two types of contacts between the α and β subunits of Hb (Perutz, 1970; Dickerson and Geis, 1983). The $\alpha_1\beta_1$ (or $\alpha_2\beta_2$) contacts, involving B, G, and H helices, and GH corners, are called packing contacts. These contacts remain unchanged and hold the dimer together even when there is a change in the ligation state of the heme. The $\alpha_1\beta_2$ (or $\alpha_2\beta_1$) contacts, mainly involving C and G helices, and FG corners, are called sliding contacts, and undergo significant motion when there is a change in the ligation state of the heme. The movement of heme iron atoms and the sliding motions of the $\alpha_1\beta_2$ (or $\alpha_2\beta_1$) subunit interface, as well as the breaking of intra- and intermolecular salt bridges and hydrogen bonds as a result of the ligation of the Hb molecule, are among the most important features of the stereochemical mechanisms for the cooperative oxygenation of Hb (Perutz, 1970). For a recent discussion including a new view on this subject, refer to Srinivasan and Rose (1994).

The Hill coefficient (n_{max}), which measures the cooperative oxygenation of Hb, provides a convenient measure of some of the allosteric properties of this protein (Dickerson and Geis, 1983). Under usual experimental conditions, Hb A has an n_{max} value of approximately 3 in its binding with O_2. Human abnormal Hbs with amino acid substitutions in the $\alpha_1\beta_2$ subunit interface generally have high oxygen affinity and reduced cooperativity in O_2 binding compared to Hb A (Dickerson and Geis, 1983; Bunn and Forget, 1986), suggesting the importance of the $\alpha_1\beta_2$ subunit interface to the functional properties of Hb. In particular, mutant human Hbs with an amino acid substitution at the β99Asp, such as Hb Kempsey (β99Asp→Asn or β:D99N) (n_{max} = 1.1) (Reed *et al.*, 1968), Hb Yakima (β99Asp→His) (n_{max} = 1.0) (Jones *et al.*, 1967), Hb Radcliff (n_{max} = 1.1) (β99Asp→Ala) (Weatherall *et al.*, 1977), Hb Hôtel Dieu (β99Asp→Gly) (Thillet *et al.*, 1981) (n_{max} = 1.3), and Hb Ypsilanti (β99Asp→Tyr) (n_{max} not reported) (Glynn *et al.*, 1968), possess greatly reduced cooperativity and increased oxygen affinity relative to those exhibited by Hb A. X-ray crystallographic studies of Hb A (Perutz, 1970; Fermi *et al.*, 1984) have shown that β99Asp is hydrogen-bonded to both α42Tyr and α97Asn in the $\alpha_1\beta_2$ subunit interface of deoxy-Hb A, suggesting that the essential role of β99Asp is to stabilize the deoxy-Hb molecule by forming intersubunit hydrogen bonds.

Recently, two recombinant Hbs (r Hbs) with an amino acid substitution at the α42Tyr, r Hb (α42Tyr→His) and r Hb (α42Tyr→Phe), have been constructed (Ishimori *et al.*, 1989; Imai *et al.*, 1991). r Hb (α42Tyr→Phe) exhibits essentially no cooperativity in binding oxygen (n = 1.2) and possesses very high oxygen affinity, whereas r Hb (α42Tyr→His) exhibits substantial cooperativity (n = 2 at pH 6.8) and moderate oxygen affinity. These investigators have attributed the differences in function between these two mutants to the presence of a weak hydrogen bond between α42His and β99Asp in the deoxy form of r Hb (α42Tyr→His), and the lack of such a bond in deoxy-r Hb (α42Tyr→Phe). Since abnormal Hbs with an amino acid substitution at either β99Asp or α42Tyr, which lose the intersubunit hydrogen bonds in the deoxy form, also lose their functional properties, it has been suggested that these hydrogen bonds are crucial for the structure and function of the Hb molecule.

Molecular dynamics (MD) simulations have been used successfully to calculate the free energy difference between native and mutant proteins. Recent studies have involved, for example, the effects on protein stability (Dang *et al.*, 1989; Prevost *et al.*, 1991, Tidor and Karplus, 1991), on ligand binding (Komeiji *et al.*, 1992; Lau and Karplus, 1994), and on cooperativity of Hb (Gao *et al.*, 1989). In most cases, MD-simulation results demonstrate excellent agreement with experimentally determined data. In particular, MD simulations of Hb Radcliff (β99Asp→Ala) show remarkable agreement between the measured thermodynamic value and the calculated cooperativity. The overall free energy change calculated is -5.5 kcal/mol, which has the same sign and is of the same order as the experimentally measured value of -3.4 kcal/mol, suggesting that MD-simulation results may be used to gain insights into specific interactions within a protein molecule (Gao *et al.*, 1989). The analysis of the free energy of simulation shows that the effect of the mutation, *i.e.*, β99Asp→Ala, is more complex than the crystal structure suggests. Some of the contributions to the difference in the free energy of cooperativity are as large as 60 kcal/mol, indicating that essential thermodynamic elements are hidden in the measured value (-3.4 kcal/mol.) (Gao *et al.*, 1989). Thus, the partly canceling individual contributions in solvent-protein interactions and electrostatic interactions, which are not evident from a crystal structure, can be exposed only by a free-energy simulation, which could provide new insights into the origin of thermodynamic values.

Up to now, there have been no completely reliable approaches to predicting protein tertiary structure from its amino acid sequence. However, if only a few amino acid side chains in a protein are mutated, the conformational changes may, in some cases, be predictable by MD simulations. Recently, the calculated structures of mutant subtilins, in which the methionine at the 222 position has been replaced, *i.e.*, 222Met→Ala, 222Met→Phe, and 222Met→Gln, have been compared with the X-ray structures of the respective mutants, and have shown good agreement between the predicted and X-ray results (Heiner *et al.*, 1993). These calculations involved gradually changing methionine at the 222 position into the replacing amino acid using a thermodynamic integration method for free-energy determination.

We have used MD simulations to design mutant Hbs which have altered $\alpha_1\beta_2$ subunit interfaces. Since known, naturally-occuring abnormal Hbs with an amino acid substitution at β99Asp not only lose the intersubunit hydrogen bonds in the deoxy form, but also lose their functional properties, our approach has been to design compensatory mutant Hbs by introducing additional mutations in the local environment of the β99 mutation in order to create new hydrogen bonds to compensate for the missing ones. Several novel Hbs have been designed in this way. In particular, for Hb Kempsey (β99Asp→Asn), which has a high oxygen affinity and exhibits essentially no cooperativity in binding oxygen (Bunn *et al.*, 1974), our computer simulations indicate that a new hydrogen bond involving β99Asn can be induced in the deoxy form by replacing α42Tyr by a stronger hydrogen-bond acceptor, such as Asp. This suggests that a new Hb with the amino acid substitutions, β99Asp→Asn and α42Tyr→Asp (or β:D99N; α:Y42D), *i.e.*, a mutant of Hb Kempsey, can regain the cooperativity lost in Hb Kempsey (Kim *et al.*, 1994).

To test the validity of the MD simulations, it is necessary to have the appropriate mutant proteins for experimental investigation. We have recently constructed a plasmid (pHE2) (Shen *et al.*, 1993) in which synthetic human α- and β-globin genes (Hoffman *et al.*, 1990) are coexpressed with the *Escherichia coli* methionine aminopeptidase (Met-AP) gene (Ben-Bassat *et al.*, 1987) under the control of separate tac promoters. In *E. coli* cells harboring the pHE2 plasmid, the N-terminal methionine residues of the expressed Hb A have been effectively cleaved by the coexpressed Met-AP, and this expressed r Hb A lacking an N-terminal methionine is identical to native Hb A in a number of structural and functional properties (Shen *et al.*, 1993). Using this *E. coli* expression system, we have applied site-directed mutagenesis to produce several new r Hbs which have replacements at the $\alpha_1\beta_2$ interface. We have determined the O_2 binding properties of these r Hbs, and have used proton nuclear magnetic resonance spectroscopy (^{1}H NMR) to investigate the tertiary structures around the heme groups and the quaternary structure. In this article, we describe r Hb (β99Asp→Asn, α42Tyr→Asp) as an example of our appoach (for details, see Kim *et al.*, 1994).

^{1}H NMR spectroscopy has proven to be a powerful technique for investigating the structure-function relationship in Hb. Owing to the presence of the unpaired electrons in the high-spin ferrous atoms in deoxy-Hb and the highly conjugated porphyrins of the Hb molecule, the proton chemical shifts of various Hb derivatives cover a wide range. Resonances vary from about 20 ppm upfield from the proton resonance of a standard, 2,2-dimethyl-2-silapentane-5-sulfonate (DSS), to about 80 ppm downfield from DSS, depending on the spin state of the iron atoms and the nature of ligands attached to the heme groups. This unusually large spread of proton chemical shifts for deoxy-Hb A provides the selectivity and the resolution necessary to investigate specific regions of the Hb molecule. For details, see Ho (1992, and the references therein).

MD Simulations

MD simulations of proteins involve treating each atom in the molecule as a particle following the classical Newtonian equations of motion. By integrating Newton's equations of motion, atomic positions and velocities can be obtained as a function of time. In a protein consisting of many atoms, the total force acting on any one atom at any given time depends on the position of all the other atoms. Thus, a high-speed computer is needed to solve the Newtonian equations in order to determine the positions and velocities of all the atoms in a protein molecule over a long period of time. Recent advances in computing technology have made it possible to investigate proteins in solution and to extend the simulation time in the range of nanoseconds. For details, see Karplus and Petsko (1990).

For the design of Hb mutants with an altered subunit interface, transformation between wild-type and mutant proteins using the thermodynamic integration method (Kirkwood, 1935) has been employed. This method has been used for the calculation of free energy, *i.e.*, by gradually changing the potential from that representing wild-type Hb to the mutant form during a simulation. Energy minimized X-ray structures of the normal Hb tetramer were used to generate the starting configurations. MD simulations were carried out by a stochastic boundary method (Brooks and Karplus, 1989) using CHARMM 22 with standard parameters for the polar hydrogen protein model (param19). The molecule was partitioned into MD and Langevin regions with radii of 10 Å and 15 Å, respectively. The inside sphere was filled with charmm-adapted pre-equilibrated TIP3P water molecules (Jorgensen *et al.*, 1983).

The transformation from wild-type to mutant proteins was achieved by using a hybrid potential function $V_\lambda = (1-\lambda)V_A + \lambda V_B$ (Gao *et al.*, 1989; Tidor and Karplus, 1991), where λ is a coupling factor between 0 and 1. V_A and V_B are potential energy functions for Hb A and mutant Hb, respectively. Simulations were done at nine values of λ_i (λ = 0.1, 0.2,........0.9), with 5 ps of equilibration followed by 5 ps of production dynamics, except at $\lambda = 0.1$ and $\lambda = 0.9$, where 10 ps of equilibration was employed. Non-bonded interactions which had been truncated to zero at 8.5 Å and a constant dielectric ($\varepsilon = 1$) were used. All bonds involving hydrogen atoms were constrained with the SHAKE algorithm (Ryckaert *et al.*, 1977).

The free energy of simulation can be obtained from the trajectory files of the MD simulations for both deoxy and oxy forms of Hb using the thermodynamic integration method (Kirkwood, 1935) with the following equation:

$$\Delta G = G_B - G \quad = \int_0^1 \langle \Delta V \rangle_\lambda d\lambda \approx \sum_i \langle \Delta V \rangle_{\lambda i} \Delta\lambda$$

where $\Delta V = V_B - V_A$, and the thermodynamic average $\langle \Delta V \rangle_\lambda$ indicates the average of V_λ over the hybrid system. The linear form of the thermodynamic equations shows that the total free energy of the simulations can be decomposed into individual additive contributions.

The change in the free energy of cooperativity resulting from the mutations can be indirectly obtained by the difference in the free energy of simulation between deoxy and oxy form from the thermodynamic cycle (Gao *et al.*,1989).

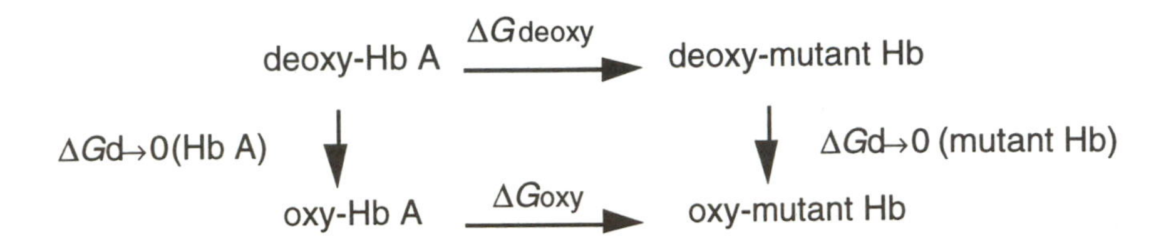

For the purpose of a structural analysis, average structures for wild-type (λ = 0) and mutant systems (λ = 1) have been calculated from the nearest states that are simulated (λ = 0.1 and λ = 0.9, respectively) by using the exponential formula (Brooks, 1986):

$$X_0 = \frac{< Xe^{+0.1\beta\Delta v} >_{\lambda = 0.1}}{< e^{+0.1\beta\Delta v} >_{\lambda = 0.1}} \quad \text{and} \quad X_1 = \frac{< Xe^{-0.1\beta\Delta v} >_{\lambda = 0.9}}{< e^{-0.1\beta\Delta v} >_{\lambda = 0.9}}$$

where X represents the Cartesian coordinates of the system, and X_0 and X_1 are the simulation-average wild-type and mutant coordinates, respectively, $\beta = 1/k_BT$, and $<>\lambda$ is an ensemble average under the potential V_λ.

Design and expression of a compensatory mutation for Hb Kempsey (β99Asp→Asn)

Design of a compensatory mutation

For a successful application of MD simulations to predict the structural changes in a mutant protein, it is essential that the local conformations of wild-type and mutant protein be similar. Deoxy-Hb Kempsey crystallizes in an oxy-like (or R-type) structure (Perutz *et al.*, 1974), and ^{1}H NMR spectroscopic results also suggest that it has an oxy-like quaternary structure (Perutz *et al.*, 1974; Takahashi *et al.*, 1982). Thus, its quaternary structure is quite different from that of deoxy-Hb A. However, if new hydrogen bond(s) can be formed in the $\alpha_1\beta_2$ interface of Hb Kempsey by compensatory mutations, it would be expected that a stable deoxy-like quaternary structure could be maintained in this mutant. With this assumption, MD simulations have been used to design compensatory mutations in Hb Kempsey (Kim *et al.*, 1994).

Our computer simulations indicate that a new hydrogen bond involving β99Asn can be induced in the deoxy form by replacing α42Tyr by a stronger hydrogen-bond acceptor, such as Asp. The average simulated structure of our mutant Hb (β99Asp→Asn, α42Tyr→Asp) in the deoxy form shows that new hydrogen bonds between β99Asn and α42Asp in the $\alpha_1\beta_2$ subunit interface can be formed (Figure 1). The sidechain angles (χ^1) of α42Asp and β99Asn in the averaged simulated structure are -77.3°, and -171.3°, respectively. These angles are the most frequently observed values in the rotamer library (48% and 33%, respectively) (Ponder and Richards, 1987), indicating that these amino acids have a stable sidechain orientation. 100-ps MD simulations of the transformed mutant Hb (β99Asp→Asn, α42Tyr→Asp) in the deoxy form show that the new hydrogen bond between β99Asn and α42Asp is very stable, *i.e.*, with an average simulated distance of 2.19 ± 0.16 Å (Kim *et al.*, 1994).

The change in the free energy of cooperativity resulting from the mutations (β99Asp→Asn, α42Tyr→Asp) can be indirectly obtained from the thermodynamic cycle. The measured thermodynamic value for Hb Kempsey is -3.4 kcal/mol per interface (Turner *et al.*, 1992). If mutant Hb (β99Asp→Asn, α42Tyr→Asp) restores some cooperativity in binding oxygen, the change in the free energy of cooperativity is expected to be somewhere between 0 and -3.4 kcal. The calculated value is -3.3 kcal/mol per interface, which is on the same order and same sign as the expected value. Considering that both mutations introduced involve charge changes, this calculated value seems to be in good agreement. The simulated results may be used to obtain information about the specific interactions which contribute to the total free energy difference.

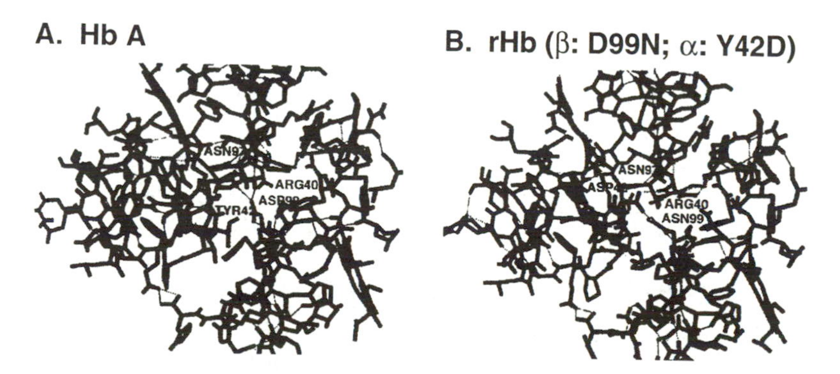

Figure 1: (A) Energy minimized X-ray structure of Hb A and (B) average simulated structure of r Hb (b:D99N; a:Y42D), both showing the $\alpha_1\beta_2$ interface region in the deoxy form. Dotted lines display the hydrogen bonds. This figure was prepared by the molecular graphic program, GRAPHX, developed at the Pittsburgh Supercomputing Center. This figure is taken from Figure 1 of Kim *et al.* (1994).

Oxygen binding properties

We have produced the new r Hb (β99Asp→Asn, α42Tyr→Asp) suggested by our computer simulations. The O_2-binding properties of r Hb (β99Asp→Asn, α42Tyr→Asp), Hb Kempsey, and Hb A are shown in Figure 2. r Hb (β99Asp→Asn, α42Tyr→Asp) exhibits about 40% of the Bohr effect of Hb A over the pH range 6.8 to 7.9. In 0.1 M sodium phosphate at 29°C, r Hb (β99Asp→Asn, α42Tyr→Asp) exhibits intermediate oxygen affinity; for example, at pH 7.2, p_{50} = 2.5 mm Hg for r Hb (β99Asp→Asn, α42Tyr→Asp) versus p_{50} = 10.5 mm Hg for Hb A, and p_{50} = 0.2 mm Hg at pH 7.2 for Hb Kempsey (Bunn *et al.*, 1974). However, our double mutant exhibits significant cooperativity in binding of oxygen, especially at low pH; for example, n_{max} = 2.0 at pH 6.8 versus n_{max} = 3.0 for Hb A, and n_{max} = 1.1 at pH 7.2 for Hb Kempsey. A marked change occurs when the allosteric effector, inositol hexaphosphate (IHP), is added to r Hb (β99Asp→Asn, α42Tyr→Asp). The oxygen affinity is reduced significantly [p_{50} = 12.9 mm Hg at pH 7.2 for r Hb (β99Asp→Asn, α42Tyr→Asp) versus p_{50} = 46.7 mm Hg for Hb A and p_{50} = 1.1 mm Hg at pH 7.2 for Hb Kempsey], and the cooperative oxygenation process for this double mutant approaches the normal value for Hb A as manifested by the Hill coefficient, with an n_{max} value of 2.3 to 2.5 over the pH range from 6.8 to 7.9.

The restoration of cooperativity in our r Hb (β99Asp→Asn, α42Tyr→Asp) relative to Hb Kempsey (β99Asp→Asn) can most likely be explained by the presence of a new hydrogen bond between α42Asp and β99Asn in the $\alpha_1\beta_2$ interface introduced by the additional mutation α42Tyr→Asp. This new interfacial hydrogen bond can stabilize a deoxy-like (or T-type) structure, and thus, could provide the necessary free energy of cooperativity in binding oxygen.

We have also expressed r Hb (α42Tyr→Asp) in *E. coli*. However, r Hb (α42Tyr→Asp) is an unstable hemoglobin which appears to lose hemes, *i.e.*, the formation of Heinz bodies (Bunn and Forget, 1986) (unpublished results of Ho, C., Ruiz-Noriega, M., Shen, T.-J., Kim, H.-W., Zou, M., and Ho, N.). Hb Kempsey (β99Asp→Asn) exhibits very high oxygen affinity and greatly reduced cooperativity (Bunn *et al.*, 1974). Thus, r Hb (β99Asp→Asn, α42Tyr→Asp) may be described as a case of "two wrongs making a right"!

Structural investigation by ^1H NMR spectroscopy

^1H NMR spectroscopy has been shown to be an excellent tool to investigate the tertiary and quaternary structural features of Hb (Ho, 1992). Low-field ^1H resonances of Hb A, Hb Kempsey, and r Hb (β99Asp→Asn, α42Tyr→Asp) are shown in Figure 3 (unpublished results of Sun, D.P., Kim, H.- W., Ho, N.T., and Ho, C.). The resonance at ~63 ppm from DSS has been assigned to the hyperfine-shifted H-exchangeable $H^{\delta 1}$ of the proximal histidine residue

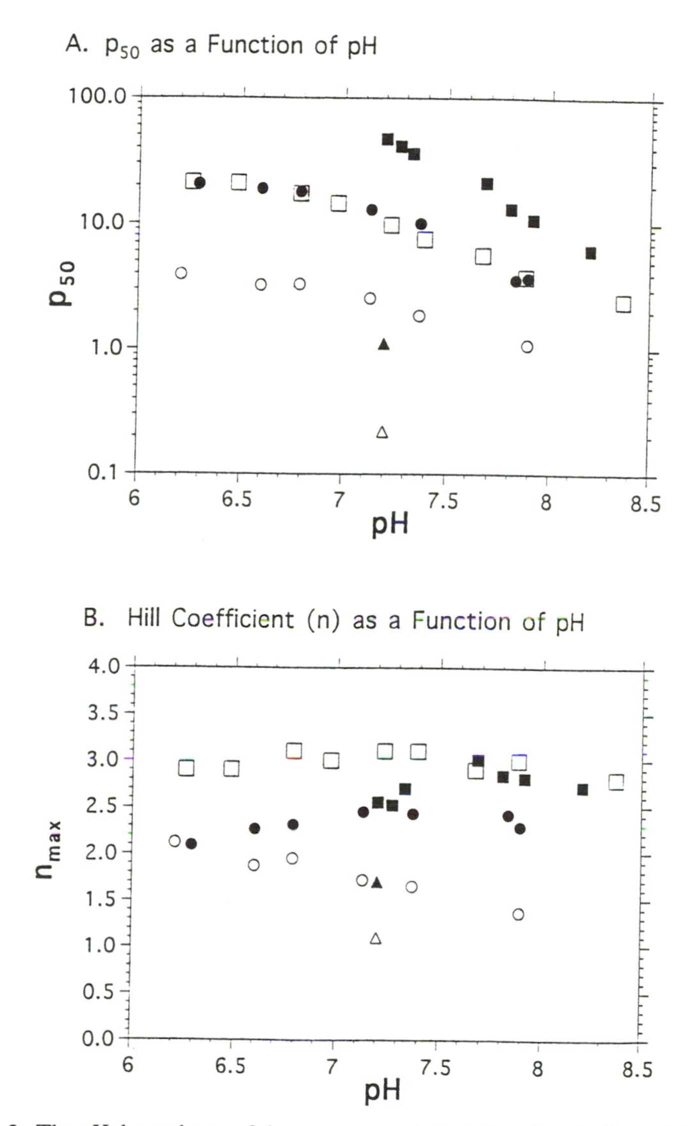

Figure 2: The pH dependence of the oxygen affinity (A) and the Hill coefficient (B): (**O**) r Hb (β:D99N; α:Y42D); (●) r Hb (β:D99N; α:Y42D) with 2 mM IHP; (□) Hb A; (n) Hb A with 2 mM IHP; (Δ) Hb Kempsey (β:D99N); and (s) Hb Kempsey (β:D99N) with 1 mM IHP. Oxygen dissociation curves were measured by a Hemox-Analyzer, and p_{50} and the Hill coefficient were determined from each curve. Oxygen dissociation data were obtained with 0.1 mM Hb in 0.1 M sodium phosphate buffer in the pH range 6.5-8.0. p_{50} is plotted on the y-axis in logarithmic scale. The data for Hb Kempsey are taken from Bunn *et al.* (1974). This figure is taken from Figure 2 of Kim *et al.* (1994).

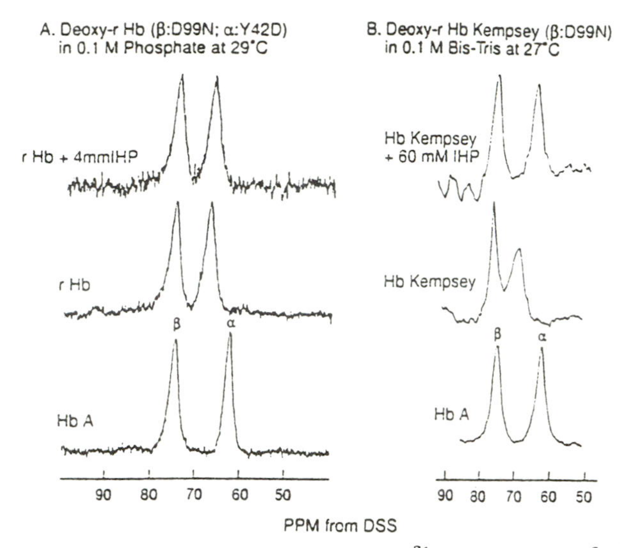

Figure 3: Hyperfine-shifted H-exchangeable $H^{\delta 1}$ resonances of the proximal histidine residues of Hb in the deoxy form: (A) r Hb (β:D99N; α:Y42D) and Hb A; and (B) Hb Kempsey (β:D99N) with and without IHP. The ^1H NMR spectra of (A) were obtained on a Bruker AM-300 spectrometer operating at 300 MHz. The experimental conditions for (A) were: [Hb] ≈ 1.8 to 3.6 mM in 0.1 M phosphate in H_2O at pH 7.0 at 29°C. The ^1H NMR spectra of (B) were obtained from a home-built 250-MHz spectrometer using correlation spectroscopy and were taken from Figures 1 and 2 of Takahashi *et al.* (1982). The experimental conditions for (B) were: [Hb] ≈ 4 mM in 0.1 M Bis-Tris with and without 60 mM IHP in H_2O at pH 6.4 at 27°C.

(α87His) of the α chain of deoxy-Hb A and the one at ~77 ppm from DSS has been assigned to the corresponding residue of the β chain (β92His) of deoxy-Hb A (Takahashi *et al.*, 1980; La Mar *et al.*, 1980). The chemical shift positions of the two proximal histidyl resonances in r Hb (β99Asp→Asn, α42Tyr→Asp) are very similar to those of Hb Kempsey, which shows a downfield resonance shift in the histidyl resonance of the α chain. In Hb Kempsey, the presence of IHP causes the proximal histidyl resonance from the α chain to be shifted upfield to a chemical shift position similar to that of deoxy-Hb A (Takahashi *et al.*, 1982). In contrast, IHP exerts no significant effect on the proximal histidyl resonances of r Hb (β99Asp→Asn, α42Tyr→Asp).

Figure 4 shows the ring-current shifted resonances of Hb A, r Hb (β99Asp→Asn, α42Tyr→Asp), and Hb Kempsey (β99Asp→Asn) in the CO form. The proton resonances over the region from 0 to -2.0 ppm from DSS arise from the protons of amino-acid residues located in the vicinity of the heme groups and from those of the heme groups (Ho, 1992). The resonances from -1.7 to -1.9 ppm from DSS, which have been assigned to the γ_1- and γ_2-methyl groups of β67E11Val (distal valine) (Lindstrom *et al.*,1972; Dalvit and Ho, 1985), merge into one peak in the spectrum of the r Hb (β99Asp→Asn, α42Tyr→Asp). This has also been observed for Hb Kempsey in the CO form (Lindstrom *et al.*, 1973).

The exchangeable and ferrous hyperfine-shifted proton resonances of Hb A, Hb Kempsey (β99Asp→Asn), and r Hb (β99Asp→Asn, α42Tyr→Asp) in the deoxy form are shown in Figure 5. As most of the proton resonances appearing from ~9 to ~14 ppm from DSS have been assigned to the exchangeable proton resonances from interfacial hydrogen bonds, the presence of an extra hydrogen bond in the $\alpha_1\beta_2$ interface of r Hb (β99Asp→Asn, α42Tyr→Asp) can, in principle, be demonstrated by the appearance of a new resonance in this region. The resonance at ~14 ppm from DSS of Hb A has been assigned to the intersubunit hydrogen bond between α42Tyr and β99Asp of Hb A (Fung and Ho, 1975), and is a key marker for the deoxy-quaternary structure of Hb A (Perutz, 1970; Fermi *et al.*, 1984). This resonance is completely absent from the spectra of Hb Kempsey and r Hb (β99Asp→Asn, α42Tyr→Asp), as expected. There are two broad, shoulder resonances around 15 and 23 ppm from DSS in the spectra of Hb Kempsey and r Hb (β99Asp→Asn, α42Tyr→Asp).

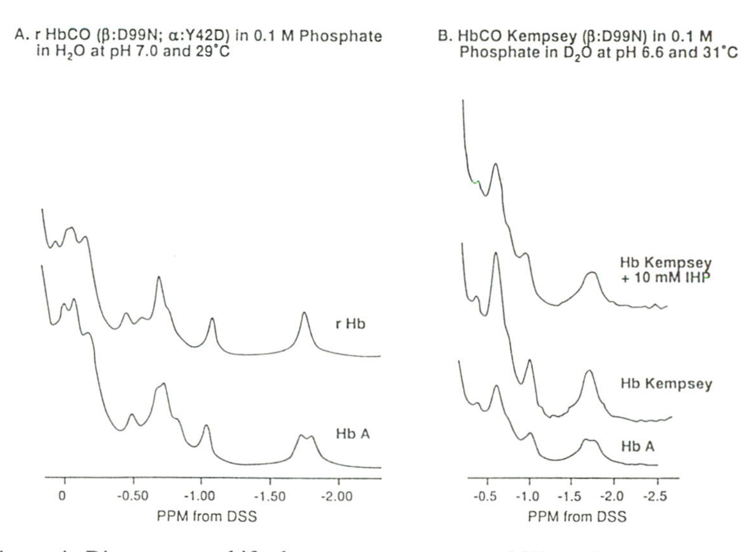

Figure 4: Ring-current shifted proton resonances of Hb in the carbonmonoxy form: (A) r Hb (β:D99N; α:Y42D) and Hb A; and (B) Hb Kempsey (β:D99N) with and without IHP. The ^{1}H NMR spectra of (A) were obtained on a Bruker AM-300 spectrometer operating at 300 MHz and were taken from Figure 3 of Kim *et al.* (1994). The experimental conditions for (A) were: [Hb] ≈ 1.8 to 3.6 mM in 0.1 M phosphate in H_2O at pH 7.0 at 29°C. The ^{1}H NMR spectra of (B) were obtained on a home-built 250-MHz spectrometer using correlation spectroscopy and were taken from Figure 2 of Lindstrom *et al.* (1973) and were reproduced from Figure 3 of Kim *et al.* (1994). The experimental conditions for (B) were: [Hb] ≈ 4 mM in 0.1 M phosphate with and without 10 mM IHP in D_2O at pH 6.6 at 31°C.

However, additional work is still needed to ascertain the origin of these two resonances. For details, see Kim *et al.* (1994). The occurrence of an exchangeable proton resonance in the spectrum of a protein molecule depends on the exchange rate between the hydrogen-bonded proton and the protons of H_2O. Thus, the absence of resonances in the exchangeable proton resonance region does not necessarily mean the absence of interfacial hydrogen bonds. One of the $\alpha_1\beta_2$ interfacial hydrogen bonds (*i.e*, the one between α97Asn and β99Asp), shown in the crystal structure of deoxy-Hb A (Fermi *et al.*, 1984), has also not been observed or identified in the exchangeable resonance region by ^1H NMR spectroscopy (Ho, 1992).

The hyperfine-shifted proton resonances of our deoxy-r Hb (β99Asp→Asn, α42Tyr→Asp) observed at 15 to 20 ppm downfield from DSS show significant changes from those of deoxy-Hb A and deoxy-Hb Kempsey (Figure 5), indicating that there are structural differences among these three Hbs. In contrast to deoxy-Hb Kempsey, where the addition of IHP can convert the spectrum to one similar to that of deoxy-Hb A in the hyperfine-shifted proton resonance

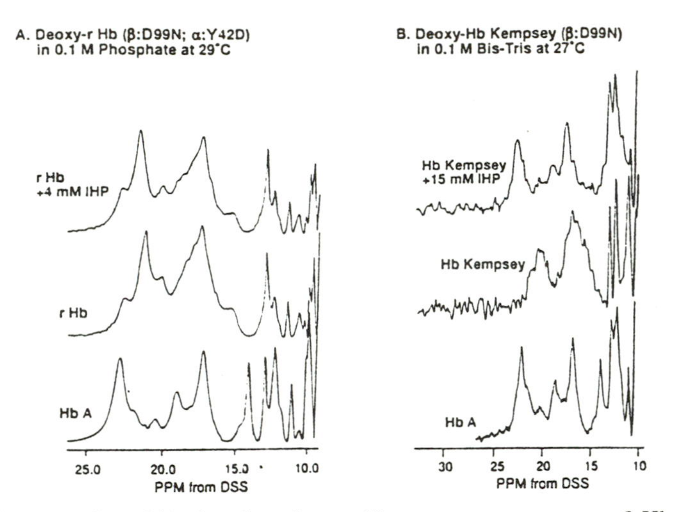

Figure 5: Hyperfine-shifted and exchangeable proton resonances of Hb in deoxy form: (A) r Hb (β:D99N; α:Y42D) in the presence and absence of 4 mM IHP, and Hb A; and (B) Hb Kempsey (β:D99N) in the presence and absence of 15 mM IHP, and Hb A. ^1H-NMR spectra of (A) were obtained on a Bruker AM-300 spectrometer operating at 300 MHz and were taken from Figure 4 of Kim *et al.* (1994). The experimental conditions for (A) were: [Hb] ≈ 1.8 to 3.6 mM in 0.1 M phosphate in H_2O at pH 7.0 and 29°C. The ^1H NMR spectra of (B) were obtained on a home-built 250-MHz spectrometer using correlation spectroscopy and were taken and modified from Figures 37 and 38 of Ho (1994) and were reproduced from Figure 4 of Kim *et al.* (1994). The experimental conditions for (B) were [Hb] ≈ 4 mM in 0.1 M Bis-Tris in H_2O at pH 7.0 at 27°C.

region, the addition of IHP to deoxy-r Hb (β99Asp→Asn, α42Tyr→Asp) does not cause any noticeable changes, as shown in Figure 5. It should also be notedthat the addition of IHP does not cause any noticeable changes in the hyperfine-shifted N_δH-exchangeable resonances of deoxy-r Hb (β99Asp→Asn, α42Tyr→Asp) (Figure 3), while the addition of IHP to Hb Kempsey causes a significant upfield shift in the histidyl resonance from the α chain (Takahashi *et al.*, 1982; Figure 3). A possible explanation is that deoxy-Hb Kempsey does not have the usual interfacial hydrogen bonds which stabilize the T-structure, *i.e.*, deoxy-Hb Kempsey exists in an oxy-like (or R-type) quaternary structure (Perutz *et al.*, 1974). Thus, the addition of a strong allosteric effector, IHP, can produce large conformational changes in the direction of the T-structure. These conformational changes are reflected in the hyperfine-shifted proton resonance region, as well as by an increase in the Hill coefficient from 1.1 to 1.7. Due to the presence of new interfacial hydrogen bonds, deoxy-r Hb (β99Asp→Asn, α42Tyr→Asp) can have a more stable deoxy-like structure, and the addition of IHP does not cause significant structural changes to affect the hyperfine-shifted proton resonances (Kim *et al.*, 1994).

The existence of significant cooperativity in r Hb (β99Asp→Asn, α42Tyr→Asp) despite large alterations in the heme environment suggests that the $\alpha_1\beta_2$ interfacial hydrogen bonds are essential for maintaining cooperativity in the oxygenation process. It is likely that alterations in the heme environment are responsible for the high oxygen affinity of this mutant Hb (Lindstrom *et al.*, 1973; Kim *et al.*, 1994). Most of the known human abnormal Hbs that have high oxygen affinity also exhibit greatly reduced cooperativity. However, the present findings on r Hb (β99Asp→Asn, α42Tyr→Asp) suggest that high oxygen affinity and reduced cooperativity may arise by different mechanisms (Kim *et al.*, 1994).

Summary

We have demonstrated that we can design mutant Hbs which have an altered $\alpha_1\beta_2$ subunit interface. A compensatory mutation for a naturally occurring abnormal human Hb, Hb Kempsey, has been designed, and this mutation allowed the molecule to regain its allosteric response. The calculated values for the difference in the free energy of cooperativity show excellent agreement with experimentally determined thermodynamic values, suggesting that the MD simulation results can be used to obtain information about the specific interactions that contribute to the total free energy of cooperativity. Our present results provide encouragement to begin a systematic investigation of the molecular basis of the subunit interactions between the α_1 and β_2 chains of Hb A by designing appropriate r Hbs. These studies could lead to the design of Hbs with desired cooperativity in the oxygenation process and to the restoration of functional properties of abnormal hemoglobins associated with hemoglobinopathies. Thus, the present results are pertinent to the use of gene therapy to treat patients with hemoglobinopathies

Acknowledgement

We thank Drs. E. Ann Pratt and Susan R. Dowd for helpful discussions. This work is supported by a grant from the National Institutes of Health (HL-24525).

References

Ben-Bassat, A., Bauer, K., Chang, S.-Y., Myambo, K., Boosman, A., and Chang, S. (1987). *J. Bacteriol.* **169**, 751.

Brooks III, C. L. (1986). *J. Phys. Chem.* **90**, 6680.

Brooks III, C. L., and Karplus, M. (1989). *J. Mol. Biol.* **208**, 159.

Bunn, H. F., Wohl, R. C., Bradley, T. B., Cooley, M., and Gibson, Q. H. (1974). *J. Biol. Chem.* **249**, 7402.

Bunn, H. F., and Forget, B. G. (1986). *Hemoglobin; Molecular, Genetic, and Clinical Aspects.* W. B. Saunders, Philadelphia.

Dalvit, C., and Ho, C. (1985). *Biochemistry* **24**, 3398.

Dang, L. X., Merz, K. M, Jr., and Kollman, P. A. (1989). *J. Am. Chem. Soc.* **111**, 8505.

Dickerson, R.E., and Geis, I. (1983). *Hemoglobin: Structure, Function, Evolution, and Pathology.* The Benjamin Cummings Publishing Co., Menlo Park, California.

Fermi, G., Perutz, M.F., Shaanan, B., and Fourme, R. (1984). *J. Mol. Biol.* **175**, 159.

Gao, J., Kuczera, K., Tidor. B., and Karplus, M. (1989). *Science* **244**, 1069.

Glynn, K. P., Penner, J. A., Smith, J. R., and Rucknagel, D. L. (1968). *Ann. Int. Med.* **69**, 769.

Heiner, A. P., Berendsen, H. J., and van Gunsteren, W. F. (1993). *Protein Eng.* **4**, 397.

Ho, C. (1992). *Advan. Protein Chem.* **43**, 153.

Ho, C., and Perussi, J. R. (1994). *Methods Enzymol.* **232**, 97.

Hoffman, S. L., Looker, R. D., Roehrich, J. M., Cozart, P. E., Durfee, S. L.,Tedesco, J., and Stetler, G. (1990). *Proc. Natl. Acad. Sci. U.S.A.* **87**, 8521.

Imai, K., Fushitani, K., Miyazaki, G., Ishimori, K., Kitagawa, T., Wads, Y., Morimoto, H., Morishima, I., Shih, D. T.-b., and Tame, J. (1991). *J. Mol. Biol.* **218**, 769.

Ishimori, K., Morishima, I., Imai, K., Fushitani, K., Miyazaki, G., Shih, D. T.-b., Tame, J., Pagnier, J., and Nagai, K. (1989). *J. Biol. Chem.* **264**, 14624.

Jones, R. T., Osgood, E. E., Brimhall, B., and Koler, R. D. (1967). *J. Clin. Invest.* **46**, 1840.

Jorgensen, W. L., Chandrasekar, J., Madura, J. D., Impey, R. W., and Klein, M. L. (1983). *J. Chem. Phys.* **83**, 3050.

Karplus, M., and Petsko, G. A. (1990). *Nature* **347**, 631.

Kim, H.-W., Shen, T.-J., Sun, D. P., Ho, N. T., Madrid, M., Tam, M. F., Zou, M., Cottam, P. F., and Ho, C. (1994). *Proc. Natl. Acad. Sci. U.S.A.* **91** 11547.

Kirkwood, J. G. (1935). *J. Chem. Phys.* **3**, 300.

Komeiji, Y., Uebayasi, M., Someya, J.-I., and Yamamonto, I. (1992). *Protein Eng.* **5**, 759.

La Mar, G. N., Nagai, K., Jue, T., Budd, D. L., Gersonde, K., Sick, H., Kagimoto, T., Hayashi, A., and Taketa, F. (1980). *Biochem. Biophys, Res. Commun.* **96**, 1172.

Lau, F. T. K., and Karplus, M. (1994). *J. Mol. Biol.* **236**, 1049.

Lindstrom, T. R., Noren, I. B. E., Charache, S., Lehman, H., and Ho, C. (1972). *Biochemistry* **11**, 1677.

Lindstrom, T. R., Baldassare, J. J., Bunn, H. F., and Ho, C. (1973). *Biochemistry* **12**, 4212.

Perutz, M. F. (1970). *Nature* (London) **228**, 726.

Perutz, M. F., Ladner, J. E., Simon, S. R., and Ho, C. (1974). *Biochemistry* **13**, 2163.

Ponder, J. W., and Richards, F. M. (1987). *J. Mol. Biol.* **193**, 775.

Prevost, M., Wodak, S. J., Tidor, B., and Karplus, M. (1991). *Proc. Natl. Acad. Sci. U.S.A.* **88**, 10880.

Reed, C. S., Hampson, R., Gordon, S., Jones, R. T., Novy, M. J., Brimhall, B., Edwards, M. J., and Koler, R. D. (1968). *Blood* **31**, 623.

Ryckaert, J.-P., Ciccotti, G., and Berendsen, H. J. C. (1977). *J. Comput. Phys.* **23**, 327.

Shen, T.-J., Ho, N. T., Simplaceanu, V., Zou, M., Green, B. N., Tam, M.F., and Ho, C. (1993). *Proc. Natl. Acad. Sci. U.S.A.* **90**, 8108.

Srinivasan, R., and Rose, J. D. (1994). *Proc. Natl. Acad. Sci. U.S.A.* **91**, 11113.

Takahashi, S., Lin, A. K.-L., and Ho, C. (1980). *Biochemistry* **19**, 5196.

Takahashi, S., Lin, A. K.-L., and Ho, C. (1982). *Biophy. J.* **39**, 33.

Thillet, J., Arons, N., and Rosa, J. B. (1981). *Biochim. Biophys. Acta* **660**, 260.

Tidor, B., and Karplus, M. (1991). *Biochemistry* **30**, 3217.

Turner, G. J., Galacteros, F., Doyle, M. L., Hedlund, B., Pettigrew, D. W., Turner, B. W., Smith, F. R., Moo-Penn, W., Rucknagel, D. L., and Ackers, G. K. (1992). *Proteins* **14**, 333.

Weatherall, D. J., Clegg, J. B., Callender, S. T., Wells, R. M. G., Gale, R E., Huehns, E. R., Perutz, M. F., Viggiano, G., and Ho, C. (1977). *British. J. Haematol.* **35**, 177.

8

The Role of NMR Spectroscopy in Understanding How Proteins Fold

C.M. Dobson

Oxford Centre for Molecular Sciences
University of Oxford
Oxford OX1 3QT U.K.

Proteins are synthesized within the cell on ribosomes. Although there is debate as to the beginnings of folding, it is clear that the major events in the folding process of a protein occur following departure from the ribosome. Folding may involve a series of auxiliary proteins, including molecular chaperones, and for extracellular proteins may occur in part following secretion from the cell itself (Ellis, 1994). Nevertheless, many proteins also fold efficiently and correctly in isolation, for example, following transfer from a denaturing medium to a medium in which the native state is thermodynamically stable (Anfinsen, 1973). It seems most unlikely, given the improbability that folding could occur in a finite time on a random search basis (Levinthal, 1968), that the principles behind the folding process differ fundamentally in the two situations (*in vivo* and *in vitro*). Studies of the molecular basis of protein folding are therefore appropriately initiated *in vitro*, where physical techniques capable of providing detailed structural information can be used most readily and where folding of molecules can be examined in isolation (Evans and Radford, 1994).

It has long been recognized that NMR spectroscopy, with its ability to define protein structure and dynamics in solution, is ideally suited as a technique for studying the structural transitions that take place during folding. The rapidity of folding of small proteins under most conditions, however, has until recently limited its direct application in 'real time' kinetic studies. Early applications of NMR in folding studies therefore included investigations of the equilibrium between folded and unfolded states, and a search for stable intermediate species (Jardetzky *et al.*, 1972). This approach has in fact become very important in recent years with the discovery that a wide range of stable partially structured states can be generated under carefully chosen conditions, and with the development of heteronuclear NMR techniques that make possible their detailed characterisation (Dobson, 1994). The most famous of these partially folded states are known as 'molten globules', compact species with extensive secondary structure but lacking persistent tertiary interactions; these are of particular interest as they appear to be closely linked to intermediates observed in kinetic refolding experiments (Ptitsyn, 1995). The use of magnetisation transfer methods and hydrogen exchange techniques to relate the structure of these intermediates to that of the native states has extended the scope of NMR in this area (Baum *et al.,* 1989). Moreover, the coupling of hydrogen exchange methods with quenched flow procedures has enabled NMR to be used to probe transient as well as stable intermediates (Roder *et al.*, 1988; Udgaonkar and Baldwin, 1988). It has been this development, in particular, that has generated major advances in kinetic folding studies, particularly when combined with recently developed mass spectrometric procedures and complementary stopped flow optical techniques (Evans and Radford, 1994).

Steps along an *in vitro* folding pathway

Overview

Proteins can be unfolded *in vitro* most readily by addition of chemical denaturants such as urea or guanidinium chloride in high concentration. The chemical basis for denaturation is not yet understood in any detail, although it seems evident that a major factor involves the increased ability of water to solvate non-polar amino acids, and hence reduce the contribution of hydrophobic interactions to the stability of the native state (Dill and Shortle, 1991). It is clear that in many cases the idea that a random coil model describes the properties of such states is only an approximation, but NMR studies have shown that the residual structure present in highly denatured states is generally rather localized, and involves clusters of hydrophobic groups and transient, rather than well-defined, elements of secondary structure or persistent long-range interactions (Wüthrich, 1994). *In vitro* folding is conventionally initiated by dilution from solutions with high denaturant concentrations in order to generate an environment in which the native state is thermodynamically more stable than the denatured state. The indication from experimental studies is that the refolding of proteins may not be critically dependent on the means by which the denatured state is generated, unless extensive residual structure remains, although it is strongly dependent on the conditions under which refolding takes place (Kotik *et al.*, 1995).

Experimental studies of protein folding are involving an increasing number of proteins and a widening range of techniques and methodologies, although relative few proteins have yet been studied by a large number of different techniques. One group of proteins that have been investigated in particular detail in our laboratory is the c-type lysozymes (Radford *et al.*, 1992; Dobson *et al.*, 1994; Radford and Dobson, 1995). This article refers particularly to results from the investigation of these proteins and primarily to the archetypal member of the family, the hen protein (Figure 1). The experimental evidence from a range of kinetic refolding studies is summarised in Figure 2, and a schematic folding pathway based on this is given in Figure 3. One of the features of the folding of this particular protein is that a variety of distinct events can be very clearly observed. In the first of these, secondary structure is formed and at least partial collapse of the structure occurs (to give the "interconverting collapsed states" in Figure 3). In the second, regions of structure assemble into persistent units or domains (to give the "protected states" in Figure 3), and in the third the final assembly and reorganisational events take place to generate persistent tertiary interactions and hence the native structure.

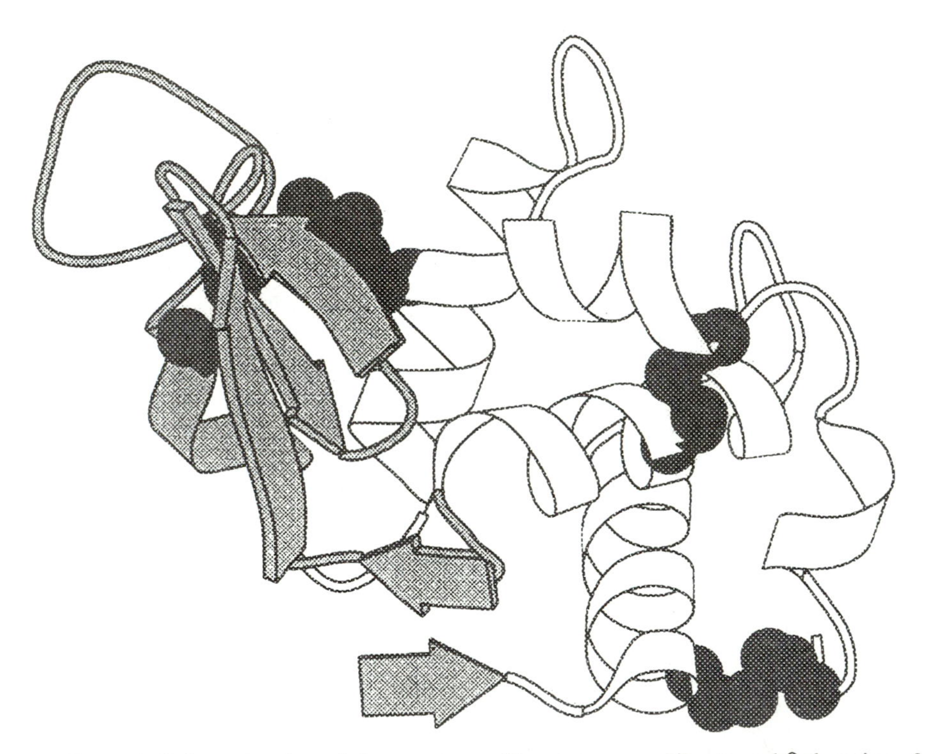

Figure 1: Schematic view of the structure of hen lysozyme. The α- and β-domains of the protein are shaded white and grey, respectively. The overall fold is the same for other c-type lysozymes and α-lactalbumins. The diagram was drawn using the program Molscript and is taken from Radford and Dobson, 1995.

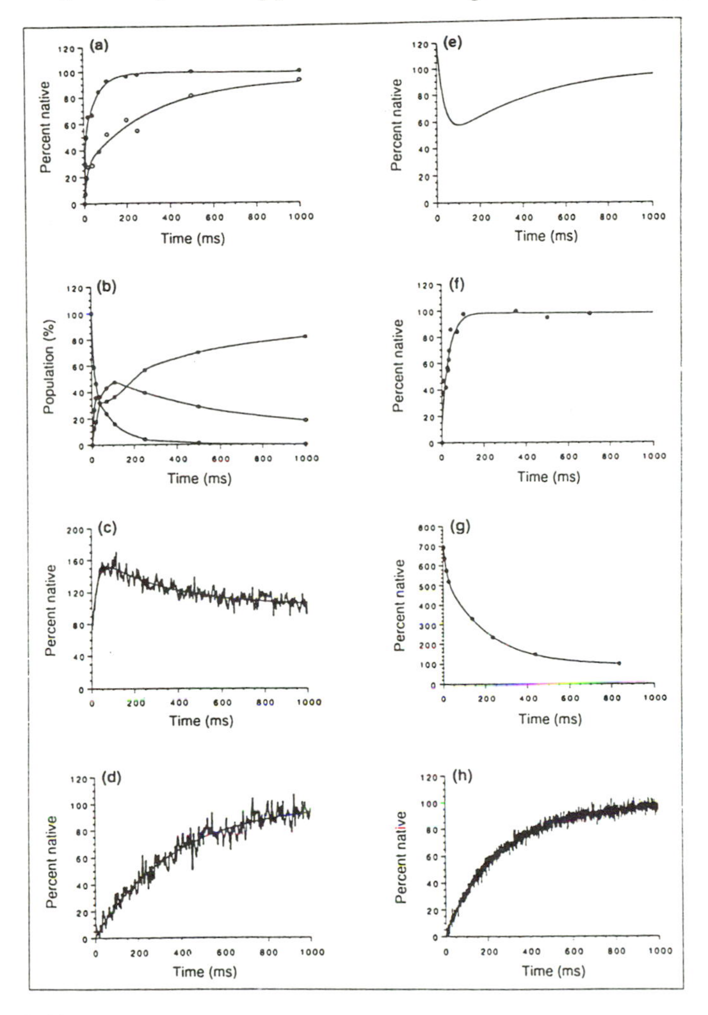

Figure 2: The folding kinetics of hen lysozyme initiated by 11-fold dilution (pH 5.5) from 6M guanidine hydrochloride at 20 °C. (a) Hydrogen exchange labeling and two-dimensional NMR; average protection of amides in the α- (•) and β-domains (o) is shown. (b) Hydrogen exchange labeling and electrospray mass spectrometry (ESMS); unprotected state (•); species substantially protected only in the a- domain (o); species with native-like protection (n). (c) Far UV CD (225nm). (d) Near UV CD (289 nm). (e) Intrinsic tryptophan fluorescence. (f) Quenching of fluorescence with iodide ions. (g) Binding of 1-anilino napthalene sulfonate (ANS). (h) Binding of a fluorescent inhibitor (4-methyl umbelliferyl-N,N' diacetyl-β-chitobiose). Values are expressed as a percentage of the value of the native state, which is arbitrarily given a value of 100, except for (b) which represents the percentage of each species, given a total population of 100%. The contribution of about 20% of very slow folding molecules, which probably arise as a consequence of *cis/trans* proline isomerism, has been ignored. Reproduced from Dobson *et al.* (1994).

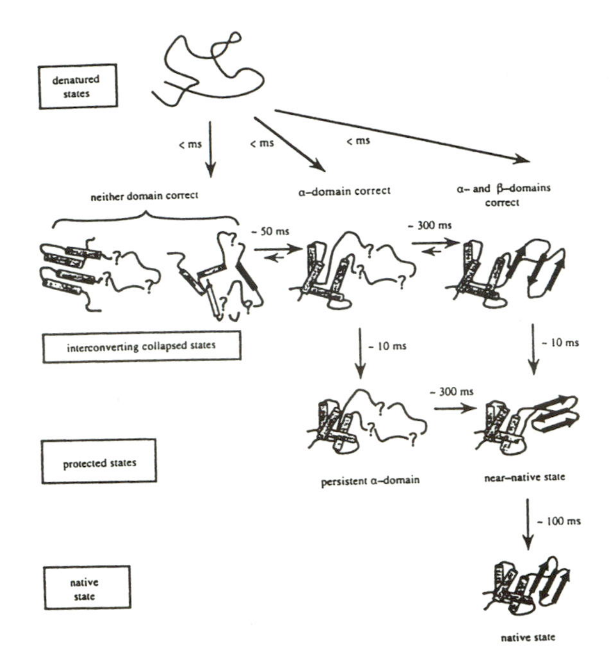

Figure 3: Schematic representation of a possible folding pathway of hen lysozyme based on the experimental data shown in Figure 2. The majority of molecules pass through the species of the left hand side of the picture, whilst a minority of molecules folds more rapidly. This kinetic heterogeneity has been attributed to the formation of both correctly and misfolded structures during the collapse process (Radford *et al.*, 1992; Dobson *et al.*, 1994; Radford and Dobson, 1995). From Radford and Dobson (1995).

Collapse and secondary structure formation

Studies using far UV CD indicate that extensive secondary structure is formed very rapidly (milliseconds or less) after the initiation of folding and prior to tertiary interactions detectable by near UV CD or fluorescence, Figure 1. A major question at once arises as to how such extensive structure can develop so rapidly, given that intrinsic preferences of one type of secondary structure over another for regions of polypeptide chains are often marginal and that the population of molecules in aqueous solution having such structure in isolation is, even in the most favorable cases, usually small (Dyson and Wright, 1993). It is possible that the earliest stages of folding involve nucleation sites, short regions of the polypeptide chain that can become structured in the absence of interactions with the remainder of the protein, which form and then generate larger structures either by propagation from these points (Wetlaufer, 1973) or by "diffusion-collision", in which independently formed labile regions of structure interact and stabilize each other (Karplus and Weaver, 1994).

A somewhat different view is that the environment generated in the folding process itself could play a key role in promoting the formation of secondary structure. The results of experimental studies of intrinsic fluorescence and of binding of hydrophobic dyes such as ANS (see Figure 2) indicate that substantial global collapse of the protein structure takes place very early in folding (Ptitsyn *et al.*, 1990) driven by the energetic advantage of removing apolar residues from an aqueous environment. On this model of folding, collapse and the formation of secondary structure are intimately linked (Chan and Dill, 1990) and that they act in a synergic manner to generate a partially structured state with at least a rudimentary hydrophobic core. The collapsed state effectively amplifies the intrinsic structural preferences of given regions of the polypeptide chain and stabilizes certain types of structure over others. For a sequence of amino acids corresponding to a native protein, this process will drive the structure efficiently toward the native fold (Dobson *et al.*, 1994).

At this stage of folding, even though the extent of secondary structure present in the collapsed state of the protein is extensive, it appears to be flexible and rather disordered. Pulse labeling studies, for example, show that hydrogens, even of those amides involved in secondary structure, are not significantly protected against solvent exchange; an almost complete complement of secondary structure appears to be formed at least an order of magnitude faster than substantial protection of amide hydrogens takes place, Figure 2. Moreover for hen lysozyme the aromatic residues have not yet become sufficiently immobilized to give intense signals in the near UV CD, although in the case of human lysozyme a fraction of the native-like signal is formed very early in folding (Hooke *et al.*, 1994). Some further indication of the nature of the protein structure at this stage of folding is emerging from studies of stable compact denatured states (molten globules) formed under mild denaturing conditions (Dobson, 1994; Ptitsyn, 1995). In the case of the lysozymes, the most studied stable, partially-folded states are those of the homologous α-lactalbumins. A variety of studies of these have provided clear evidence for native-like secondary structure and for the presence of at least a rudimentary native-like fold (Chyan *et al.*, 1993; Peng and Kim, 1994) suggesting that these partly folded species are related to intermediates on kinetic pathways. Such a conclusion has been drawn particularly clearly for apo-myoglobin, where the pattern of hydrogen exchange protection in a stable intermediate has been shown to be similar to that deduced for kinetic intermediates in pulse labeling studies (Jennings and Wright, 1993).

One extremely important general feature emerging from the study of the stable partly folded states of α-lactalbumins is their dynamic nature. Even though they have a compactness comparable to the native protein, their hydrogen exchange protectional, though measurable (and indeed valuable for defining aspects of their structure) is much less than for the native proteins or for species detected in pulse labeling experiments. In addition, NMR experiments show that their sidechains are extremely disordered, and persistent tertiary contacts appear to be limited (Baum *et al.*, 1989; Chyan *et al.*, 1993), Figure 4. Interestingly, denatured states of proteins with a high degree of sequence and structural homology may differ significantly in their ability to form a collapsed state stable enough to exist at equilibrium in preference to highly unfolded states. The c-type lysozymes and α-lactalbumins are classic examples of this, as the stability of molten globules formed at equilibrium appears in general to be substantially greater for the latter than the former (Kuwajima, 1989). Stabilization of a collapsed state relative to more unfolded states under different conditions appears to depend significantly on the amino

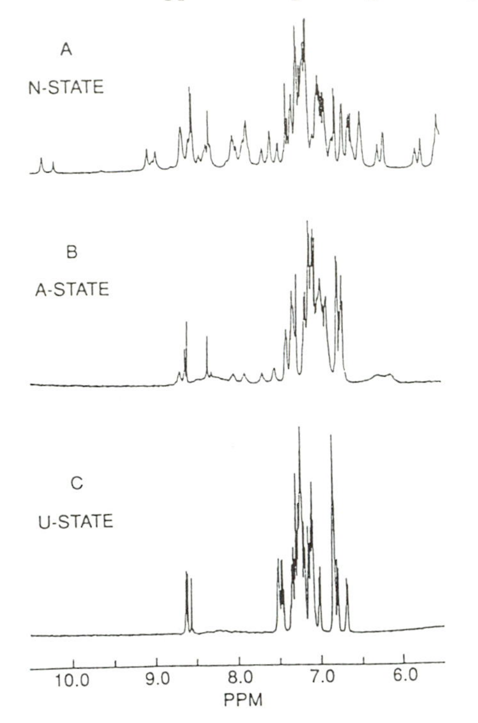

Figure 4: Part of the NMR spectra of guinea pig α-lactalbumin at (A) pH 5.4 (the native state), (B) pH 2.0 (the molten globule or A-state) and (C) pH 2.0 in the presence of 9 M urea (the unfolded state). In each case the protein was dissolved in ^2H_2O for 6 h prior to recording the spectrum. The resonances in the spectra shown corresponds to protons from aromatic residues and from amide protons that have not exchanged with deuterons from the ^2H_2O solvent. In the case of the A-state the most protected of these have been identified as arising from two of the regions which in the native state form helices (B and C) in the α-domain of the protein. The overall similarity between the spectra of the A-state and the unfolded state indicates that the ordered environment for sidechains characteristic of the native state is lost in the partly folded protein. (From Baum *et al.*, 1989).

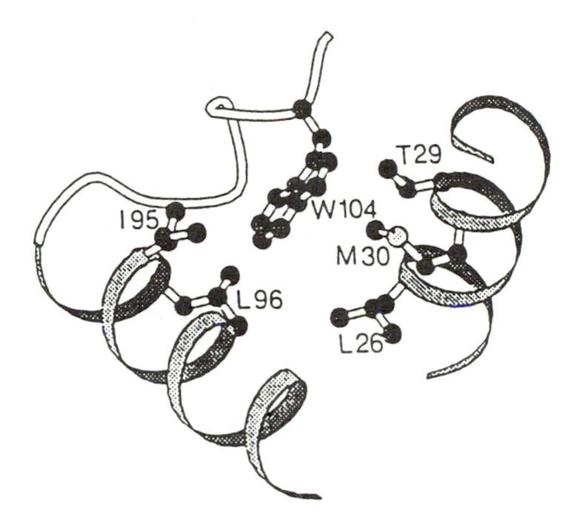

Figure 5: Tertiary interactions involving some of the residues forming the B- and C-helices of the native state of α-lactalbumin. That the most slowly exchanging amides of the A-state are located in this region of the protein suggests that similar, although less persistent, interactions could be maintained in a rudimentary hydrophobic core of this partially folded state. From Chyan *et al.*, 1993.

acid composition and sequence, and on the existence of stabilizing factors such as cross-links between distinct regions of the polypeptide chain. In the case of lysozyme, the formation of disulfide bonds during oxidative refolding is likely to be an important factor in the stabilization of intermediates, although evidence suggests that such bonds are not the driving force in structure formation. Even a rather disordered rudimentary collapsed state formed rapidly can, however, be a key intermediate if it is the means by which the unfolded protein is directed efficiently toward more stable species on a folding pathway, Figure 5 (Dobson *et al.*, 1994; Morozova *et al.*, 1995).

Stabilization of structural units

The second major stage of folding involves the conversion of the disordered collapsed state into more persistent structural units, and the coming together and mutual stabilization of distinct elements of secondary structure, Figure 3. A particularly intriguing feature of the proposed pathway for hen lysozyme is the indication that key steps at this stage of folding appear to involve the reorganization of initially formed species to those with sufficient native-character to proceed with folding (Radford *et al.*, 1992; Dobson *et al.*, 1994; Radford and Dobson, 1995). Such rearrangement processes could occur even within substantially collapsed states, provided they have the fluidity discussed above. For small proteins stabilization of the secondary structure may well require all of the elements of the native state to come together in order for a persistent stable folding unit to form (Kragelund *et al.*, 1995). For larger proteins, however, distinct regions of the protein may become stabilized prior to others; that this occurs in hen lysozyme is indicated by pulse labeling hydrogen exchange experiments (Figure 2) showing that the two structural domains of the native structure (Figure 1) fold to a highly protective state with very different kinetics (Radford *et al.*, 1992; Dobson *et al.*, 1994; Radford and Dobson, 1995). Interestingly in the case of human lysozyme three such regions can be identified, the α-domain forming in two distinct steps (Hooke *et al.*, 1994). There is, however, clear evidence for a high degree of cooperativity in the formation of such domains, and evidence that their emergence stabilizes elements of otherwise flexible structure. More generally, for all the proteins studied by pulse labeling methods substantial hydrogen exchange protection seems to occur only when a number of elements of secondary structure (helices or strands) come together (Kragelund *et al.*, 1995). It is at this stage of folding that the overall topologies of proteins are likely to become firmly established.

An important feature of species formed at this stage of folding is that any structure formed, even in well-defined persistent domains, is not fully native-like in character. For example, although the presence of a native-like secondary complement of structure in hen lysozyme is clearly indicated by far UV CD and hydrogen exchange measurements, the sidechain packing as indicated by near UV CD or fluorescence measurements is not organized to the extent characteristic of native globular proteins (Figure 2). The intermediate states appearing at this second stage of folding therefore again manifest characteristics of molten globules, albeit ones having greater order, as reflected for example in much higher protection against hydrogen exchange, than those formed in the first stage of folding. Although stable analogues of these states have not yet been found for lysozyme or its homologues, two structures of stable, partly-folded states that might have characteristics of such late folding intermediates have recently been determined by NMR methods (Redfield *et al.*, 1994; Feng *et al.*, 1994). The structures of both of these show well defined regions of secondary structure linked by regions of significant disorder, and support the view that the critical interactions for stability involve largely hydrophobic contacts between secondary structure units. The formation of these contacts in a stabilizing manner is likely to be crucial in maintaining the direction of folding toward the native state.

Acquisition of the native structure

The final stage in folding involves the transition of partially folded intermediates to the compact native state and is likely to involve a number of events that may or may not be correlated. In the case of the lysozymes, with well-defined folding domains, a crucial event appears to be the interaction or docking of these to form the overall topology of the protein, Figure 2. In all cases two other events are likely to occur at this stage. One is the specific packing of sidechains and the loss of conformational freedom, at least for the interior residues. Even a small expansion from the highly compact native state is likely to generate substantial disorder in sidechain interactions. This is indicated particularly clearly in NMR studies of the dynamics of the native states themselves where distinct correlations between rigidity and incorporation in the close packed interior of the protein have been found (Buck *et al.*, 1995). The second event involves the expulsion of residual solvent molecules from the interior of the protein. Interestingly, experimental studies indicate that these final steps in folding, about which rather little is known, may well be relatively slow for some proteins, despite the fact that only relatively small conformational rearrangements appear be involved. This could be because highly specific interactions involving many groups in the protein need to be correctly established in states sufficiently closely packed that significant barriers to reorganisation may exist. In the case of small proteins with a single folding domain, however, it appears that these events may be very efficient, and that the rate limiting steps occur earlier in the folding process(Kragelund *et al.*, 1995; Jackson and Fersht, 1991). Although much remains to be learned about this aspect of folding, considerable insight into the nature of transition states for folding is now being gained for several proteins, notably barnase, from studies of the influence of site-specific mutagenesis on protein stability and on folding and unfolding kinetics (Fersht, 1995). The evidence for barnase is that the transition state occurs late in folding and has a very high degree of native-like structure.

Concluding remarks

In recent years progress both in experimental and theoretical aspects of folding has been remarkable. In the case of *in vitro* folding, the existence of different stages in folding is now well established for a number of proteins (Matthews, 1993). In the early stages of folding it seems likely that distinctive properties of individual polypeptide chains, such as the pattern of hydrophobic and hydrophilic residues and secondary structure propensities, are crucial in allowing rapid and efficient formation of a collapsed state with a substantial degree of secondary structure and the correct overall topology. Such a state substantially restricts the volume of conformational space that needs to be searched to find the correct fold, thereby substantially solving the Levinthal paradox. In the later stages of folding, partially structured near native states need to become organised into the close packed structures characteristic of fully folded proteins. It is possible that significant barriers to conformational rearrangements could develop to become rate determining in the folding of some proteins, and that misfolding events may be responsible for slow folding events in others (Radford *et al.*, 1992; Dobson *et al.*, 1994; Radford and Dobson, 1995; Sosnik *et al.,* 1994). The further development of NMR techniques to permit the structural details of these events to be defined in greater detail will be crucial to enhance our understanding of these events.

Even though proteins fold *in vivo* in the presence of catalytic and protective devices, these are likely to serve to control and modulate the various stages of the folding process rather than to determine its fundamental character (Ellis, 1994). It is, however, of great interest to investigate how the various auxiliary factors might moderate or control the process in the light of our understanding of the refolding of proteins in isolation. Efforts to extend the use of physical techniques to enable these objectives to be achieved are increasingly being made. For example, initial experiments in which hydrogen exchange behaviour is monitored by NMR or by mass spectrometry have been used to suggest that the molecular chaperone GroEL binds to species analogous to those formed in the very early stages of *in vitro* folding (Zahn *et al.,* 1994; Robinson *et al.*, 1994).

As protein folding is usually fast, extremely accurate, and often rather insensitive to changes in conditions and even amino acid sequence, it seems likely that it must depend on rather simple universal characteristics of protein sequences and structures, and theoretical models are being developed to explore the nature of these (Karplus and Sali, 1995). Experiments such as those discussed here are beginning to give insights into at least some of the determinants of folding. NMR spectroscopy, as anticipated in the earliest studies of proteins, has been the key technique in this progress. The full potential of this technique is, however, only just beginning to be realized. For example, coupling of rapid mixing techniques directly with NMR spectroscopy is beginning to enable real-time experiments to be carried out rapidly enough to detect even early folding intermediates. This approach in particular promises to allow the full power of NMR in determining structures and dynamics at the atomic level to be applied to study protein folding (Balbach *et al.*, 1995).

Acknowledgements

I should like to acknowledge the invaluable contribution to the ideas described in the article by the many students, post-doctoral fellows and colleagues whose names appear in the published articles on which this paper is based. This is a contribution from the Oxford Centre for Molecular Sciences that is supported by the EPSRC, and BBSRC. The research of CMD is also supported in part by an International Research Scholars award from the Howard Hughes Medical Institute.

References

Anfinsen, C.B. (1973). *Science* **181**, 223.

Balbach, J., Forge, V., van Nuland, N.A.J., Winder, S., Hore, P.J., and Dobson, C.M. *Following Protein Folding in Real Time using NMR Spectroscopy,*

submitted for publication.
Baum, J., Dobson, C.M., Evans, P.A., and Hanley, C. (1989). *Biochemistry* **28**, 7.
Buck, M., Boyd, J., Redfield, C., MacKenzie, D.A., Jeenes, D.J., Archer, D.B., and Dobson, C.M. (1995). *Biochemistry* **34**, 4041.
Chan, H.S., and Dill, K.A. (1990). *Proc. Natl. Acad. Sci. U.S.A.* **87**, 6388.
Chyan, C.L., Wormald, C., Dobson, C.M., Evans, P.A., and Baum, J. (1993). *Biochemistry* **32**, 5681.
Dill, K.A., and Shortle, D. (1991). *Annu. Rev. Biochem.* **60**, 795.
Dobson, C.M. (1994). *Curr. Biol.* **4**, 636.
Dobson, C.M., Radford, S.E., and Evans, P.A. (1994). *Trends Biochem. Sci.* **19**, 31.
Dyson, H.J., and Wright, P.E. (1993). *Curr. Op. Struct. Biol.* **3**, 60.
Ellis, R.J. (1994). *Curr. Op. Struct. Biol.* **4**, 117.
Evans, P.A., and Radford, S.E. (1994). *Curr. Op. Struct. Biol.* **4**, 100.
Feng, Y., Sligar, S.G., and Wand, A.J. (1994). *Nature Struct. Biol.*, **1**, 30.
Fersht, A.R. (1995). *Phil. Trans. R. Soc. Lond. B* **348**, 11.
Hooke, S.D., Radford, S.E., and Dobson, C.M. (1994). *Biochemistry* **33**, 5867.
Jackson, S.E., and Fersht, A.R. (1991). *Biochemistry* **30**, 10428.
Jardetzky, O., Thielman, H., Arata, Y., Markley, J.L., and Williams, M.N. (1972). *Cold Spring Harb. Symp. Quant. Biol.* **36**, 257.
Jennings, P.A., and Wright, P.E. (1993). *Science* **262**, 892.
Karplus, M., and Sali, A. (1995). *Curr. Op. Struct. Biol.* **5**, 58.
Karplus, M., and Weaver, D.L. (1994). *Protein Sci.* **3**, 650.
Kotik, M., Radford, S.E., and Dobson, C.M. (1995). *Biochemistry* **34**, 1714.
Kragelund, B.B., Robinson, C.V., Knudsen, J., Dobson, C.M., and Poulsen, F.M. (1995). *Biochemistry* **34**, 7217.
Kuwajima, K. (1989). *Proteins* **6**, 87.
Levinthal, C. (1968). *Chim. Phys.* **65**, 44.
Matthews, C.R. (1993). *Annu. Rev. Biochem.* **62**, 653.
Morozova, L.A., Haynie, D.T., Arico-Muendel, C., Van Dael, H., and Dobson, C.M. (1995). *Nature Struct. Biol., in press.*
Peng, Z.Y., and Kim, P.S. (1994). *Biochemistry* **33**, 2136.
Ptitsyn O.B. (1995). *Curr. Op. Struct. Biol.* **5**, 74.
Ptitsyn, O.B., Pain, R.H., Semisotnov, G.V., Zerovnik, E., and Razgulyaer O.I. (1990). *FEBS Lett.* **262**, 20.
Radford, S.E., and Dobson, C.M. (1995). *Phil. Trans. R. Soc. Lond. B.* **348**, 17.
Radford, S.E., Dobson, C.M., and Evans, P.A. (1992). *Nature Lond.* **358**, 302.
Redfield, C., Smith, R.A.G., and Dobson, C.M. (1994). *Nature Struct. Biol.* **1**, 23.
Robinson, C.V., Grob, M., Eyles, S.J., Mayhew, M., Hartl, F.U., Dobson, C.M., and Radford, S.E. (1994). *Nature Lond.* **372**, 646.
Roder, H., Elöve, G.A., and Englander, S.W. (1988). *Nature Lond.* **335**, 700.
Sosnik, T.R., Mayne, L., Hiller, R., and Englander, S.W. (1994). *Nature Struct. Biol.* **1**, 149.
Udgaonkar J.B., and Baldwin R.L. (1988). *Nature* **335**, 694.
Wetlaufer, D.B. (1973). *Proc. Natl. Acad. Sci. U.S.A.* **70**, 697.
Wüthrich, K. (1994). *Curr. Op. Struct. Biol.* **4**, 93.
Zahn, R., Spitzfaden, C., Ottiger, M., Wüthrich, K., and Plückthun, A. (1994). *Nature* **368**, 261.

9

NMR Approaches To Understanding Protein Specificity

G.C. K. Roberts, L.-Y. Lian, I.L. Barsukov, S. Modi, and W.U. Primrose

Biological NMR Centre and Department of Biochemistry
University of Leicester
Leicester LE1 9HN U.K.

The biological functions of proteins all depend on their highly specific interactions with other molecules, and the understanding of the molecular basis of the specificity of these interactions is an important part of the effort to understand protein structure-function relationships. NMR spectroscopy can provide information on many different aspects of protein-ligand interactions, ranging from the determination of the complete structure of a protein-ligand complex to focussing on selected features of the interactions between the ligand and protein by using "reporter groups" on the ligand or the protein. It has two particular advantages: the ability to study the complex in solution, and the ability to provide not only structural, but also dynamic, kinetic and thermodynamic information on ligand binding. Early analyses of ligand binding (Jardetzky and Roberts, 1981) focused on measurements of relaxation times, chemical shifts and coupling constants, which gave relatively limited, although valuable, structural information. More recently, it has become possible to obtain much more detailed information, due to the extensive use of nuclear Overhauser effect measurements and isotope-labeled proteins and ligands; a number of reviews of this area are available (Feeney and Birdsall, 1993; Lian *et al.*, 1994; Wand and Short, 1994; Petros and Fesik, 1994; Wemmer and Williams, 1994). In this article, we describe some recent work from our laboratory which illustrates the use of NMR spectroscopy to obtain structural and mechanistic information on relatively large enzyme-substrate and protein-protein complexes.

Protein-protein interactions of bacterial antibody-binding proteins: identification of the interaction surface by chemical shift changes

A number of species of pathogenic bacteria, notably *Streptococci* and *Staphylococci*, have proteins on their surface that bind immunoglobulins (reviewed in Boyle (1990)). Protein A from *S. aureus* and protein G from species of *Streptococci* are widely used as immunological tools and are the most extensively studied of these antibody-binding proteins. A detailed understanding of the binding mechanisms of these proteins is important, not only for providing us with the structural basis for their functions, but also as a contribution toward understanding the general rules of protein-protein interactions.

Protein A contains five highly homologous Fc binding domains, each of about 60 amino acid residues (Langone, 1982; Moks *et al.*, 1986), which bind to the Fc portion of immunoglobulin G (IgG) with an affinity which varies with the species and subclass of IgG. The C terminal half of protein G contains three IgG binding domains, referred to as domains I, II, and III, each consisting of 55 residues and separated from the others by short linker sequences. These domains are closely homologous to one another, but they show no sequence similarity to those of protein A (Guss *et al.*, 1986; Olsson *et al.*, 1987). Protein G has a broader specificity than protein A for IgGs from different sources, and its IgG-binding domains are able to bind to both the Fab and the Fc portions of the anti body molecule (Eliasson *et al.*, 1991; Derrick and Wigley, 1992).

The IgG-binding domains from protein A and protein G have different three-dimensional structures. The solution structure of the B domain of protein A determined by NMR (Gouda *et al.*, 1992; Torigoe *et al.*, 1990) shows it to be a three-helix bundle (Figure 1). We have previously reported the solution structure determination by ^{1}H NMR of domains II and III from protein G of *Streptococcus* strain G148 (Lian *et al.*, 1991; Lian *et al.*, 1992), and the crystal structure of domain III has recently been determined (Derrick and Wigley, 1994). Each of the IgG-binding domains was found to consist of an α-helix packed against a four-stranded antiparallel-parallel-antiparallel β-sheet (Figure 1). Essentially identical structures have been determined by others for a different IgG binding domain of protein G from another *Streptococcus* strain by NMR (Gronenborn *et al.*, 1991; Orban *et al.*, 1992) and X-ray crystallography (Achari *et al.*, 1992; Gallagher *et al.*, 1994).

The binding of protein G to Fc fragments is competitive with respect to protein A (Stone *et al.*, 1989; Frick *et al.*, 1992), suggesting that the binding sites for protein A and protein G on Fc overlap, notwithstanding the fact that they lack sequence or structural similarity. At the same time, the small antibody-binding domains of protein G are able to bind to both Fab and Fc

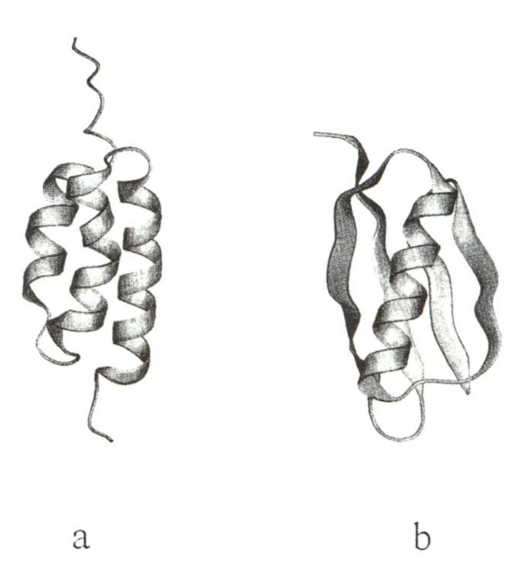

Figure 1: Comparison of the solution structures of IgG-binding domains of (a) protein A (Langone, 1982; Moks *et al.*, 1986) and (b) protein G (Guss *et al.*, 1986; Olsson *et al.*, 1987).

fragments of IgG. There are thus interesting questions about the structural origins of the specificity of protein-protein interactions in this system. The molecular mass of the complexes between a domain of protein G or protein A and an Fab or an Fc fragment of IgG is approximately 58,000, so that a complete structure determination by NMR is not feasible. However, since the structure of each component of the complex is known, it is possible to obtain a medium-resolution model of the complex if the residues in each component involved in the interaction can be identified, at a simple level by using appropriately isotope-labeled proteins to identify residues whose chemical shifts are affected by complex formation.

Residues of domain II of protein G involved in binding Fab and Fc

Complete ^{1}H, ^{15}N and almost complete ^{13}C resonance assignments of domain II of protein G are available (Lian *et al.*, 1992; Lian, L.-Y., Barsukov, V., Derrick, J.R., and Roberts, G.C.K., unpublished work), and ^{1}H-^{15}N and ^{1}H-^{13}C heteronuclear correlation spectra thus provide a convenient means of identifying resonances affected by the binding of protein G to IgG fragments.

Figure 2a compares the ^{1}H-^{15}N correlation spectra of protein G domain II, alone and in its complex with an Fab fragment of mouse IgG (Lian *et al.*, 1994). The resonances of rather more than a third of the residues either undergo

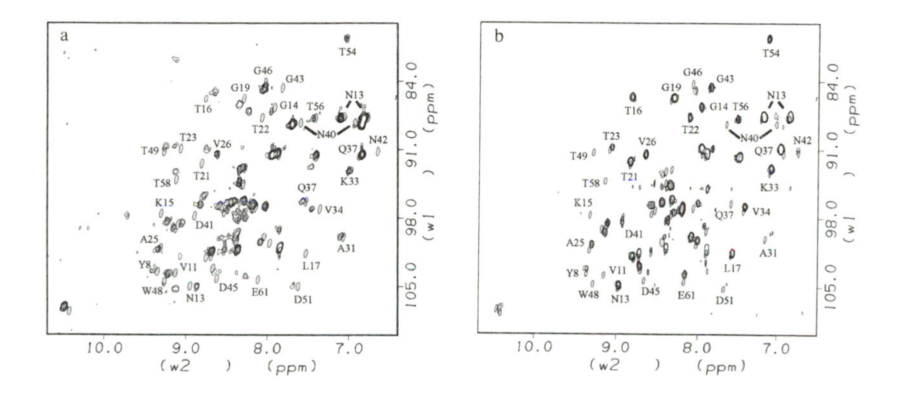

Figure 2: The effects on the ^{15}N-^{1}H HSQC spectra of domain II of protein G (uniformly labeled with ^{15}N and ^{13}C) of binding to (a) the Fab fragment of mouse IgG (Gronenborn *et al.*, 1991), and (b) the Fc fragment of mouse IgG (Orban *et al.*, 1992). In each case, spectra are shown for the domain alone (solid lines) and in its complex (dashed lines). Assignments (in the spectra of the domain alone) are indicated by the single-letter code.

significant chemical shift changes (greater than the linewidth of the crosspeak) or are so broadened as to be undetectable. This marked line-broadening is specific to these residues, and distinct from the smaller increase in linewidth seen for all resonances due to the increase in correlation time on complex formation, and probably arises from the fact that the chemical shift changes of these resonances are such that the exchange rate between the free and bound forms of the domain is in the intermediate range on the NMR timescale.

These specific changes may reflect direct protein-protein contacts and/or changes in conformation on complex formation - although the crystal structure of this complex (Derrick and Wigley, 1992; Derrick and Wigley, 1994) suggests that any conformational change is slight. The *unaffected* resonances, on the other hand, can be unambiguously assigned and must arise from residues that are *not* involved in intermolecular contacts. The unaffected backbone amide resonances in the ^{1}H-^{15}N correlation spectrum of the complex between domain II and the Fab fragment are those of residues 1-11, 13, 23, 25-40, 44, 46, 47, 49-57 and 59-64. The regions of the protein G molecule that are involved in Fab binding can thus be identified as the turn between the strands 1 and 2 of the β-sheet and the first two-thirds of β-strand 2, and the loop following the helix, but not the helix itself. As discussed previously (Lian *et al.*, 1994), the changes observed in the NMR spectrum on complex formation in solution are entirely consistent with the crystal structure (Derrick and Wigley, 1992, 1994); it is clear that the crystal structure does correspond to the structure of the complex in solution.

Figure 2b similarly compares the ^{1}H-^{15}N correlation spectra of protein G domain II, alone and in its complex with an Fc fragment of mouse IgG (Kato *et al.*, 1995). In this case, the unaffected backbone amide resonances are those of residues 1-10, 12-14, 16-27, 33-34, 38, 41-44, 51-57 and 59-64. Affected resonances arise primarily from residues in the α-helix and in the third strand of the β-sheet. This pattern of affected amide resonances is almost identical to that observed for the binding of a slightly different protein G domain (from *Streptococcus* strain GX7809) to a human Fc fragment (Gronenborn and Clore, 1993). It is clear that the same part of the surface of protein G is involved in binding to the two Fc fragments, although it binds them with markedly different affinity; equally, it is clear that a different region of domain II is involved in binding to the Fab and the Fc fragment of IgG (Lian *et al.*, 1994; Kato *et al.*, 1995).

The ^{1}H-^{15}N spectra give information primarily about the backbone of the protein, while ^{1}H-^{13}C correlation spectra give more information on sidechain contacts. Changes on binding to Fab and Fc are also observed, for example, in the methyl region of the ^{1}H-^{13}C spectrum, and clear differences are again observed in the effects of the two antibody fragments (Lian *et al.*, 1994; Kato *et al.*, 1995).

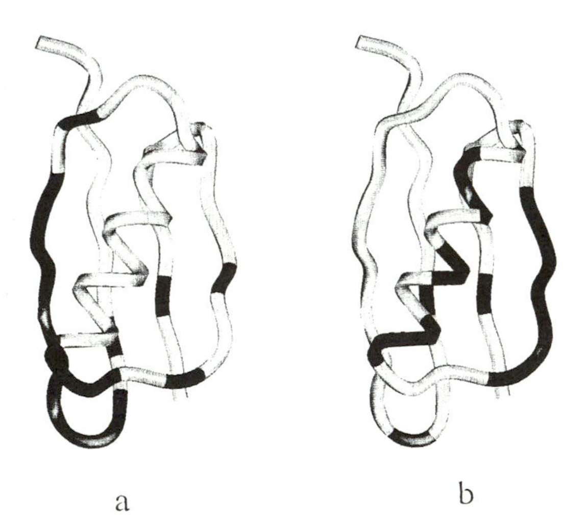

Figure 3: Comparison of the residues of the IgG-binding domains of protein G affected by binding to (a) Fab and (b) Fc fragments of IgG. Those residues whose amide $^{1}H/^{15}N$ or methyl $^{1}H/^{13}C$ chemical shifts and/or linewidths are altered on formation of the respective complexes are shaded (Gronenborn *et al.*, 1991; Orban *et al.*, 1992). The protein G domain is oriented with its N-terminus at the top.

The regions of the protein G domain that form the "contact surface" in the complexes with the Fab and Fc fragments are compared schematically in Figure 3. In the complex with Fab, the *second* strand of the β-sheet plays a major role, together with the loop at the C-terminal end of the helix, while in the complex with Fc, the regions most affected are the helix and the *third* strand of the β-sheet, but not the intervening loop. The IgG-binding domains of protein G thus interact with the Fab and Fc regions of the antibody in quite different ways (Lian *et al.*, 1994; Kato *et al.*, 1995).

Residues of Fc involved in binding protein G

To locate the binding site for protein G on Fc, we have used ^{13}C resonances originating from the carbonyl carbons of histidine, leucine, methionine, tryptophan, and tyrosine residues as "probes" (Kato *et al.*, 1995). The assignments of the histidine, methionine, tryptophan and tyrosine resonances to individual residues, most of which were made by the ^{13}C-^{15}N double labeling method developed by Kainosho and Tsuji (1982), have been reported (Kato *et al.*, 1993; Kato *et al.*, 1991; Kato *et al.*, 1991; Kato *et al.*, 1989). The resonances of the leucine residues have not yet been specifically assigned.

These experiments gave a total of 35 residues whose behaviour could be monitored. Of these, only Met-252, His-433, His-435 and His-436 were affected by the binding of domain II of protein G (Kato *et al.*, 1995). These four residues lie in the 'groove' between the C_H2 and C_H3 domains of Fc, indicating that this region is primarily responsible for the binding of protein G. A comparison with the results of similar experiments with domain B of protein A reveals that all these residues, 252, 433, 435 and 436, are also perturbed by the binding of protein A (Kato *et al.*, 1993). However, the binding of protein A also affects the chemical shifts of residues 310, 314 and 429, which are not affected by protein G. Thus, protein A and protein G bind to overlapping but not identical sites on Fc. The differences between the effects of protein G and protein A suggest that the former interacts more with residues from the C_H3 domain than with those from the C_H2 domain.

A model for the protein G - Fc complex

Although crystal structures of the complexes between a protein A domain and Fc (Deisenhofer, 1981) and between a protein G domain and Fab are available (Derrick and Wigley, 1992, 1994), there is no published structure of a protein G domain - Fc complex. The information on the residues in the two proteins which are involved in the interaction, together with the structures of the IgG-binding domains of protein G and of the Fc fragment, allows us to construct an approximate model for the structure of this complex. This was done by Monte Carlo minimization (details are given in Kato *et al.* (1995)). The NMR information was introduced into the calculations by defining the interacting region on each protein as including all surface residues within 8 Å of one or more residues whose chemical shift was affected by complex formation and using a pseudo-potential which constrained the affected residues in either partner to lie close to one or more residues of the other partner which were within this interacting region. In Figure 4, the model calculated in this way is compared with that of the complex between Fc and domain B of protein A (Deisenhofer, 1981). In this model the protein G domain is located in the 'groove' between the C_H2 and C_H3 domains, with the helix lying more or less in the groove, and the third strand of the sheet making contact with the C_H3 domain. Although no information on the structure of the protein A - Fc complex was used in arriving at this model, the helix of domain II of protein G is found to lie in an essentially identical position to that occupied by helix 1 of domain B of protein A in its complex with human Fc (Deisenhofer, 1981). However, the orientation of these two helices differs by 180°, so that the third strand of the β-sheet of

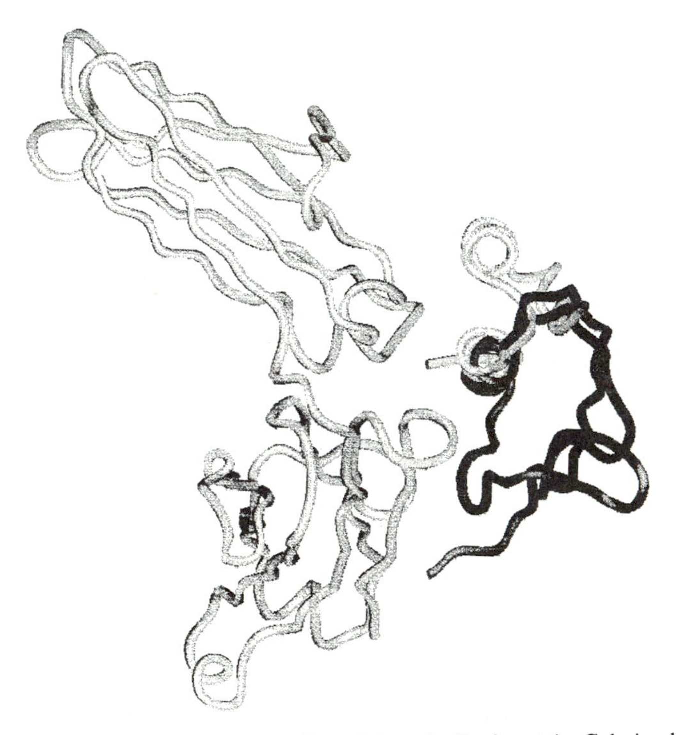

Figure 4: A model of the complex between Fc and domain II of protein G derived from the chemical shift changes observed in solution (Orban *et al.*, 1992), superimposed on the crystal structure of the complex between Fc and domain B of protein A (Gronenborn and Clore, 1993). The two structures were superimposed by least-squares superposition of the backbone atoms of the Fc fragments only; for

protein G interacts only with the C_H3 domain, while protein A in its complex has more extensive interactions with the C_H2 domain. This accounts for the observed differences in the residues of Fc affected by the binding of protein G and protein A. The procedure used to arrive at this model depended on the assumption that there is no substantial change in the conformation of either partner on formation of the complex. It therefore led only to an approximate, medium-resolution model, but one which can be tested by, for example, site-directed mutagenesis, and these experiments are in progress.

The molecular bases of the specificity of the interactions of these antibody-binding proteins with their 'target' immunoglobulins revealed by these experiments show two interesting features. First, although the IgG-binding domains of protein A and protein G have quite different three-dimensional structures, they bind to closely similar sites on the Fc fragment of IgG. In both cases an α-helix plays an important role in the recognition, but in a very different orientation - two different structural solutions to the recognition of the same region of a protein surface (Kato *et al.*, 1995). Secondly, although the constant domains in Fab and Fc are of course structurally related, protein G binds quite differently to them (Lian *et al.*, 1994; Kato *et al.*, 1995). This small domain is able to recognize specifically two quite different protein surfaces, by employing an almost completely different set of residues on its surface.

Substrate binding to cytochrome P450: paramagnetic relaxation studies of substrate binding

The cytochrome P_{450}s are a family of versatile oxygen-activating heme enzymes, currently with more than 200 members, which catalyze the insertion of one atom of molecular oxygen into a wide variety of endogenous and exogenous compounds. In mammals, members of this family play a major role in determining the response of the organism to exogenous chemicals, and an understanding of the structural basis of their specificity is important in predicting the metabolism, and hence the possible toxic effects, of such chemicals.

It has proved very difficult to obtain eukaroytic cytochromes P_{450} in a soluble form suitable for structural studies. A good deal of structural and mechanistic work has therefore been done with soluble bacterial P_{450}s (Sariaslani, 1991; Guengerich, 1991). Until recently, the only cytochrome P_{450} for which high-resolution structural information was available was cytochrome $P_{450\ CAM}$ from *Pseudomonas putida*, which catalyses the hydroxylation of camphor (Poulos *et al.*, 1987; Raag *et al.*, 1993; and references therein). Studies of this enzyme have provided valuable insight into the mechanism of cytochrome P_{450}s, but its limited sequence similarity to the enzymes of the mammalian endoplasmic reticulum limits its usefulness as a model for these.

cytochrome P_{450}s, but its limited sequence similarity to the enzymes of the mammalian endoplasmic reticulum limits its usefulness as a model for these. We have begun to study cytochrome $P_{450\ BM3}$ from *Bacillus megaterium*, which catalyzes hydroxylation at the ω-1, ω-2 and ω-3 positions of fatty acids (Fulco, 1991). This is a unique catalytically self-sufficient P_{450}, containing both a cytochrome P_{450} domain and a flavoprotein NADPH-cytochrome P_{450} reductase domain, which both show clear sequence similarity to the corresponding mammalian microsomal enzymes (Ruettinger *et al.*, 1989). The crystal structure of the P_{450} domain of $P_{450\ BM3}$ has recently been determined (Ravichandran et al., 1993), and comparison with the structure of cytochrome $P_{450\ CAM}$ (Ravichandran *et al.*, 1993; Li and Poulos, 1994) shows that, while the overall topology is broadly similar, there are substantial differences between the two enzymes around the substrate binding pocket. The substrate-binding site in $P_{450\ BM3}$ is a large, almost entirely hydrophobic, channel leading from the surface to the deeply buried heme. Toward the outer end of this channel is a single charged residue, Arg47, which is postulated to bind the carboxylate of the fatty acid substrates (Ravichandran *et al.*, 1993); support for this role of Arg47 has very recently been obtained from mutagenesis experiments (Gibson, C.F. *et al.,* unpublished). Detailed information on substrate and inhibitor binding to the restricted active site of cytochrome $P_{450\ CAM}$ is available, but as yet there is little direct structural information on the binding of substrates to the much larger active site of cytochrome $P_{450\ BM3}$.

As a first step toward understanding the specificity of this enzyme, we have used the paramagnetic relaxation effect of the unpaired electron of the heme iron to obtain information on the location of the substrate lauric acid in the binding pocket of the P_{450} domain (Modi *et al.*, 1995). The results of these initial experiments provide an interesting contrast to observations on camphor binding to $P_{450\ CAM}$. In the "resting" enzyme, which contains low-spin Fe(III), the relaxation experiments provide an estimate of 2.7 Å for the distance between the iron and the protons of the nearest water molecule, consistent with the observation in the crystal structure (Li and Poulos, 1994) of a water molecule directly coordinated to the iron, and with NMR experiments on $P_{450\ CAM}$ and $P_{450\ scc}$ (Philson *et al.*, 1979; Jacobs *et al.*, 1987). In all P_{450}s, the formation of the initial enzyme-substrate complex (before transfer of electrons to the reductase) leads to conversion of the heme iron to the high-spin state; this can be followed by characteristic changes in the visible spectrum, as shown for cytochrome $P_{450\ BM3}$ in Figure 5. In $P_{450\ CAM}$ the substrate binds in close proximity to the iron in this initial complex (Poulos *et al.*, 1987; Raag *et al.*, 1993), and the nearest water molecule is 8-10 Å from the iron (Jacobs *et al.*, 1987, Griffin and Peterson, 1975). In the case of $P_{450\ BM3}$, our relaxation measurements show that the water molecule remains only 5.2 Å from the iron when the substrate lauric acid is bound, and at the same time lauric acid binds at

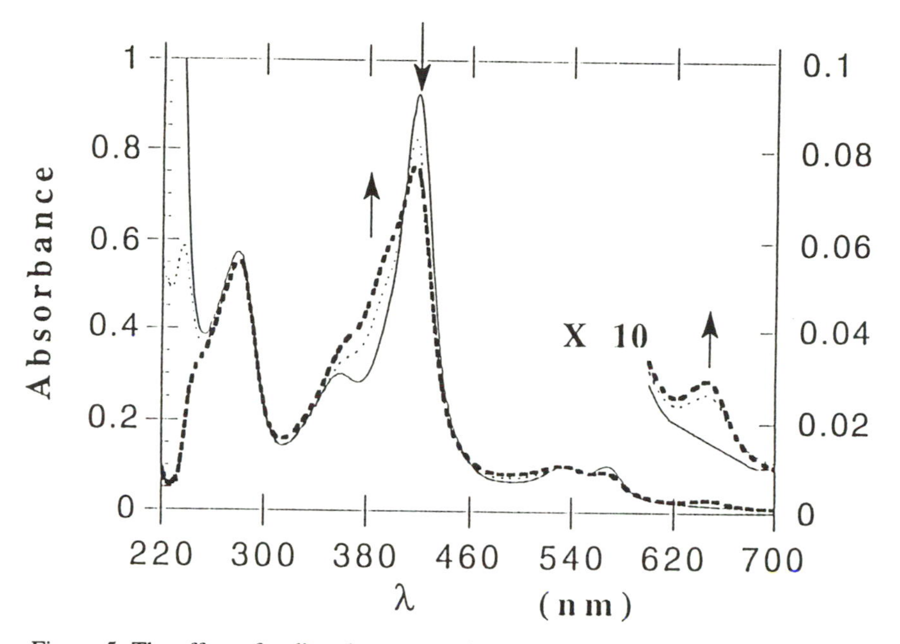

Figure 5: The effect of sodium laurate on the UV-visible absorption spectrum of the P_{450} domain of cytochrome $P_{450\ BM3}$. The spectra are those of the P_{450} domain (11.9 mM) alone (----) and in the presence of 150 mM (- - - -) and 300 mM (- - - -) sodium laurate (Modi *et al.*, 1994). The right-hand scale refers to the ten-fold vertical expansion of the region 600-700 nm.

protons of the methylene group adjacent to the carboxyl group are 16.4 Å from the iron (a distance consistent with an interaction of the carboxylate with Arg47) (Ravichandran *et al.*, 1993). This demonstrates a clear difference in the mode of substrate binding between cytochromes $P_{450\ CAM}$ and $P_{450\ BM3}$. In the initial substrate complex of $P_{450\ CAM}$ the substrate binds in close proximity to the iron, displacing the water molecule coordinated to the iron in the resting enzyme, and in an orientation appropriate for hydroxylation. By contrast, in $P_{450\ BM3}$, while the water molecule is displaced from the iron, it remains closer to the iron than the substrate, demonstrating that the displacement (and presumably the low-spin to high-spin conversion) must arise from a conformational change produced by substrate binding rather than from a simple steric clash. The observation that the atoms of the substrate which are hydroxylated are approximately 9 Å from the iron (Modi *et al.*, 1995) further implies that a rearrangement of the complex must occur after electron transfer from the reductase. The structural basis of the specificity of cytochrome $P_{450\ BM3}$, and presumably of the homologous mammalian enzymes, thus appears to be more complex than that of $P_{450\ CAM}$.

Chloramphenicol acetyltransferase: transferred nuclear Overhauser effect studies of the conformation of bound substrate

Chloramphenicol acetyl transferase (CAT), a trimer of total Mr 75,000, is the primary effector of resistance of bacteria to the antibiotic chloramphenicol. It catalyzes acetyl transfer from acetyl-CoA to chloramphenicol to produce 3-acetyl-chloramphenicol. This can then undergo a non-enzymatic rearrangement to form 1-acetylchloramphenicol, which is the substrate for a second cycle of acetylation by CAT, finally yielding 1,3-diacetyl chloramphenicol. The structures of the two binary complexes of CAT with chloramphenicol and CoA have been determined by crystallography (Shaw and Leslie, 1990). However, attempts to obtain satisfactory crystals of the CAT-CoA-chloramphenicol ternary complex have so far proved unsuccessful.

This protein is very large for study by NMR, and we have used it as a system for the development of NMR methods suitable for large proteins. Two general approaches to this can be envisaged - selective isotope labeling in conjunction with heteronuclear "editing" experiments, and magnetization transfer methods such as the transferred NOE. In CAT, we have used selective ^{13}C-labeling of both the protein (with [imidazole-2-^{13}C]-histidine) and the ligand (Derrick *et al.*, 1991; Derrick *et al.*, 1992). ^{13}C-edited ^{1}H-^{1}H NOESY experiments with diacetyl-[^{13}C]-chloramphenicol revealed a number of clear intermolecular NOEs, in particular to aromatic protons of the protein. Candidate aromatic residues were identified by model-building on the basis of the crystal structure, and were replaced in turn by isoleucine residues. ^{13}C-edited NOESY spectra of the complexes with these mutants allowed the two aromatic residues with which the acetyl groups of the ligand made contact to be identified unambiguously, thus allowing the orientation of the product in the binding site to be defined (Derrick *et al.*, 1992).

More recently, we have used transferred NOEs and ROEs to determine the conformation of CoA bound to the enzyme in both binary and ternary complexes. In these experiments, the exchange rate must be faster than the spin-lattice relaxation rate of the bound ligand; if this condition is satisfied, changes in magnetization of a bound ligand proton resulting from intra- or intermolecular NOEs are transferred to the free ligand by the exchange between bound and free ligand. For small ligands, cross-relaxation in the bound state is large and negative while that in the free state is small and positive. The intramolecular cross-relaxation of the ligand is hence governed by the relaxation in the bound state scaled down by the fraction of bound ligand. Thus, these experiments rely upon *indirect* characterization of ligand-protein interactions and of the conformation of the bound ligand *via* the averaged NMR spectrum of the free and bound ligand. Early one-dimensional transferred NOE experiments were interpreted only qualitatively, although in several cases this sufficed to give useful information about the conformation of bound ligands [*e.g.*, Albrand *et al.*, 1979; Feeney *et al.*, 1983; Lian *et al.*, 1994]. Quantitative interpretation involves a rather large system of coupled differential equations which prevents the derivation of simple analytical results, and this led to the use of the "two spin approximation", which can only give semi-quantitative information. The success of the relaxation matrix approach in treating cross-relaxation phenomena in the context of protein structure determination stimulated its application to the analysis of the combination of cross-relaxation and chemical exchange involved in the transferred NOE, and a general method of analysis is now available (Lian *et al.*, 1994, and references therein).

The first step in a study of the conformation of a bound ligand by means of transferred NOEs is to establish whether the exchange of the ligand on and off the enzyme is sufficiently fast on the NMR timescale (Lian and Roberts, 1993); in the case of CoA binding to CAT, analysis of the chemical shifts and lineshape of ligand resonances (together with, subsequently, detailed fitting of transferred NOE data at two temperatures) showed that this is indeed the case (Barsukov *et al.*, 1995). Transferred NOEs between protons of CoA can then readily be measured, since an excess of ligand over protein can be used. The initial rate of build-up of the transferred NOE is independent of the ratio of ligand to protein concentrations (Lian *et al.*, 1994), and in these experiments we used a ratio of 15:1 as a compromise between increased NOESY cross-peak intensity and the problems of t_1-noise and possible nonspecific interactions at high ligand concentrations.

As with all NOE measurements, interpretation of these in terms of molecular structure requires that we differentiate direct cross-relaxation effects from spin diffusion. In a protein as large as CAT, spin diffusion is a problem that cannot be ignored, and there are a number of different approaches to minimizing the errors it can introduce. When measuring NOEs between protons of a ligand bound to a protein, spin diffusion can occur via either protons of the ligand itself or protons of the protein. The latter route involves the intermolecular relaxation processes which give rise to intermolecular NOEs which are often readily observed in protein-ligand complexes, but it has hitherto largely been ignored in the context of transferred NOE studies of ligand conformation; its significance can be established by the use of perdeuterated protein. In the case of the CoA-CAT system, any effect of deuteration of the enzyme on the intensities of transferred NOEs was clearly less than a factor of two (Barsukov *et al.*, 1995), although accurate measurement of these intensities was much easier with the deuterated enzyme. The magnitude of the error produced by intermolecular relaxation processes will obviously vary considerably from one system to another, and it would seem a wise precaution to use perdeuterated protein in relaxation processes will obviously vary considerably from one system to another, and it would seem a wise precaution to use perdeuterated protein in experiments of this type; deuteration is almost always feasible for proteins expressed in a bacterial system.

We have employed a combination of two approaches to deal with the problems introduced by *intra*molecular spin diffusion within the bound ligand: the use of both NOESY and ROESY data, and the analysis of the data using a complete combined relaxation and kinetic matrix (Lian *et al.*, 1994). In the ROESY experiment, cross-peaks arising from a direct cross-relaxation effect have the opposite sign to those arising from a two-step spin-diffusion. Thus cross-peaks in the ROESY spectrum which have opposite phase to the diagonal peaks correspond predominantly to direct magnetization transfer, and cross-peaks which are in-phase with diagonal peaks arise from indirect transfer; zero cross-peak intensity indicates either the absence of magnetization transfer due to a large interproton distance, or that direct and indirect transfer are equally effective. Figure 6 shows the part of the NOESY/ ROESY spectra obtained from the binary CAT-CoA complex, which contains cross-peaks between the protons of the two methyl groups of CoA (labeled HP8 and HP9) and the other protons of the ligand. In the NOESY spectrum obtained with a mixing time of 400 ms (Figure 6, top), both methyl groups exhibit strong cross-peaks to all of the other CoA protons, reflecting the presence of strong spin diffusion. A mixing time of 100 ms gives somewhat better distinction between spin-diffusion and direct effects, but only with a mixing time of 30 ms can cross-peaks arising predominantly from direct interactions be reliably identified for most of the proton pairs (Figure 6). Even at this mixing time, the problem of spin-diffusion still remains for the geminal protons of a methylene group, and the intensities of the NOE cross-peaks between the protons of these two methyl groups and the two protons of the neighbouring methylene (labeled HPB2 and HPB3) are as a result very similar. Part of the ROESY spectrum obtained with a mixing time of 100ms, is also shown in Figure 6. All cross-peaks in this spectrum have phase opposite to that of the diagonal, and thus correspond to predominantly direct interactions. Some cross-peaks clearly seen in the NOESY spectrum are absent in the ROESY spectrum, indicating equal contributions from direct and indirect transfer. In the ROESY spectrum, there is a clear differentiation between the intensities of the cross-peaks arises from the two methylene protons: HPB3 gives rise to the cross-peaks to both of the methyl groups, while HPB2 shows a cross-peak only to HP8.

This information allows a qualitative determination of the conformation about the central bond in the $-CH_2-C(CH_3)_2-$ fragment of the pantetheine moiety of CoA, shown as a Newman projection in Figure 6, which is the same as that observed in the crystal structure of the binary CAT-CoA complex (Leslie *et al.*, 1986). The intensity of the cross-peak between the methine proton HP7

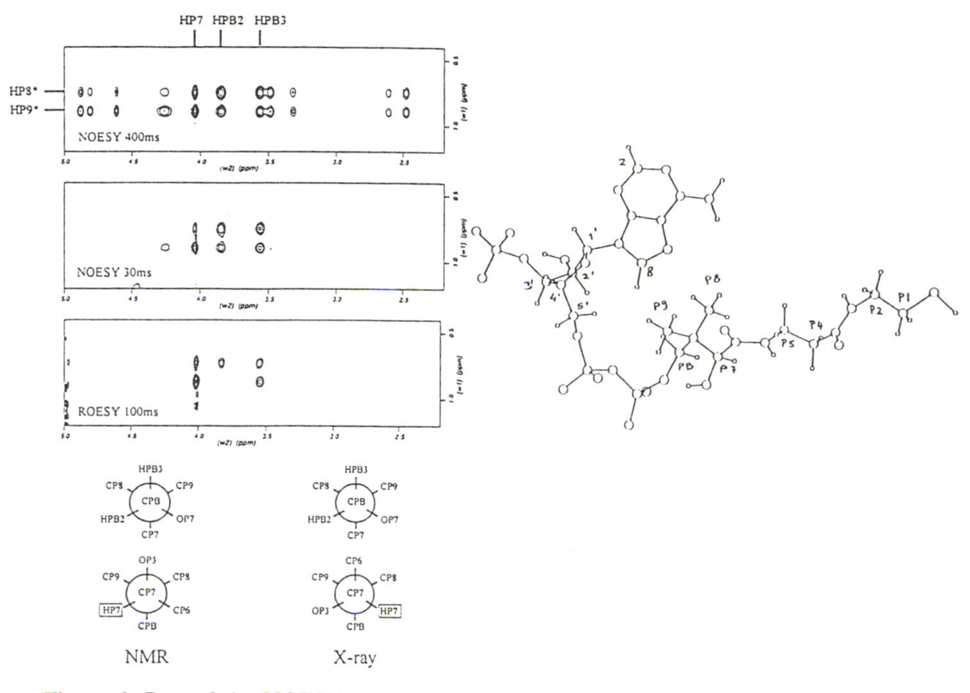

Figure 6: Part of the NOESY and ROESY spectra of coenzyme A in the presence of chloramphenicol acetyltransferase (ligand:protein 15:1), showing transferred NOEs/ROEs to the CP8 and CP9 methyl protons (Barsukov *et al.*, 1994). From the top, the spectra are NOESY (τm 400 ms); NOESY (τm 30 ms) and ROESY (τm 100 ms). At the bottom, conformations about two bonds in the pantetheine moiety of coenzyme A are shown in Newman projections, as deduced from the NMR data and as seen in the crystal structure; top, the -CH_2-$C(CH_3)_2$- fragment and bottom, the -$C(CH_3)_2$-CH(OH)- fragment. The structure of coenzyme A and the nomenclature used for its protons is on the right.

conformation about the central bond in the -$C(CH_3)_2$-CH(OH)- fragment of the pantetheine, as shown in Figure 6. It is clear that in this case the conformation derived from a qualitative analysis of the NMR data is not the same as that observed in the X-ray structure.

A quantitative analysis of a total of 71 NOESY cross-peaks and 33 ROESY cross-peaks was used to determine the conformation of CoA bound to CAT, in an iterative procedure using distance geometry and simulated annealing calculations. The inclusion of the ROESY data was found to be important in obtaining internally consistent results, with different starting structures leading to essentially the same final structure. In the final "family" of structures, none of the upper limit distance restraints were violated by more the 0.3 Å and none of the experimental cross-peak intensities differ from the calculated ones by more than a factor of two.

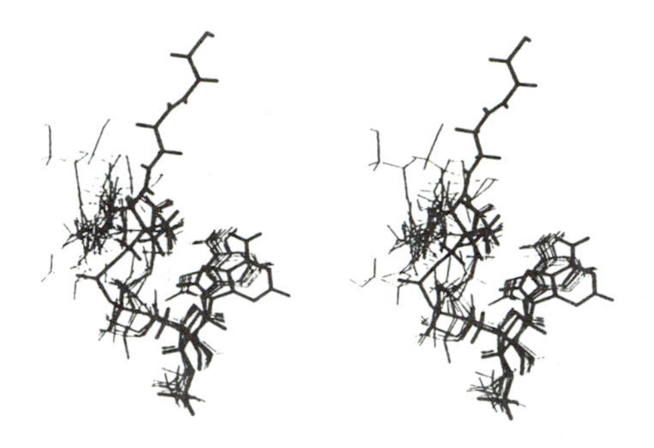

Figure 7: Stereo pair comparing the conformation of coenzyme A in its binary complex with chloramphenicol acetyltransferase calculated from distance constraints derived from transferred NOEs (Basrsukov *et al.*, 1994; thin lines) with that observed in the crystal structure (Leslie *et al.*, 1986; thick line).

The calculated structures (Barsukov *et al.*, 1995) are compared with the X-ray structure (Leslie *et al.*, 1986) in Figure 7. The precision of the NMR structure and its similarity to the X-ray structure differ in different parts of the CoA molecule. The adenosine part of the molecule is very well defined and its conformation is close to that in the crystal. The conformation of the pyrophosphate link is not defined as there are no protons and hence no NOEs in this part of the molecule. The pantetheine "arm" of CoA shows a clear difference between the NMR and X-ray structures. In the crystal structure, this arm has an extended conformation which makes some of the distances between the β-mercaptoethylamine methylene protons of the pantetheine and the adenosine protons as large as 10 Å. However, relatively intense NOE cross-peaks can be observed for these proton pairs, and the constraints derived from these NOEs lead to a more "folded" conformation of the pantetheine arm in the NMR structure.

Several factors may account for the difference between NMR and X-ray conformations of CoA. First, there might be a genuine difference between the solution and crystal structures; there is, however, no indication in the crystal structure that crystal contacts might influence the conformation of CoA. Second, because an excess of ligand is used in the transferred NOE experiments, the discrepancy might arise from secondary binding sites for CoA. Fluorescence experiments give no indication of more than one binding site per subunit, and the changes in chemical shift and linewidth of a number of CoA proton resonances as a function of ligand concentration are all in a good agreement with this simple model.

The most likely explanation for the discrepancy seems to be that the binding process is more complex than a simple two-site exchange. The equilibrium and kinetic data on the binary complex can be explained by postulating a two-step binding of CoA to the enzyme:

$$\text{CAT} + \text{CoA} \leftrightarrow \text{CAT-CoA} \leftrightarrow \text{CAT-CoA}^*$$

where CAT-CoA represents an intermediate complex and CAT-CoA* is the final complex after a conformational readjustment of some kind. In a system described by a model of this kind, what one observes will depend crucially on the technique used, and on the equilibrium constant of the second step. In X-ray diffraction, one would see a superposition of all conformations of the complex; the fact that a single clear-cut structure was observed indicates that the equilibrium constant of the second step is well over to the right - by a factor of ten or more. However, if the rate of interconversion between CAT-CoA and CAT-CoA* is sufficiently fast, the measured NMR parameters will be an average of the parameters of the two states. Furthermore, since these parameters, such as the NOE, are non-linear functions of distances and angle, this average will be a non-linear one (Jardetzky and Roberts, 1981; Jardetzky, 1980). Suppose that in the intermediate CAT-CoA complex the conformation of the CoA molecule has the pantotheine arm in the folded state, while in the final CAT-CoA* complex it is, as shown by the crystal structure, in the extended state. Since in the extended conformation most of the distances to the end of the pantetheine arm from the adenine part of the ligand are too large to give rise to NOEs, the averaged NMR information will be dominated by NOEs from the folded intermediate state. This will be true even if this intermediate state is only populated to a minor extent at equilibrium. Support for this two-step binding model comes from a comparison of the estimates of the dissociation rate constant of CoA from the binary complex made by fluoresence stopped-flow methods and by NMR methods (Barsukov *et al.*, 1995). Stopped-flow measurements could be made over the range 5-20 °C; at higher temperatures, the rate was too fast to be determined accurately. At 7 °C, stopped-flow gave k_{off} = 200 s^{-1} while analysis of the transferred NOE data gave an estimate of at least 500 s^{-1}; at 25 °C, NMR lineshape analysis gave a value of 3500 s^{-1}, while extrapolation of the stopped-flow data gave an estimate of 650 s^{-1}. These differences can readily be explained if the stopped-flow experiments measure dissociation from the predominant CAT-CoA* complex, while the NMR data refer to an average of the CAT-CoA and CAT-CoA* complexes.

Similar transferred NOE experiments have been carried out on the ternary CAT-CoA-chloramphenicol complex, for which no crystallographic information is available, again using deuterated enzyme to eliminate spin diffusion through the protein. Strikingly, *interligand* NOEs were observed in the complex, between the pantothenate protons of CoA and protons of chloramphenicol; NOEs involving the C2 and C3 protons of chloramphenicol are shown in Figure 8. This provides the first experimental evidence for the relative orientation of the two ligands when both are bound simultaneously to CAT. Furthermore, calculation of the structure of the ternary complex, using the NOE and ROE data together with the steric constraints imposed by the structure of the binding site, revealed that in the ternary complex the pantetheine arm of CoA

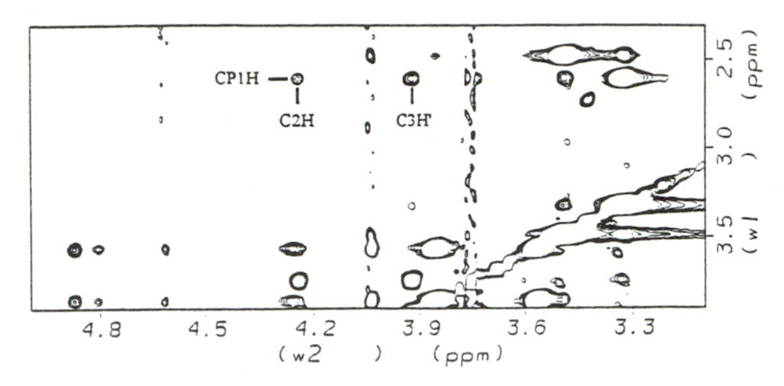

Figure 8: Part of the NOESY spectrum from a sample containing coenzyme A, chloramphenicol and perdeuterated chloramphenicol acetyltransferase (ratio 15:7:1), showing transferred NOEs between the CP1 proton of coenzyme A and the C_2H and C_3H' protons of chloramphenicol (Barsukov *et al.*, 1994).

adopts a conformation somewhat more extended than that in the binary complex. This is consistent with the two-step model for CoA binding outlined above, suggesting that in the ternary complex the equilibrium between the two states of the complex is far to the right, favoring a more extended conformation of CoA, which is necessary for the end of the pantetheine arm to be sufficiently close to chloramphenicol for acyl transfer to occur.

These experiments emphasize the care that must be taken in the interpretation of transferred NOE experiments - or indeed any NMR experiment involving rapid exchange - but also show how the combination of NMR and crystallography can provide not only structural but mechanistic information on enzyme-substrate interactions.

Acknowledgements

We are most grateful to our collaborators in the work summarized here for their invaluable practical and intellectual contributions: for protein G Drs. J.P. Derrick, K. Kato, I. Shimada and Prof. Y. Arata; for cytochrome P_{450} Dr. J. Miles and Prof. C.R. Wolf; and for CAT Dr. J. Ellis and Prof. W.V. Shaw. We are also very grateful to A. Prescott and J. Boyle for excellent technical assistance. The work in Leicester was supported by the Biotechnology and Biological Sciences Research Council and the Medical Research Council. G.C.K.R. expresses his considerable gratitude to Oleg Jardetzky for introducing him to protein NMR in its infancy, and for being a continuing source of inspiration and encouragement - and argument - over the ensuing years.

References

Achari, A., Hale, S.P., Howard, A.J., Clore, G.M., Gronenborn, A.M., Hardman, K.D., and Whitlow, M. (1992). *Biochemistry* **31**, 10449.

Albrand, J.P., Birdsall, B., Feeney, J., Roberts, G.C.K., and Burgen, A.S.V. (1979). *Int. J. Biol. Macromol.* **1**, 37.

Barsukov, I.L., Lian, L.-Y., Ellis, J., Sze, K.-H., Roberts, G.C.K., and Shaw,W.V. (1995). *in preparation.*

Boyle, M.D.P. (1990). Ed: *Bacterial Immunoglobulin Binding Proteins*, Academic Press, San Diego.

Deisenhofer, J. (1981). *Biochemistry* **20**, 2361.

Derrick, J.P., Lian, L.-Y., Roberts, G.C.K., and Shaw, W.V. (1991). *FEBS Lett.* **280**, 125.

Derrick, J.P., Lian, L.-Y., Roberts, G.C.K., and Shaw, W.V. (1992). *Biochemistry* **31**, 8191.

Derrick, J.P., and Wigley, D.B. (1992). *Nature* **359**, 752.

Derrick, J.P., and Wigley, D.B. (1994). *J. Mol. Biol.* **243**, 906.

Eliasson, M., Anderson, R., Nygren, P.A., and Uhlen, M. (1991). *Mol. Immunol.* **28**, 1055.

Feeney, J., Birdsall, B., Roberts, G.C.K., and Burgen, A.S.V. (1983). *Biochemistry* **22**, 628.

Feeney, J., and Birdsall, B. (1993). in *NMR of Biological Macromolecules* (Roberts, G.C.K. ed.). IRL Press at Oxford University Press. pp.183-216.

Frick, I.M., Wikström, M., Forsén, S., Drakenberg, T., Gomi, H., Sjöbring, U., and Björck, L. (1992). *Proc. Natl. Acad. Sci. U.S.A.* **89**, 8532.

Fulco, A.J. (1991). *Ann. Rev. Pharmacol. Toxicol.* **31**, 177.

Gallagher, T., Alexander, P., Bryan, P., and Gilliland, G.L. (1994). *Biochemistry* **33**, 4721.

Gibson, C.F., Modi, S., Primrose, W.U., Lian L.-Y., and Roberts, G.C.K. (1994). *unpublished work.*

Gouda, H., Torigoe, H., Saito, A., Sato, M., Arata, Y., and Shimada, I. (1992). *Biochemistry* **31**, 9665.

Griffin, B.W., and Peterson, J.A. (1975). *J. Biol. Chem.* **250**, 6445.

Gronenborn, A.M., Filpula, D.R., Essig, N.Z., Achari, A., Whitlow, M., Wingfield, P.T., and Clore, G.M. (1991). *Science* **253**, 657.

Gronenborn, A.M., and Clore, G.M. (1993). *J. Mol. Biol.* **233**, 331.

Guengerich, F.P. (1991). *J. Biol. Chem.* **266**, 10019.

Guss, B., Eliasson, M., Olsson, A., Uhlen, M., Frej, A.-K., Jornvall, H., Flock, J.-I., and Lindberg, M. (1986). *EMBO J.* **5**, 1567.

Jacobs, R.E., Singh, J., and Vickery, L.E. (1987). *Biochemistry* **26**, 4541.

Jardetzky, O. (1980). *Biochim. Biophys. Acta* **612**, 227.

Jardetzky, O., and Roberts, G.C.K. (1981). NMR in Molecular Biology. New York, Academic Press.

Kainosho, M., and Tsuji, T. (1982). *Biochemistry* **21**, 6273.

Kato, K., Matsunaga, C., Nishimura, Y., Waelchli, M., Kainosho, M., and Arata, Y. (1989). *J. Biochem. (Tokyo).* **105**, 867.

Kato, K., Matsunaga, C., Igarashi, T., Kim, H., Odaka, A., Shimada, I., and Arata, Y. (1991). *Biochemistry* **30**, 270.

Kato, K., Matsunaga, C., Odaka, A., Yamato, S., Takaha, W., Shimada, I., and Arata, Y. (1991). *Biochemistry* **30**, 6604.

Kato, K., Gouda, H., Takaha, W., Yoshino, A., Matsunaga, C., and Arata, Y. (1993). *FEBS Lett.* **328**, 49

Kato, K., Lian, L.-Y., Barsukov, I.L., Derrick, J.P., Kim, H., Tanaka, R., Yoshino, A., Shiraishi, M., Shimada, I., Arata Y., and Roberts, G.C.K. (1995). *Structure*

3, 79.

Langone, J.J. (1982). *Adv. Immunol.* **32**, 157.

Leslie, A.G.W., Liddell, J.M., and Shaw, W.V. (1986). *J. Mol. Biol.* **188**, 283.

Li, H., and Poulos, T.L. (1994). *Structure* **2**, 461.

Lian, L.-Y., Yang, J.C., Derrick, J.P., Sutcliffe, M.J., Roberts, G.C.K., Murphy, J.P., Goward, C.R., and Atkinson, T. (1991). *Biochemistry* **30**, 5335.

Lian, L.-Y., Derrick, J.P., Sutcliffe, M.J., Yang, J.C., and Roberts, G.C.K. (1992). *J. Mol. Biol.* **228**, 1219.

Lian, L.-Y., and Roberts, G.C.K. (1993). in *NMR of Biological Macromolecules* (Roberts, G.C.K. ed.). IRL Press at Oxford University Press, pp. 153-182.

Lian, L.-Y., Barsukov, I.L., Derrick, J. P., and Roberts, G.C.K. (1994). *Nature Structural Biology* **1**, 355.

Lian, L.-Y., Barsukov, I.L., Sutcliffe, M.J., Sze, K.H., and Roberts, G.C.K. (1994). *Methods in Enzymology* **239**, 657.

Modi, S., Primrose, W.U., Boyle, J., Gibson, C.F., Lian L.-Y., and Roberts, G.C.K. (1995). *Biochemistry, in press.*

Moks, T., Abrahmsen, L., Nilsson, B., Hellman, U., Sjöquist, J., and Uhlen, M. (1986). *Eur. J. Biochem.* **156**, 637.

Olsson, A., Eliasson, M., Guss, B., Nilsson, B., Hellman, U., Lindberg, M., and Uhlen, M. (1987). *Eur. J. Biochem.* **168**, 319.

Orban, J., Alexander, P., and Bryan, P. (1992). *Biochemistry* **31**, 3604.

Petros, A.M., and Fesik, S.W. (1994). *Methods in Enzymology* **239**, 717.

Philson, S.B., Debrunner, P.G., Schimidt, P.G., and Gunsalus, I.C. (1979). *J. Biol. Chem.* **254**, 10173.

Poulos, T.L., Finzel, B.C., and Howard, A.J. (1987). *J. Mol. Biol.* **195**, 687.

Raag, R., Li, H., Jones, B.C., and Poulos, T.L. (1993). *Biochemistry* **32**, 4571.

Ravichandran, K.G., Boddupalli, S.S., Hasemann, C.A., Peterson, J.A., and Deisenhofer, J. (1993). *Science* **261**, 731.

Ruettinger, R.T., Wen, L.-P., and Fulco, A.J. (1989). *J. Biol. Chem.* **264**, 10987.

Sariaslani, F.S. (1991). *Adv. Appl. Microbiol.* **36**, 133.

Shaw, W.V., and Leslie, A.G.W. (1990). *Ann. Rev. Biophys. Biophys. Chem.* **20**, 363.

Stone, G.C., Sjöbring, U., Björck, L., Sjöquist, J., Barber, C.V., and Nardella, F.A. (1989). *J. Immunol.* **143**, 565.

Torigoe, H., Shimada, I., Saito, A., Sato, M., and Arata, Y. (1990). *Biochemistry* **29**, 8787.

Wand, A.J., and Short, J.H. (1994). *Methods in Enzymology* **239**, 700.

Wemmer, D.E., and Williams, P.E. (1994). *Methods in Enzymology* **239**, 739.

10

NMR and Mutagenesis Investigations of a Model *Cis:Trans* Peptide Isomerization Reaction: Xaa^{116}-Pro^{117} of Staphylococcal Nuclease and its Role in Protein Stability and Folding

A.P. Hinck, W.F. Walkenhorst, D.M. Truckses, and J.L. Markley

Department of Biochemistry,
College of Agricultural and Life Sciences
University of Wisconsin, Madison
Madison, WI 53706 USA

The slow rates of peptide bond isomerization in imino acids and the substantial population of the *cis* peptide bond isomer in Xaa-Pro linkages in peptides were first recognized in NMR studies of proline-containing model compounds (Maia *et al.*, 1971). The important role of this isomerization in protein stability and folding (reviewed by Kim and Baldwin, 1982, 1990; Schmid, 1993) were recognized several years later (Brandts *et al.*, 1975) and the biological relevance of this process was substantiated by the discovery of a ubiquitous enzyme that catalyzes Xaa-Pro peptide bond isomerization (Fischer *et al.*, 1984, 1989; Takahashi *et al.*, 1989). The strict evolutionary conservation of some prolyl residues and the observation that the kinetics of interconversion between alternative functional forms of some systems is consistent with the time scale of proline isomerization suggest that proline isomerization may play a wide role in protein structure and function. Suggestive examples include the sodium pump of *Escherichia coli*, the disulfide isomerase/thioredoxin class of enzymes, concanavalin A, and bovine prothrombin fragment I (Brown *et al.*, 1977; Marsh *et al.*, 1979; Dunker, 1982; Brandl and Deber, 1986; Langsetmo *et al.*, 1989). NMR spectroscopy is one of the most suitable tools for studying this isomerization reaction. The rates generally are slow on the time scale of NMR chemical shifts but, in favorable cases, are comparable to longitudinal relaxation rates so that the isomerization process can be investigated by chemical exchange spectroscopy. NMR data obtained on calbindin $D9_k$ (Chazin *et al.*, 1989), insulin (Higgins *et al.*, 1988), and staphylococcal nuclease (nuclease) as discussed below have shown that each exists in solution under native conditions as a mixture of slowly exchanging conformers.

The fact that dynamic molecular heterogeneity in nuclease was first observed in the laboratory of Oleg Jardetzky, as manifested by splitting of the histidyl $^1H^{\varepsilon 1}$ resonance from His^{46} in one-dimensional ^1H NMR spectra recorded at 100 MHz (Markley *et al.*, 1970), makes this topic particularly appropriate to a volume celebrating his scientific contributions. The topic was taken up again later by Evans and coworkers (1987) who used magnetization transfer to demonstrate that heterogeneity observed among the histidyl $^1H^{\varepsilon 1}$ resonances arises from conformational exchange between different folded forms and site-directed mutagenesis of Pro^{117} to Gly to establish the role of this residue in the conformational equilibrium. Elegant one- and two-dimensional transfer studies of interconversion rates in partially-folded nuclease (Evans *et al.*, 1989) showed that the system could be modeled by four states: a folded state with a *cis* peptide bond, a folded state with a *trans* peptide bond, an unfolded state with a *cis* peptide bond, and an unfolded state with a *trans* peptide bond. More recent site-directed mutagenesis and magnetization-transfer studies (Loh *et al.*, 1991) have shown that isomerization of the His^{46}-Pro^{47} peptide bond is responsible for the splitting of the $^1H^{\varepsilon 1}$ signal of His^{46}. This latter equilibrium is independent of the isomerization at the Lys^{116}-Pro^{117} peptide bond, which is responsible for splitting of the $^1H^{\varepsilon 1}$ signals of His^8, His^{121}, and His^{124}. The net result is that the folded protein in solution consists of four, slowly-interconverting conformational species: F_{tc} (72%), F_{cc} (18%), F_{tt} (8%), and F_{ct} (2%), where the first subscripted letter (c, *cis*; t, *trans*) refers to the configuration of the His^{46}-Pro^{47} peptide bond, and the second letter to that of the Lys^{116}-Pro^{117} peptide bond (Alexandrescu *et al.*, 1990).

On simple thermodynamic grounds, it is likely that configurational heterogeneity at Xaa-Pro peptide bonds, at a level detectable by NMR spectroscopy, is fairly common in proteins. A basic question concerns the possible functional benefit of conformational heterogeneity in a given system. Is the heterogeneity simply the consequence of the limited amount of folding energy that can be devoted to shifting the equilibrium, or does it confer some essential benefit. For example, does it serve as a poised conformational switch, or does it allow for increased flexibility for ligand binding?

Recent NMR spectroscopic, X-ray crystallographic, and thermodynamic studies of staphylococcal nuclease (nuclease), primarily in our laboratory and those of Robert O. Fox at Yale University and Christopher M. Dobson at Oxford University, have focused on the molecular basis for the observed structural heterogeneity at Pro^{47} and Pro^{117} and the roles these conformational equilibria play in protein stability, catalytic activity, and protein folding. Topics of interest have included the three-dimensional structures of protein variants that display different populations of the substates (Hodel *et al.*, 1993; 1994; Wang *et al.*, 1995; Truckses *et al.*, 1995), the configurational states and interconversion rates of the Pro^{47} and Pro^{117} peptide bonds in the unfolded protein (Evans *et al.*, 1989; Rayleigh *et al.*, 1992), the influence of residues close to and distant from the prolines on the conformational equilibria (Hodel *et al.*, 1994; Alexandrescu *et al.*, 1990; Hinck, 1993; Royer *et al.*, 1993), effects of engineered disulfide bonds on the positions of the conformational equilibria (Hinck, 1993; Hinck *et al.*, 1995a), correlations between the equilibrium constants for the isomerization and the overall stability of the protein (Alexandrescu *et al.*, 1990; Hinck, 1993), and the role of these conformational equilibria in the folding kinetics (Evans *et al.*, 1989; Chen, *et al.*, 1991; Kuwajima *et al.*, 1991; Sugawara *et al.*, 1991; Shalongo *et al.*, 1992; Nakano *et al.*, 1993; Walkenhorst *et al.*, 1995ab).

This review covers six topics: (1) findings from studies of peptide bond isomerizations in proline-containing model peptides and proteins, (2) NMR strategies for the identification and quantification of prolyl peptide bond configurations and their application to nuclease, (3) models proposed to explain the position of the *cis:trans* equilibria at the His^{46}-Pro^{47} and Lys^{116}-Pro^{117} peptide bonds, (4) non-covalent interactions responsible for stabilization of *cis* Lys^{116}-Pro^{117} peptide bond in nuclease, (5) perturbation of the peptide bond equilibrium at Lys^{116}-Pro^{117} by the introduction of disulfide bridges, and (6) the role of peptide bond isomerization in the folding of nuclease.

Peptide bond isomerizations in proline-containing model peptides and proteins

High-resolution NMR studies have shown that the solution conformations of short flexible proline-containing peptides can be characterized by two populations that differ only with respect to the configuration of the Xaa-Pro peptide bond. The population of the *cis* isoform varies from peptide to peptide, but generally falls within the range of 5% to 40%, depending largely on the identity of the residue preceding proline (Grathwohl and Wüthrich, 1976a, 1976b, 1981; Cheng and Bovey, 1977) and to a lesser extent on the residue following proline (Dyson *et al.*, 1988). In contrast, the population of the *cis* isoform for peptides that lack proline is considerably lower, 0.01 to 0.1% (MacArthur and Thornton, 1991). This difference is attributed to steric interactions in Xaa-Pro peptides between the pyrrolidine ring δ-position and the H^{α} and sidechain of the preceding residue. This raises the free energy of the *trans* form and effectively diminishes the net free energy difference between the *cis* and *trans* configurations.

Table I.

Perturbation Energies of the *Trans* ⇄ *Cis* Equilbrium at 298K (kcal/mol)[a]

Unfolded Configuration	Folded Configuration			
	1.0 % *Cis*	5.0 % *Cis*	95.0 % *Cis*	99.0 % *Cis*
5 % *cis*	-0.98	0.00	3.49	4.46
10 % *cis*	-1.42	-0.44	3.04	4.02
20 % *cis*	-1.90	-0.92	2.56	3.54
30 % *cis*	-2.12	-1.24	2.25	3.22
40 % *cis*	-2.48	-1.50	1.98	2.96

[a]Perturbation energies given by $\Delta G_U - \Delta G_F$

As of 1991, only seven examples of *cis* peptide bonds for residues other than proline were found in the protein data bank (Hertzberg and Moult, 1991). In these cases, the special geometry appears to be intimately related to function. By contrast, 5.7% of the Xaa-Pro linkages in a set of X-ray structures of non-homologous proteins under 2.5 Å resolution deposited in the Protein Data Bank adopted the *cis* configuration (MacArthur and Thornton, 1991). Structurally, *cis* Xaa-Pro peptide bonds, which are essential to the formation of type VI reverse turns, provide an important means of reversing the chain direction and allowing the backbone to pack onto itself.

The perturbation of a given Xaa-Pro isomerization equilibrium that accompanies protein folding can be translated into a perturbation free energy: $\Delta\Delta G^{Folding}_{Cis/Trans}$. If one assumes that the *cis:trans* equilibrium in the unfolded protein is equivalent to that in a model peptide (*i.e.*, that the equilibrium is not perturbed by the structure of the unfolded state), then one can estimate the value of the perturbation energy from the position of the equilibrium in the folded protein relative to that in a suitable model peptide (Table I).

NMR strategies for the identification and quantification of prolyl peptide bond configurations and their application to nuclease

Prolyl peptide bond configurations in peptides and proteins can be detected either by X-ray crystallography or NMR spectroscopy. The X-ray technique has the advantage of simplicity and is applicable to crystallized proteins of any size; the limitations of this method for the detection of conformational equilibria stem from difficulties in detecting minor forms in electron density maps and from the possibility that crystal packing forces perturb conformational equilibria (Kricheldorf *et al.*, 1985). NMR spectroscopy is applicable to peptides and proteins in solution in both folded and unfolded states, but the method generally is limited to proteins under 30 kD. Because of the slow rate of isomerization, separate NMR signals are expected from molecules with different Xaa-Pro peptide bond configurations; also it may be possible by changing the temperature of the sample to adjust the lifetimes of the states to values comparable with longitudinal NMR relaxation. Thus with NMR it is possible, in principle, to detect separate signals from each of the substates that arise as a consequence of isomerization of prolyl peptide bonds in folded and unfolded proteins and to study their rates of interconversion (Evans *et al.*, 1987; 1989; Loh *et al.*, 1991).

A number of indirect measures for following prolyl peptide bond configuration in solution have been reported in the literature. The most common of these is the observation of chemical shift heterogeneity for amino acid residues close in space to a prolyl residue undergoing isomerization. In nuclease, isomerization of Pro^{117} leads to heterogeneity among the well-resolved histidyl $^{1}H^{\varepsilon 1}$ resonances as well as doubling of a number of resonances from residues elsewhere in the protein (Wang *et al.*, 1990; Hinck, 1993). In calbindin $D9_{k}$, isomerization of Pro^{43} leads to structural heterogeneity affecting a number of residues in the loop that bridges its two domains, and this effect could be monitored easily by one- and two-dimensional NMR spectroscopy (Chazin *et al.*, 1989).

In the early studies of nuclease and its mutants, the individual resonances of doubled NMR signals were assigned to either the *cis* or *trans* conformer on the basis of comparison with the X-ray crystal structure (on the assumption that the major species in solution would be that seen in the X-ray structure) or on the basis of perturbations in their relative intensity when the inhibitor pdTp was added (on the assumption that the *cis* form at Lys^{116}-Pro^{117} is stabilized in the complex).

As noted above, assignment of the prolyl residue responsible for signal doublings can be established by site-directed mutagenesis of the proline to another residue, usually glycine (Evans *et al.*, 1987; Loh *et al.*, 1990; Hinck, 1993). This approach has been applied to various prolyl residues in nuclease, as illustrated in Figure 1, and to a prolyl residue of calbindin $D9_{k}$ (Chazin *et al.*, 1989).

Two different criteria have been used to distinguish more directly whether a set of NMR signals corresponds to the *cis* or *trans* Xaa-Pro configuration. The first stems from the observation that the ^{13}C NMR chemical shift of a prolyl $^{13}C^{\gamma}$ in a *cis* peptide bond is approximately 3.0 ppm upfield from that in a *trans* peptide bond. This approach has been used to characterize a number of proline-containing peptides, including poly(L-proline) (Dorman *et al.*, 1972); it has been applied as well to wild-type nuclease selectively enriched with [$^{13}C^{\gamma}$]proline (Stanczyk *et al.*, 1989). The second approach relies on the

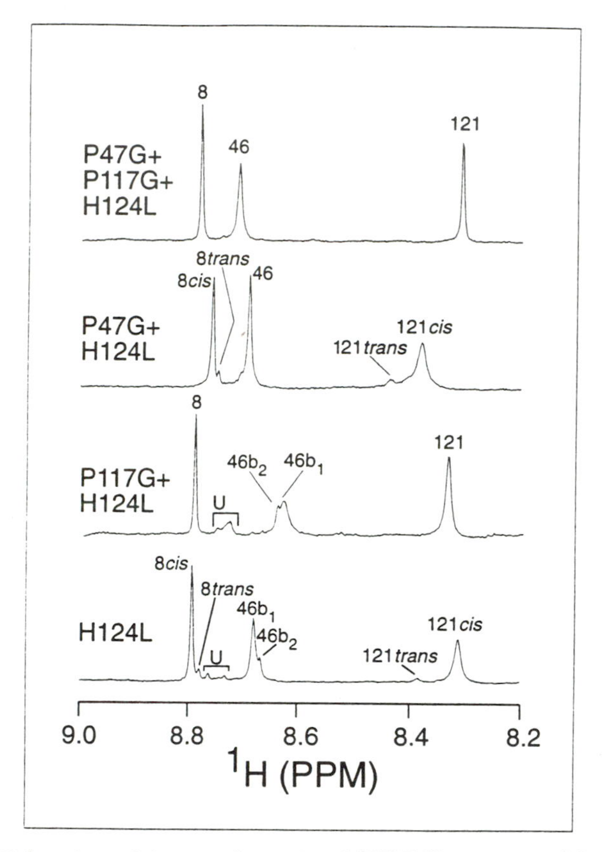

Figure 1: The $^1H^{\epsilon1}$ region of the one-dimensional ^1H NMR spectrum of four nuclease mutants recorded in ^2H_2O at 500 MHz. The histidyl $^1H^{\epsilon1}$ peaks are labeled by the residue number (8, 46, 121) and conformational form (*cis* or *trans*) to which they have been assigned.

observation of sequential nuclear Overhauser enhancement (NOE) patterns that are characteristic of *cis* or *trans* Xaa-Pro linkages. When the configuration is *trans* ($\omega = \pm 180°$), short (d < 2.7 Å) Xaa H^{α}-Pro $H^{\delta1/\delta2}$ and long (d > 4.5 Å) Xaa H^{α}-Pro H^{α} interproton distances are anticipated (Figure 2). Upon isomerization to the *cis* configuration (w = 0°), the peptide configuration is characterized by short (d < 2.7 Å) Xaa H^{α}-Pro H^{α} and long (d > 4.5 Å) Xaa H^{α}-Pro $H^{\delta1/\delta2}$ interproton distances.

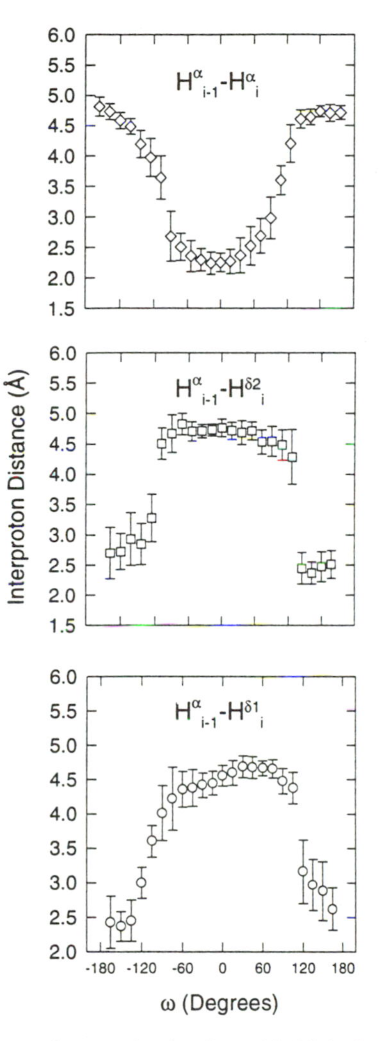

Figure 2: Three interproton distance in the dipeptide N-Ac-Lys-Pro-OMe as a function of rotation of the peptide bond dihedral angle, ω. The calcualted distances were obtained by fixing ω with a force and by performing energy minimization and 15 ps of molecular dynamics. The calculated distances and associated errors were calcualated from the mean and the standard deviation of the interproton vector sampled every 0.1 ps during the last 10 ps of molecular dynamics.

The latter approach was applied by Torchia and coworkers (Torchia *et al.*, 1989) to the ternary complex of nuclease with the activator ion Ca^{2+} and the mononucleotide inhibitor, thymidine-3′,5′-bisphosphate (pdTp). Each of the prolyl residues in nuclease, except Pro^{117}, showed strong sequential NOEs between the prolyl H^{δ} protons and H^{α} of the preceding residue. Furthermore, selective labeling of nuclease with [$^{13}C^{\alpha}$ Pro] and isotope-edited NOE spectroscopy assisted in assigning the resonances of Pro^{117} and in the identification of a strong NOE cross peak between the H^{α} resonances of Pro^{117} and Lys^{116}. These results were consistent with the prolyl peptide bond configuration identified in the X-ray structure of the ternary complex and confirmed that the conformation of nuclease in solution does not differ significantly from that in the crystalline environment.

Hinck and coworkers (1993) subsequently refined the NOE method by introducing a double labeling approach that permits unambiguous identification of NMR signals from a given Xaa-Pro linkage in any state of the protein (folded or unfolded) along with the diagnosis of its configuration as either *cis* or *trans*. In this approach, a labeled analogue is prepared by incorporating [$^{13}C^{\alpha}$]Xaa along with either [$^{13}C^{\alpha}$]Pro or [13C^{δ}]Pro. The sample containing [$^{13}C^{\alpha}$]Pro will yield an isotope-edited Xaa $^{1}H^{\alpha}$ - Pro $^{1}H^{\alpha}$ NOE if the Xaa-Pro linkage is *cis*; by contrast, the sample containing [$^{13}C^{\delta}$]Pro will yield an isotope-edited Xaa $^{1}H^{\alpha}$ - Pro $^{1}H^{\delta}$ NOE if the Xaa-Pro linkage is *trans*. This method was used to diagnose the predominant configuration of the Pro^{117} peptide bond in nuclease H124L[2] in the absence of active site ligands as *cis* and those in two single amino acid variants of nuclease as *trans* (Figure 3). These results provided the first direct evidence that the alternative population of protein molecules has a *trans* Lys^{116}-Pro^{117} peptide bond and qualitatively demonstrated that the heterogeneity observed among the histidyl residues is directly correlated with *cis:trans* isomerization of the Lys^{116}-Pro^{117} peptide bond. Because of the limited sensitivity of the method, it is better used for the configurational assessment of the predominant form present in solution rather than for quantitative determination of the equilibrium constant for the process. The method, however, is general and provides an attractive approach for determining the dominant form of an Xaa-Pro peptide bond under various conditions because the sequential assignment is encoded by the labeling pattern provided that the labeled dipeptide is unique.

Models proposed to explain the positions of the *cis:trans* equilibria at the His^{46}-Pro^{47} and Lys^{116}-Pro^{117} peptide bonds

The indirect effect of Lys^{116}-Pro^{117} *cis:trans* isomerization on the chemical shifts of the His^{121} $^{1}H^{\varepsilon 1}$ resonances in partially heat denatured nuclease has been used to ascertain the population of the *cis* isoform in the unfolded

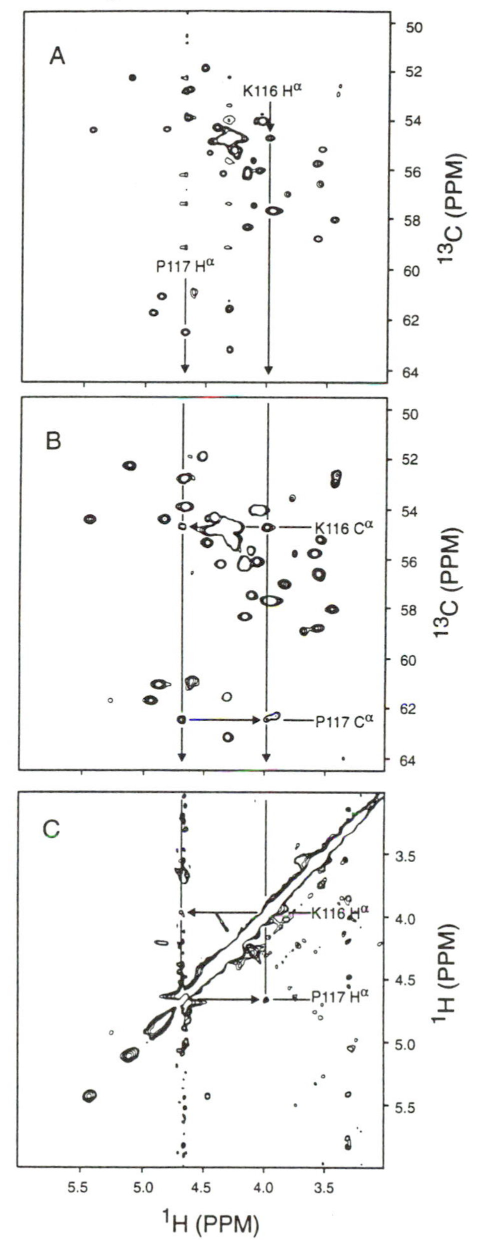

Figure 3: HSMQC (A), HMQC-NOE (B), and ω_1 IE-NOESY (C) spectra of [$^{13}C^{\alpha}Lys^{13}C^{\beta,\delta}$ Pro] G79S+H124L in the absence of active site ligands. The data were collected at 37 °C on a Bruker AM500 spectrometer equipped with a $^1H\{^{13}C\}$ variable temperature probe. The NOE mixing time was 80 ms in the HMQC-NOE and IE-NOESY experiments. Symmetric "off-diagonal" noesy cross-peak linking $Lys^{116}H^{\alpha}$ and Pro^{117} $H^{\delta 1/\delta 2}$ are diagnostic of a *trans* peptide bond.

population. This estimate, 4.7%, which was obtained by observing magnetization transfer cross peaks between the previously assigned $^1H^{\varepsilon 1}$ peaks corresponding to *cis* and *trans* forms of the folded protein and the *cis* and *trans* forms of the unfolded protein (Evans *et al.*, 1989, Alexandrescu, *et al.*, 1989), is in good agreement with the value of 4 ± 2% obtained from NMR studies of a synthetic linear peptide corresponding to residues 113 - 122 of nuclease WT (Raleigh *et al.*, 1992). Similarly, the most accurate estimates for the proportion of the *cis* isoform of the Lys^{116}-Pro^{117} peptide bond in the folded population has been obtained by recording peak areas of the histidyl $^1H^{\varepsilon 1}$ resonances corresponding to the *cis* and *trans* forms of the folded protein. This equilibrium has been found to depend on a number of factors including amino acid sequence (Alexandrescu *et al.*, 1990; Hodel, 1993; Hinck, 1993), temperature (Evans *et al.*, 1989), pH (Alexandrescu *et al.*, 1989), and the state of ligation by Ca^{2+} and pdTp (Evans *et al.*, 1989; Alexandrescu *et al.*, 1989). In ^2H_2O at pH*[3] 5.5 and 37 °C, wild type nuclease contains between 83 and 88% of the *cis* isoform (Alexandrescu *et al.*, 1990; Evans *et al.*, 1989), whereas the more thermostable naturally occurring variant, nuclease H124L, contains 95% of the *cis* isoform (Alexandrescu *et al.*, 1990). Thus, the Lys^{116}-Pro^{117} peptide bond in folded staphylococcal nuclease exists predominantly in the intrinsically unfavorable *cis* configuration. Estimates of the energy required for the perturbation of the *cis:trans* equilibrium of this peptide bond upon protein folding, $\Delta\Delta G^{Folding}_{Cis/Trans}$, range between 2.9 and 3.6 kcal/mol (Table II).

Estimates for $\Delta\Delta G^{Folding}_{Cis/Trans}$ associated with the His^{46}-Pro^{47} peptide bond have been more difficult to obtain, owing to the somewhat larger linewidths for the His^{46} $^1H^{\varepsilon 1}$ resonances and the fact that they partially overlap with histidyl $^1H^{\varepsilon 1}$ resonances of the unfolded protein. ^{15}N relaxation studies suggest that this broadening is due to conformational exchange on the millisecond time scale (Kay *et al.*, 1989). Rather than resulting from isomerization of either the His^{46}-Pro^{47} or Thr^{41}-Pro^{42} peptide bond, which would be much slower, the lineshape effect may be due to changes in proline ring puckering at Pro^{42} as suggested by Loh and coworkers (1991). Interestingly, the omega loop

Table II

Strain Energies for the *Trans* ⇄ *Cis* Equilibrium of the Lys^{116}-Pro^{117} Peptide Bond in Staphylcoccal Nuclease (kcal/mol)[a]

Mutant	% *cis*, Unfolded	% *cis*, Folded	$\Delta\Delta G^{Folding}_{Trans \Leftrightarrow Cis}$
Wild Type	4 ± 2	85 ± 3	2.9
H124L	4 ± 2	95 ± 3	3.6

[a]Strain energies given by ΔG_U - ΔG_F

containing residues 42 and 47 is disordered in the X-ray structures of nuclease (Hynes and Fox, 1991) and in that of the ternary complex o nuclease with Ca^{2+} and pdTp (Loll and Lattman, 1989; Cotton and Hazen, 1979). Although NMR signals from His^{46} in nuclease H124L were found to sharpen when either Pro^{47} or Pro^{42} was mutated to Gly (Figure 1), the omega loop region, which is disordered in the X-ray structure of nuclease H124L, is similarly disordered in the X-ray structure of nuclease H124L+P47G+P117G (Truckses *et al.*, 1995).

Attempts have been made to determine the molecular basis for the preferred *cis* configuration at the Lys^{116}-Pro^{117} peptide bond. Single amino acid substitutions at residue positions remote from Pro^{117} have large effects on the equilibrium of *cis* and *trans* forms at the $Xaa^{116}Pro^{117}$ peptide bond ($K_{Cis/Trans}$) (Alexandrescu *et al.*, 1990; Hinck, 1993). The perturbation energy for this equilibrium is negatively correlated with the overall free energy for folding, ΔG_F (Figure 4). These results are consistent with the notion that the stability and packing of the overall fold are important for restricting the Lys^{116}-Pro^{117} peptide bond to the intrinsically unfavorable *cis* configuration.

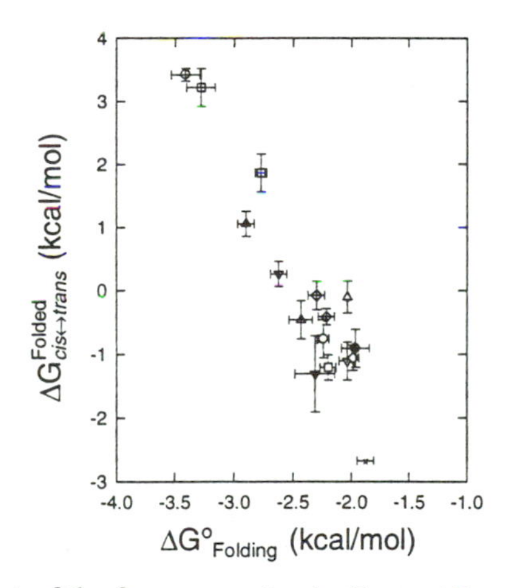

Figure 4: Correlation plot of the free energy for the $F_{cis} \rightleftharpoons F_{trans}$ equilibrium and the global stability ($\Delta G_{Folding}$) as measured by extrapolating $\ln(K_{app})$ to zero denaturant concentration in GdmCl titration experiments. Each point corresponds to a single mutant: (○) H124L; (□) WT; (△) G79Y+H124L; (▽) G79D+H124L; (◇) G79A+H124L; (●) G79P+H124L; (■) G79I+H124L; (▲) G79F+H124L; (▼) G79L+H124L; (◆) G79S+H124L; (○) G79V+H124L; (□) K78I+H124L; (△) T120A+H124L; (▽) N119A+H124L; (◇) N118D+H124L;(x) D77A+H124L.

Analysis of the overall stability of nuclease and free energies for the *cis:trans* equilibrium involving Pro^{117} in terms of a four-state model (Eq. 1) revealed that $\Delta G^{Folded}_{Trans \Leftrightarrow Cis}$ and $\Delta G^{Cis}_{Folding}$ are coupled, but that $\Delta G^{Folded}_{Trans \Leftrightarrow Cis}$ and $\Delta G^{Trans}_{Unfolding}$ are not (Alexandrescu, 1990).

$$\begin{array}{ccccc} & & \Delta G_{cis \rightleftharpoons trans}^{Folded} & & \\ & F_{cis} & \rightleftharpoons & F_{trans} & \\ \Delta G_{Folding}^{Cis} & \updownarrow & \circlearrowleft & \updownarrow & \Delta G_{Unfolding}^{Trans} \\ & U_{cis} & \rightleftharpoons & U_{trans} & \\ & & \Delta G_{trans \rightleftharpoons cis}^{Unfolded} & & \end{array} \qquad (1)$$

These results were explained in terms of a model in which the *cis* to *trans* transition involves disruption of intramolecular interactions normally present in the *cis* folded form. This model suggests that amino acid substitutions that promote the *cis* to *trans* transition destabilize molecular interactions within F_{cis} that are unique to the folded *cis* form. This model explains the finding that mutants that affect the *cis:trans* equilibrium destabilize the *cis* folded form much more than the *trans* folded form: such mutations are assumed to disrupt favorable interactions present in the *cis* folded form of wild-type nuclease that must be broken upon unfolding; since they are already disrupted in the *trans* form of both wild-type and mutant nuclease, their absence in the mutant has little effect on the stability of the *trans* form. The observed temperature dependence of $\Delta G^{Folded}_{Trans \Leftrightarrow Cis}$ suggests that F_{cis} is stabilized primarily by enthalpic contributions, whereas, the higher entropy of F_{trans} is consistent with the idea that this form of the protein is more disordered (Evans et al., 1989; Alexandrescu et al., 1990; Hinck, 1993) and less compact (Royer et al., 1993). The model also is consistent with kinetic parameters for folding and unfolding obtained from a series of magnetization transfer studies (Evans et al., 1989) which suggested that the folding and unfolding transitions between F_{cis} and U_{trans} occur more readily by the $F_{cis} \Leftrightarrow F_{trans} \Leftrightarrow U_{trans}$ pathway rather than by the $F_{cis} \Leftrightarrow U_{cis} \Leftrightarrow U_{trans}$ pathway. These data suggest that the *cis* peptide bond is stabilized only at a late stage in the folding process. A positive correlation was found between the rate of $F_{trans} \rightarrow F_{cis}$ and the population of the F_{cis} state (Hinck and Loh, 1992); this suggests that the interactions that stabilize the *cis* Lys^{116}-Pro^{117} peptide bond promote the interconversion, possibly by stabilizing the transition state.

Hodel and coworkers (1993) reexamined the energetics of these processes and discussed two mechanisms to account for the preferred *cis* configuration at the Lys^{116}-Pro^{117} peptide bond in the folded form of nuclease: (1) stabilization of *cis* Lys^{116}-Pro^{117} by favorable interactions that exist only in the *cis* configuration, and (2) destabilization of the *trans* form by restraining the β-turn containing Pro^{117} at the loop ends (V^{111}and N^{118}, respectively) in such a way that the local backbone interactions of all accessible conformations with a *trans* Lys^{116}-Pro^{117} peptide bond are more strained and energetically less favorable than those of the observed type VIa β-turn containing a *cis* peptide bond (which the authors designate as the "loop anchorage" mechanism). In an effort to distinguish between these two mechanisms, mutant proteins were investigated in which Lys^{116} was changed, respectively, to Gly or Ala (Hodel *et al.*, 1993). X-ray, NMR, and thermodynamic analyses revealed that the K116A mutant maintains a predominantly *cis* configuration (92%) and has a thermal stability similar to that of wild type nuclease, whereas the K116G mutant possesses a significantly lower population of the *cis* configuration (20%) and has enhanced thermal stability. The X-ray structure of K116A is nearly identical to that of wild type nuclease, whereas that of K116G differs mainly in residues 112 through 117. In this structure, the loop formed by these residues is somewhat bowed out, and Gly^{116} occupies a "forbidden" region of ϕ,ψ space. These results were interpreted as evidence for the loop-anchorage mechanism. The Lys to Gly substitution eliminates the coupling between loop anchorage and destabilization of the *trans* peptide bond at Pro^{117}. The Gly residue can adopt a combination of ϕ,ψ dihedral angles that is "forbidden" to the Lys residue and in doing so relieves the backbone strain that would normally function to destabilize the *trans* configuration of the Lys^{116}-Pro^{117} peptide bond. The enhanced thermostability of K116G was attributed to the decreased strain energy present in the folded form of the protein containing a *trans* Gly^{116}-Pro^{117} peptide bond. It should be noted that the effect of this glycine substitution in nuclease differs from the usual destabilizing effect of such substitutions in other proteins, such as human lysozyme. The usual destabilizing effect of a Pro to Gly substitution has been explained in terms of an increase in the configurational entropy of the unfolded state (Herning *et al.*, 1992; Nemethy *et al.*, 1966). Thus the stabilization of the native state resulting from relief of strain must be considerably greater than the stabilization of the unfolded form as a result of its increased entropy.

The loop anchorage hypothesis was investigated further by considering Ala and Gly substitutions for Asn^{118} (Hodel *et al.*, 1994). Asn^{118} follows *cis* Pro^{117} and is believed to function as the C-terminal loop anchor. Unlike the sidechain of Lys^{116}, which is completely solvated, that of Asn^{118} shows complicating interactions with the protein backbone. The Asn^{118} sidechain amide and δ-oxygen participate in hydrogen-bond interactions as a donor and acceptor, respectively, to the backbone oxygen of Gln^{80} and the amide proton of Gly^{79}. Ala and Gly substitutions at positions 116 and 118 have similar overall effects on the *cis:trans* equilibrium at the 116-117 peptide bond (Hodel *et al.*, 1993): both mutations to Gly lead to major shifts away from the 90% *cis* isoform of wild-type nuclease (N118G has 28% *cis*; K116G has 20% *cis*), whereas both mutations to Ala have lesser effects on the equilibrium (N118A has 45% *cis*; K116A has 92% *cis*). Whereas changes in $K_{Cis/Trans}$ resulting from mutations at residue 116 were ascribed (as described above) to N-terminal loop anchorage effects, those resulting from mutations at 118 were attributed to disruption of interactions between the sidechain of Asn^{118} and the surrounding protein context as well as C-terminal loop anchorage effects which may extend beyond residue 118.

Non-covalent interactions responsible for stabilization of the *cis* Lys^{116}-Pro^{117} peptide bond in nuclease

The two models outlined above are not mutually exclusive but describe essential elements of conformational control. The first model was developed in order to explain why mutations at positions sequentially remote from Pro^{117} lead to changes $K_{Cis/Trans}$ at Pro^{117} (Alexandrescu *et al.*, 1990; Hinck, 1993). The second model was developed to explain the effects of mutations at or near residue 117; mutations that introduce flexibility into the backbone of the loop were found to favor the *trans* configuration of the 116-117 peptide bond. The major point of disagreement concerns whether strain exists within the loop in the wild-type protein when the Lys^{116}-Pro^{117} peptide bond is *trans*: Fox and coworkers have assumed that this state is strained whereas Markley and coworkers have assumed that it is not. The best evidence on this point are findings that the stabililty of the *trans* state is not affected by those mutations that alter the *cis:trans* ratio as the result of changes in tertiary interactions (Alexandrescu *et al.* 1990). On the other hand, such mutations do affect the rate of $F_{trans} \rightarrow F_{cis}$ (Hinck *et al.*, 1995b); this suggests that tertiary interactions present in F_{cis} but absent in F_{trans} are at least partially present in the transition state whose energy may be lowered as the result of strain-induced distortion of the peptide bond. Mutations that introduce flexibility into the loop serve to decouple the state of the 116-117 peptide bond from the state of these critical tertiary interactions. The NMR and structural data obtained on the K116A, K116G, N118A, and N118G mutants clearly established the important interplay between limited local backbone flexibility and the conformational preference of the prolyl peptide bond (Hodel *et al.*, 1993; 1994).

Specific interactions responsible for loop anchorage can be identified in the X-ray crystal structures of wild type nuclease (Hynes and Fox, 1992) and its ternary complex with Ca^{2+} and pdTp (Loll and Lattman, 1989; Cotton, 1979). A ball and stick representation of the positions of the heavy atoms of the residues that constitute the type VIa reverse turn containing *cis* Pro^{117} is presented in Figure 5. Salient features of the model include the absence of molecular interactions between atoms of three of the four-residues in the β-turn

(Tyr[115], Lys[116], and Pro[117]) and the remainder of the protein context, the presence of two hydrogen-bonding interactions between the sidechain of the fourth residue in the β-turn (Asn[118]) and the backbone of the loop region spanning Asp[77] to Ala[87], and the presence of a network of hydrogen bonds linking the sidechains of Asp[77] and Lys[78] with the sidechain and amide proton of Thr[120].

The roles of hydrogen-bond interactions between the sidechain amide O^{δ} of Asn[118] and the backbone amide proton of Gly[79] and between one of the sidechain amide protons of Asn[118] and the backbone oxygen of Gln[80], have been investigated by considering properties of the N118D, N118A, and N118G mutants described above (Hodel *et al.*, 1994). As discussed above, the N118G and N118A mutations helped elucidate the effects of backbone flexibility on the configurational preference of Pro[117]. Here we consider the N118A and N118D mutants. Since the backbone flexibilities of aspartate and alanine are expected to

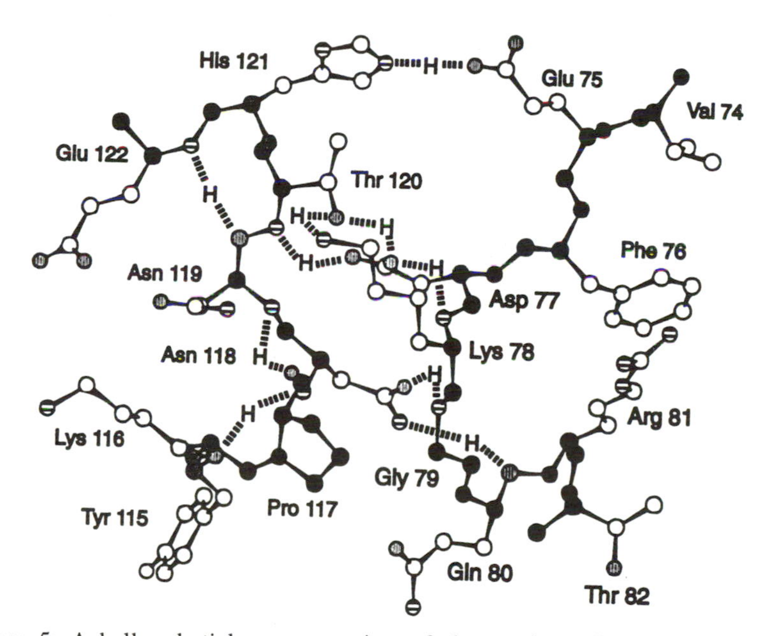

Figure 5: A ball-and-stick representation of the portion of the X-ray model of nuclease WT (Hynes and Fox, 1992) showing the type VI_a reverse turn containing *cis* Pro[117] and the adjacent loop containing Gly[79]. Except for those atoms involved in hydrogen bonding, all of the backbone atoms of residues 74-82 and 115-122 and the sidechain atoms of Pro[117] are drawn with heavy shading. Oxygen atoms are indicated by fine vertical hatching and nitrogen atoms are indicated by thick horizontal hatching. The hydrogen bonds indicated are those reported in the X-ray structure of nuclease WT (Hynes and Fox, 1992) in absence of active site ligands.

be similar to that of asparagine (present in the wild-type), changes in the population of the *cis* conformer should reflect differences in hydrogen-bonding potential between the sidechain and neighboring groups. NMR studies (Hodel *et al.*, 1994) showed that the proportion of the *cis* isoform present in solution at pH* 5.3 is reduced from 88% in the wild type to 32% in N118D and to 45% in N118A. The authors interpreted these results, along with their observation that the *cis* population in the N118D mutant increases as the pH is lowered, as evidence for disruption in the mutants of hydrogen-bonding interactions important for C-terminal loop anchorage; these hydrogen bonds link the sidechain of Asn^{118} with the amide proton of Gly^{79} and the backbone carbonyl oxygen of Asn^{118} (Hodel *et al.*, 1994).

Single-site mutations at three other positions have been found to cause significant perturbations of the *cis:trans* equilibrium at Pro^{117}: these are mutations of Asp^{77}, Gly^{79}, and His^{124}. The first to be considered is the single amino acid mutant, D77A. One-dimensional NMR studies, inhibitor titrations, and selective lysine/proline labeling, in conjunction with results from isotope-edited two-dimensional NMR spectroscopy, have shown that the proportion of the *cis* isoform present in solution at pH* 5.5 for this mutant is 1% or less (Hinck *et al.*, 1993). In addition, the D77A mutation leads to a striking decrease in the stability of the protein: over 3.0 kcal/mol at pH 7.0 and 295° K (Hinck, 1993). In the crystal structure of wild-type nuclease, the sidechain $O^{\delta 1}$ and $O^{\delta 2}$ of Asp^{77} are within hydrogen-bonding distance to the backbone nitrogen and O^{γ} of Thr^{120}, respectively. The significance of a hydrogen bond between the Thr^{120} O^{γ} and Asp^{77} $O^{\delta 2}$, is called into question by results from the T120A mutation which was found to have little effect on either $K_{Cis/Trans}$ or overall protein stability (Hinck, 1993). On the other hand, experimental evidence supports the notion that a strong hydrogen bond links Asp^{77} $O^{\delta 1}$ and the amide nitrogen of Thr^{120} and that this hydrogen bond is important to the overall stability of the protein and to the conformational preference of *cis* Pro^{117} (Hinck, 1993). Unlike those of other sidechain chains in this portion of the structure, sidechain atoms of Asp^{77} are totally inaccessible to solvent. Moreover, the backbone amide of Thr^{120} is characterized an extremely low amide D/H fractionation factor (ϕ): 0.28 and 0.39 in the presence and absence of bound ligands, respectively (Loh and Markley, 1994). A growing body of evidence, from studies of model compounds (Kreevoy and Liang, 1980), enzyme-substrate complexes (Weiss *et al.*, 1987a, 1987b), and *ab initio* calculations (Edison *et al.*, 1995), suggests that low fractionation factors ($\phi \leq 0.6$) are indicative of strong hydrogen bonding. The Asp^{77}-Thr^{120} amide hydrogen bond clearly plays an important role in the stability of nuclease and its coupling to the position of $K_{Cis/Trans}$. Studies of mutations at position 119 (for example, N119A or N119G) would help clarify the molecular mechanism for the latter effect.

Another interesting class of mutants comprises those in which Gly^{79} was substituted by a diverse selection of other amino acid types (Alexandrescu *et al.*, 1990; Hinck *et al.*, 1990; Hinck *et al.*, 1993, Hinck, 1993). The two conspicuous effects common to each of these single amino acid replacements were a dramatic shift in the *cis:trans* equilibrium in favor of the *trans* isoform and a large reduction in the overall stability of the protein (Table III). The ϕ and ψ backbone torsion angles for Gly^{79} are -102° and -145°, respectively, in the X-ray structure of unligated nuclease (Hynes and Fox, 1991). This region of ϕ-ψ space is accessible only to Gly residues. Thus, the universal effect of the substituted residues probably arises from conformational distortion of the loop segment bridging Asp^{77} and Ala^{87}. The consequences of this distortion have not been investigated directly, but the large decrease in overall stability that accompanies these substitutions and the enhanced configurational entropy characteristic of these mutants (Alexandrescu, 1990) suggests this effect interferes with the ability of the protein to form a compact folded structure. Several mechanisms can be envisioned by which the substitution of Gly^{79} could alter the configurational preference of Pro^{117}. Hodel and coworkers (1993) have suggested that mutations such as G79S lead to a destabilization of the overall fold, which in turn leads to diminished loop anchorage. A more specific mechanism, yet to be investigated, might involve steric interactions between backbone and sidechain residues of the loop bridging Asp^{77} and Ala^{87} with the type VIa reverse turn containing Pro^{117}. In the crystal structure of wild-type nuclease (Hynes and Fox, 1991), the C^{α} atoms of Gly^{79} and Pro^{117} are separated by only 5.5Å, and the pro-S hydrogen of Gly^{79} is oriented directly toward the pyrrolidine ring of Pro^{117}. Modulation of steric interactions between these two elements of secondary structure might provide an additional means of influencing the local backbone conformation of the β-turn containing Pro^{117} in

Table III.
Gly^{79} Nuclease Mutants: Global Stability and Population of the *Cis* Isoform

Mutant	$\Delta G_{Unfolding}$ (kcal/mol)[a]	$\Delta\Delta G_{Unfolding}$ (kcal/mol)[b]	% *Cis*[c]
H124L	6.78	0.00	94.5
Wild Type	5.51	1.27	82.0
G79Y+H124L	4.83	1.95	50.0
G79D+H124L	4.59	2.19	31.3
G79A+H124L	4.57	2.21	43.1
G79P+H124L	4.45	2.33	35.3
G79I+H124L	4.37	2.41	27.7
G79F+H124L	4.04	2.74	29.9
G79L+H124L	4.04	2.74	33.6
G79S+H124L	3.94	2.84	51.3
G79V+H124L	3.90	2.88	38.4

[a]As determined by monitoring the intrinsic fluorescence of Trp^{140} as a function of increasing GdmCl concentrations at pH 7.0 and 294.5 K.
[b]Calculated from ΔG_{H124L} - ΔG_{Mutant}
[c]As detetermined by integrating the intensity of the $^1H^{\epsilon 1}$ signals corresponding to *cis* and *trans* folded forms at pH* 5.2 and 313 K.

a manner analogous to that proposed by the loop anchorage hypothesis (Hodel *et al.*, 1993). Structural studies of mutants containing Gly^{79} substitutions may shed light on this question.

His^{124} is a third amino acid position at which single-site substitutions have been found to affect protein stability as well as the *cis:trans* equilibrium of the Lys^{116}-Pro^{117} peptide bond. Located on the solvent exposed face of helix 3 (residues 121-135), the sidechain of His^{124} makes contacts only with other residues in the helix. As shown in Table IV, replacement of this residue with isoleucine, alanine, or leucine leads to enhanced stability that correlates with an increased population containing a *cis* Lys^{116}-Pro^{117} peptide bond (Truckses *et al.*, 1995). Because this residue is not involved directly in tertiary interactions, the increased global stability of the protein most likely results from the helix-stabilizing effects of these substitutions which are coupled indirectly to more stable tertiary interactions as described below. Studies of amino acid substitutions in model peptides (O'Neil and DeGrado, 1990; Lyu *et al.*, 1990; Padmanabhan *et al.*, 1990; Creamer and Rose, 1994) and proteins (Horowitz *et al.*, 1992; Blaber *et al.*, 1994) suggest an increase in helical propensity for these amino acids in the following order: histidine, isoleucine, leucine, alanine. The relative stabilities of the His^{124} mutants follows this order, except for a reversed order for the alanine and leucine substitutions. In the nuclease context, leucine may be able to make additional stabilizing contacts (DeGrado and Lear, 1985) that are not possible for alanine.

Structural changes responsible for coupling between substitutions at His^{124} and the *cis:trans* equilibrium of the Lys^{116}-Pro^{117} peptide are not apparent from comparison of their crystal structures (Truckses *et al.*, 1995). Such coupling may arise from subtle changes in packing of the N-terminal side of the loop containing Pro^{117} with the sidechains of residues that reside in helix 3.

Table IV

His^{124} Nuclease Mutants: Global Stability and Population of the *Cis* Isoform

Mutant	$\Delta G_{Unfolding}$[a]	% *Cis*[b]
WT	5.5	88.0
H124I	5.9	91.0
H124A	6.0	91.5
H124L	6.6	93.5

[a]As determined by monitoring the intrinsic fluorescence of Trp^{140} as a function of increasing GdmCl concentrations at pH 7.0 and 293 K

[b]As determined by integrating the intensity of the $^{1}H^{\varepsilon 1}$ signal of His^{121} corresponding to the *cis* and *trans* folded forms at 310 K in buffer containing 50 mM [^{2}H] succinate , pH* 5.3.

More specifically, Val^{111} is within van der Waals contact distance of three residues in helix 3: Leu^{125}, Arg^{126}, and Glu^{129} Also, the sidechain of Glu^{129} hydrogen bonds to the amide proton of Val^{111}. These interactions are part of the proposed anchoring of the loop containing Pro^{117} (Hodel *et al.*, 1993). Stabilization of helix 3 would stabilize the positions of the helical residues and thus their interactions with Val^{111}. This stronger anchoring of the loop would then serve to stabilize the *cis* population of the Lys^{116}-Pro^{117} peptide bond (Hodel *et al.*, 1993). In other words, we propose that the helix-stabilizing effect of a His^{124} mutation propagates to stabilized tertiary interactions that, in turn, are stronger in the conformational state with a *cis* peptide bond than in the state with a *trans* peptide bond at Lys^{116}-Pro^{117}. The series of mutations at His^{124} illustrate how the replacement of an individual amino acid can lead to structural adjustments quite far away in the structure. Such long-range interactions can be rather difficult to uncover from structural information alone as shown by the near identity of the X-ray structures of H124L and wild-type nuclease (Truckses *et al.*, 1995).

Perturbation of the peptide bond equilibrium at Lys^{116}-Pro^{117} by the introduction of disulfide bridges

It should be possible to alter the configuration adopted by a particular Xaa-Pro peptide bond in a protein in a rational way. One approach, which follows logically from the results outlined above, would be the introduction of novel disulfide bridges designed to link the type VI_a reverse turn containing *cis* Pro^{117} and the extended loop segment bridging Asp^{77}-Ala^{87}. Disulfide bridges that stabilize this structure, which is present in the *cis* isoform but disrupted in the *trans* isoform, are expected to shift $K_{Cis/Trans}$ toward the *cis* state. The effect of the bridge can be evaluated by comparing $K_{Cis/Trans}$ determined for the oxidized (disulfide bridge intact) and reduced forms of the protein. In order to evaluate these ideas, four nuclease variants, each containing two cysteines, were designed and constructed:[5] $H124L^{C79-C118}$, $H124L^{C77-C118}$, $H124L^{C80-C116}$, and $G79S{+}H124L^{C80-C116}$ (Hinck *et al.*, 1995a), where the subscripts indicate the positions of the inserted cysteines. The locations of the residues replaced by cysteine in the protein structure are indicated in Figure 5.

The consequences of disulfide bond formation on $K_{Cis/Trans}$ and overall protein stability were characterized by one-dimensional ^1H and two-dimensional $^1H\{^{13}C\}$ isotope-edited NOE spectroscopy of selectively ^{13}C enriched samples. This analysis has revealed the presence of coupling between $K_{Cis/Trans}$ and disulfide formation in all the disulfide mutants except $H124L^{C77-C118}$. The expected increase in protein stability was found to be modulated precisely by the observed perturbation in $\Delta\Delta G^{Folding}_{Cis/Trans}$ at the Lys^{116}-Pro^{117} peptide bond (Hinck *et al.*, 1995a).

Role of peptide bond isomerization in the folding of nuclease

The means by which an unfolded polypeptide chain adopts its proper three-dimensional structure has been a topic of intense research interest. Early thermodynamic results (Privalov and Khechinashvilli, 1974) could be modeled adequately by considering only two macroscopic states: folded (F) and unfolded (U). Recent advances in structural biology along with newer calorimetric techniques have revealed that the folding of many proteins cannot be accounted for by the simple two-state mechanism (e.g., see Carra *et al.*, 1994; Carra and Privalov, 1995). Most proteins do not undergo simple monophasic folding kinetics. Instead, their folding is often described by complex pathways having fast and slow phases of varying amplitudes.

Cis:trans isomerizations at Xaa-Pro peptide bonds provided the first clear mechanism for slow steps in protein folding. Peptide bond isomerization was inferred from comparisons of the rates of slow steps in the folding of proline-containing proteins with those of prolyl *cis:trans* isomerization in proline-containing peptides (Brandts *et al.*, 1975). More direct evidence in support of this model has come from studies of isomer-specific proteolysis (Lin and Brandts, 1986) and effects of mutations of specific prolyl residues on folding kinetics (Kiefhaber *et al.*, 1990). A prototypical kinetic mechanism for the folding of a protein containing a single *cis* prolyl residue is indicated by Eq. 2.

$$U_T \rightleftarrows I_T \rightleftarrows F_C \qquad (2)$$

In this mechanism, the protein passes rapidly from the unstructured, unfolded state (U_T) to an intermediate folded state (I_T) with conservation of the *trans* isomer of the prolyl peptide bond. Final folding, concomitant with isomerization of the prolyl peptide bond from *trans* to *cis*, leads to the final folded form (F_C). The nature of I_T will vary depending on whether the *cis* form is stabilized early or late in the folding pathway of the protein under study. In some proteins, I_T will be very native-like, while in others, I_T will be largely unfolded, and only the *cis* form (U_C) will be able to fold efficiently to the native state.

Nuclease has long been a model system for kinetic studies of protein folding. As early as 1970, stopped flow fluorescence techniques revealed the existence of two kinetic phases in the refolding of nuclease from acidic pH (Schecter *et al.*, 1970). Subsequent studies (Chen *et al.*, 1991; Sugawara *et al.*, 1991; Chen and Tsong, 1994; Walkenhorst *et al.*, 1995b) have revealed at least five kinetic phases in the folding of nuclease. Several of the slowest phases have been attributed to prolyl *cis:trans* isomerization (Kuwajima *et al.*, 1991; Shalongo *et al.*, 1992; Nakano *et al.*, 1993), most notably to isomerization at P117. More recent studies, however, have shown that the nuclease mutant P117G + P47G +

H124L, which lacks the two prolines responsible for the conformational heterogeneity observed by NMR in solution (Evans *et al.*, 1989, Wang *et al.*, 1990, Loh *et al.*, 1991), retains multiphasic refolding kinetics (Walkenhorst *et al.*, 1995b). Similarly, the mutant P117G + P42G + H124L has been found to exhibit multiphasic refolding behavior. Previously, Nakano *et al.* (1993) had shown that the folding kinetics of the P31A mutant (in the wild-type background) are very similar to those of the wild-type protein. In each of the two double proline mutants described above, the amplitudes of the two slowest phases are decreased by a factor of two relative to H124L, whereas the rates of the next-to-slowest phase (~40 second lifetime) are accelerated in each case (Walkenhorst *et al.*, 1995b). It remains to be seen whether this phase will disappear completely in mutants with replacements for both Pro47 and Pro42 or if these substitutions will simply result in a faster proline-independent conformational step.

Staphylococcal nuclease exhibits apparent two-state folding behavior when measured by many different probes of native structure (Schecter *et al.*, 1970; Anfinsen *et al.*, 1972; Vidugiris *et al.*, 1995). However, when stability is measured by GdmCl denaturation experiments, different single point mutants of nuclease exhibit wide variation in their *m* values.[6] In order to account for these, Shortle proposed that the GdmCl denaturation data could be explained by residual structure in the unfolded state (Shortle, 1986; Shortle and Meeker, 1986). However more recently, Privalov and coworkers (Carra *et al.*, 1994; Carra and Privalov, 1995) have shown that nuclease mutants with *m*- denaturation profiles exhibit three-state unfolding when monitored by calorimetry. Therefore, they suggest that the variation in *m* value reflects variation in the relative stability of a folding intermediate: *m*+ mutants destabilize the intermediate, whereas *m*- mutants stabilize the intermediate. In other words, changes in the *m* value arise from changes in the cooperativity of folding. In calorimetry, a typical scan speed is 1 K/minute, so that *cis:trans* isomerizations at Xaa-Pro peptide bonds may be assumed to be at equilibrium. Several other studies have provided evidence for three-state unfolding for nuclease (Gittis *et al.*, 1993; Creighton and Shortle, 1994). Comparison of results from different probes for denaturation (Carra *et al.*, 1994) reveals that the equilibrium unfolding intermediate contains the β-sheet region but no α-helix. Pulsed hydrogen exchange studies of protein refolding have indicated extensive early protection of backbone amide in the β-sheet region of nuclease, but not in α-helices (Jacobs and Fox, 1994, Walkenhorst *et al.*, 1995ab). It is possible, therefore, that the structure of the kinetic folding intermediate corresponds to that of the hypothesized equilibrium folding intermediate.

In light of their recent data, Carra *et al.* (1994) proposed that nuclease be regarded as a two-subdomain protein divided along the active-site cleft. These two subdomains consist of a β-sheet domain and an α-helical domain, and these are connected across the active-site cleft at one end by the docking of two loops. Electrostatic interactions between residues on the two loops, such as those involving Asp^{77} described above, appear to contribute to the cooperativity of the overall unfolding reaction. Removal of these interactions, either by subjecting the protein to low pH and high salt, or by modification of the Asp^{77} sidechain by mutagenesis, results in three-state denaturation (Carra *et al.*, 1994). Three-state behavior was found to correlate strongly with *m*- behavior (Carra *et al.*, 1994; Carra and Privalov, 1995) as does the *cis:trans* ratio in many mutants of nuclease (Hinck, 1993). Thus, the D77A mutation in nuclease, which exists predominantly in the *trans* configuration, exhibits three-state denaturation by calorimetry even at neutral pH (Carra *et al.*, 1994). In nuclease then, the state of the Xaa^{116}-Pro^{117} peptide bond appears to control the docking of the two subdomains: the *cis* form is more compatible and the *trans* form is less compatible with docking. It is of interrest to note that mutations in which Pro^{117}, Pro^{47}, or Pro^{42} are replaced by Gly or Ala are nearly all *m*- (Green *et al.*, 1992) and that all three of these residues lie between the two subdomains of the protein. One indication of the functional relevance of these effects is the finding that mutations of Pro^{47} and Pro^{117} lower the enzymatic activity of nuclease (Truckses *et al.*, 1995).

A final topic, introduced in the preceding sections, concerns the kinetic mechanism for docking of the two subdomains of nuclease. Since the unfolded protein starts out largely as the *trans* isomer, an intriguing question is whether the *trans* form can dock directly, or must first isomerize to *cis*. As discussed in Section 3, the model suggested by Hodel *et al.* (1993) postulates that docking of the *trans* form should lead to strain in the 111-118 loop. However, the observed lack of influence of mutations that destabilize the *cis* isoform on the stability of the *trans* folded form (Alexandrescu *et al.*, 1990) suggests that such strain does not exist in the ground state. On the other hand, there is evidence that stabilizing interactions generated in the transition state between the *cis* and *trans* forms serves to lower the energy of that transition state (Hinck *et al.*, 1995b). This analysis suggests that unfolded molecules that have a *cis* peptide bond (species U_{cis}, Eq. 2) fold directly to F_{cis}; by contrast, those that have a *trans* peptide bond (U_{trans}) fold to F_{trans} which accumulates and then isomerizes to F_{cis} to establish the equilibrium between the two folded forms as proposed originally by Evans *et al.* (1989).

Acknowledgments

This research was supported by grant GM35976 from the National Institutes of Health. NMR studies were carried out at the National Magnetic Resonance Facility at Madison whose operation is subsidized by grant RR02301 from the NIH National Center for Research Resources and whose instrumentation was purchased with funds from the National Science Foundation, the University of Wisconsin-Madison, the NIH, and the U.S. Department of Agriculture.

References

Alexandrescu, A. T., Ulrich, E. L., and Markley, J. L. (1989). *Biochemistry* **28**, 204.

Alexandrescu, A. T., Hinck, A. P., and Markley, J. L. (1990). *Biochemistry* **29**, 4516.

Anfinsen, C. B., Schecter, A. N., and Taniuchi, H. (1972) *Cold Spring Harbor Symp. Quant. Biol.* **36**, 249.

Blaber, M., Zhang, X., Lindstrom, J.D., Pepiot, S.D., Baase, W.A., and Matthews, B.W. (1994). *J. Mol. Biol.* **235**, 600.

Brandl, C. J., and Deber, C. M. (1986). *Proc. Natl. Acad. Sci., U.S.A.* **83**, 917.

Brandts, J. F., Halvorsen, H. R., and Brennan, M.(1975). *Biochemistry* **14**, 4953.

Brown III, R.D., Brewer, C.F., and Koenig, S.H. (1977). *Biochemistry* **16**, 3883.

Carra, J. H., Anderson, E. A., and Privalov, P. L. (1994). *Biochemistry* **33**, 10842.

Carra, J. H., and Privalov, P. L. (1995). *Biochemistry* **34**, 2034.

Chazin, W. J., Kördel, J., Drakenberg, T., Thulin, E., Brodin, P., Grundström, and Forsén, S. (1989). *Proc. Natl. Acad. Sci., U.S.A.* **86**, 2195.

Chen, H. M., You, J. L., Markin, V. S., and Tsong, T. Y. (1991). *J. Mol. Biol.* **220**, 771.

Chen, H. M., and Tsong, T. Y. (1994). *Biophys. J.* **66**, 40.

Cotton, F. A., Hazen, E. E., Jr., and Legg, M. J. (1979). *Proc. Natl. Acad. Sci., U.S.A.* **76**, 2551.

Creamer, T.P., and Rose, G.D. (1994). *Proteins: Struct., Funct., and Genet.* **19**, 85.

Creighton, T. E., and Shortle, D. (1994). *J. Mol. Biol.* **242**, 670.

DeGrado, W.F., and Lear, J.D. (1985). *J. Am. Chem. Soc.* **107**, 7684.

Dorman, D. E., Torchia, D. A., and Bovey, F. A. (1972) *Macromolecules* **6**, 80.

Dunker, A. K. (1982). *J. Theor. Biol.*, **97**, 95.

Dyson, H. J., Rance, M., Houghton, R. A., Lerner, R. A., and Wright, P. E. (1988). *J. Mol. Biol.* **201**, 161.

Edison, A. S., Weinhold, F., and Markley, J. L. (1995). *J. Am. Chem. Soc., in press.*

Evans, P. A., Dobson, C. M., Kautz, R. A., Hatfull, G., and Fox, R. O. (1987). *Nature* **329**, 266.

Evans, P. A., Kautz, R. A., Fox, R. O., and Dobson, C. M. (1989). *Biochemistry* **28**, 362.

Fisher, G., Bang, H., and Mech, C. (1984). *Biomed. Biochim. Acta* **43**, 1101.

Fisher, G., Wittmann-Leibold, B., Lang, K., Kiefhaber, T., and Schmid, F.X. (1989). *Nature* **340**, 351.

Grathwohl, C., and Wüthrich, K. (1976a). *Biopolymers* **15**, 2025.

Grathwohl, C., and Wüthrich, K. (1976b). *Biopolymers* **15**, 2043.

Grathwohl, C., and Wüthrich, K. (1981). *Biopolymers* **20**, 2623.

Gittis, A. G., Stites, W. E., and Lattman, E. E. (1993). *J. Mol. Biol.* **232**, 718.

Green, S. M., Meeker, A. K., and Shortle, D. (1992). *Biochemistry* **31**, 5717.

Green, S. M., and Shortle D. (1993). *Biochemistry* **32**, 10131.

Herning, T., Yutani, K., Inaka, K., Kuroki, R., Matsushima, M., and Kikuchi, M. (1992). *Biochemistry* **31**, 7077.

Hertzberg, O., and Moult, J. (1991). *Proteins: Struct., Func., Genet.* **11**, 223.

Hinck, A. P., Loh, S. N., Wang, J., and Markley, J. L. (1990). *J. Am. Chem. Soc.*

1 1 2, 9031.

Hinck, A. P., Eberhardt, E. S., and Markley, J. L. (1993). *Biochemistry* **3 2**, 11810.

Hinck, A. P. (1993) NMR Investigations of a Model Proline *cis:trans* Isomerization Reaction in Staphylococcal Nuclease, Ph.D. Thesis, University of Wisconsin-Madison, Madison, WI USA.

Hinck, A. P., Truckses, D., and Markley, J. L. (1995a). *submitted.*

Hinck, A. P. , Loh, S. N., and Markley, J. L. (1995b). *unpublished results.*

Hodel, A., Kautz, R. A., Jacobs, M. D., and Fox, R. O. (1993). *Protein Sci.* **2**, 838.

Hodel, A., Kautz, R. A., Adelman, D. M., and Fox, R. O. (1994). *Protein Sci.* **3**, 549.

Horovitz, A., Matthews, J.M., and Fersht, A.R. (1992). *J. Mol. Biol.* **2 2 7**, 560.

Hynes, T. R., and Fox, R. O. (1991). *Proteins: Struct. Func. Genet.* **1 0**, 92.

Jacobs, M. D., and Fox, R. O. (1994). *Proc. Natl. Acad. Sci., U.S.A.* **9 1**, 449.

Kay, L. E., Torchia, D. A., and Bax, A. (1989). *Biochemistry* **2 8**, 8972.

Kiefhaber, T., Grunert, H, Hahn, U., and Schmid, F. X. (1990). *Biochemistry* **2 9**, 6475.

Kim, P.S., and Baldwin, R.L. (1982). *Annu. Rev. Biochem.* **5 1**, 459.

Kim, P.S., and Baldwin, R.L. (1990). *Annu. Rev. Biochem.* **5 9**, 631.

Kricheldorf, H.R., Haupt, E.T.K., and Müller, D. (1986). *Mag. Reson. Chem.* **2 4**, 41.

Kreevoy, M. M., and Liang, T. M. (1980). *J. Am. Chem. Soc.* **1 0 2**, 3315.

Kuwajima, K., Okayama, N., Yamamoto, K., Ishihara, T., and Sugai, S. (1991). *FEBS Lett.* **2 9 0**, 135.

Langsetmo, K., Fuchs, J., and Woodward, C. (1989). *Biochemistry* **2 8**, 3211.

Lin, L., and Brandts, J. F. (1984). *Biochemistry* **2 3**, 5713.

Loh, S. N., McNemar, C. W., and Markley, J. L. (1991). in Techniques in Protein Chemistry II (J. J. Villafranca, ed.) Academic Press, New York, p. 275.

Loh, S. N., and Markley, S. N. (1994). *Biochemistry* **3 3**, 1029.

Loh, S. N., Prehoda, K. E., Wang, J., and Markley, J. L. (1993). *Biochemistry* **3 2**, 11022.

Loll, P. J., and Lattman, E. E. (1989). *Proteins: Struct. Func. Genet.* **5**, 183.

Lyu, P.C., Liff, M.I., Marky, L.A., and Kallenbach, N.R. (1990). *Science* **2 5 0**, 669.

MacArthur, M. W., and Thornton, J. M. (1991). *J. Mol. Biol.* **2 1 8**, 397.

Markley, J. L., Williams, M. N., and Jardetzky, O. (1970). *Proc. Natl. Acad. Sci. U.S.A.* **6 5**, 645.

Marsh, H.C., Scott, M.E., Hiskey, R.G., and Koehler, K.A. (1979). *Biochem. J.* **1 8 3**, 513.

Maia, H. L., Orrlell, and Rydon, H. N. (1971). *Chem. Commun.* **1 9 7 1**, 1209.

Nakano, T., Antonino, L. C., Fox, R. O., and Fink, A. L. (1993). *Biochemistry* **3 2**, 2534.

Nemethy, G., Leach, S. J., and Scheraga, H. A. (1966). *J. Phys. Chem.* **7 0**, 998.

O'Neil, K.T., and DeGrado, W.F. (1990). *Science* **2 5 0**, 646.

Padmanabhan, S., Marqusee, S., Ridgeway, T., Laue, T.M., and Baldwin, R.L. (1990). *Nature* **3 4 4**, 268.

Privalov, P. L., and Kechinashvilli, N. N. (1974). *J. Mol. Biol.* **8 6**, 665.

Raleigh, D. P., Evans, P. A., Pitkeathly, M., and Dobson, C. M. (1992). *J. Mol. Biol.* **2 2 8**, 338.

Royer, C. A., Hinck, A. P., Loh, S. N., Prehoda, K. E., Peng, X., Jonas, J., and Markley, J. L. (1993). *Biochemistry* **3 2**, 5222.

Schechter, A. N., Chen, R. F., and Anfinsen, C. B. (1970). *Science* **1 6 7**, 886.
Schmid, F.X., Mayr, L.M., Mücke, M., and Schönbrunner, E.R. (1993). *Adv. Prot. Chem.* **4 4**, 25.
Shalongo, W., Jagannadham, M. V., Heid, P., and Stellwagen, E. (1992). *Biochemistry* **3 1**, 11390.
Shortle, D. (1986). *J. Cell. Biochem.* **3 0**, 281.
Shortle, D., and Meeker, A. K. (1986). *Proteins: Struct., Funct., Genet.* **1**, 81.
Shortle, D., and Meeker, A. K. (1989). *Biochemistry* **2 8**, 936.
Shortle, D., Stites, W. E., and Meeker, A. K. (1990). *Biochemistry* **2 9**, 8033.
Shortle, D., and Abeygunawardana, C. (1993). *STRUCTURE* **1**, 121.
Stanczyk, S. M., Bolton, P. H., DellíAcqua, M., and Gerlt, J. A. (1989). *J. Am. Chem. Soc.* **1 1 1**, 8317.
Sugawara, T., Kuwajima, K., and Sugai, S. (1991). *Biochemistry* **3 0**, 2698.
Takahashi, N., Hayano, T., and Suzuki, M. (1989). *Nature* **3 3 7**, 473.
Torchia, D. A., Sparks, S. W., and Bax, A. (1989a). *Biochemistry* **2 8**, 5509.
Torchia, D. A., Sparks, S. W., and Bax, A. (1989b). *J. Am. Chem. Soc.* **1 1 1**, 8315.
Truckses, D. M., Somoza, J. R., Prehoda, K. E., and Markley, J. L. (1995). *manuscript in preparation.*
Wang, J. F., Hinck, A. P., Loh, S. N., and Markley, J. L. (1990). *Biochemistry* **2 9**, 4242.
Wang, J., Dzakula, Z., Zolnai, Z, and Markley, J. L. (1995). *manuscript in preparation.*
Walkenhorst, W. F., Markley, J. L., and Roder, H. (1995a). *J. Cell. Biochem., Abstr. Suppl.* **2 1 B**, 24th Annual Keystone Symposia, Abstract D2-238, p. 51.
Walkenhorst, W. F., Markley, J. L., and Roder, H. (1995b). *manuscript in preparation.*
Weiss, P. M., Cook, P. F., Hermes, J. D., and Cleland, W. W. (1987a). *Biochemistry* **2 6**, 7378.
Weiss, P. M., Boerner, R. J., and Cleland, W. W. (1987b). *J. Am. Chem. Soc.* **1 0 9**, 7201.

Endnotes

[1]In this review, the term "nuclease" will be used to denote the recombinant protein produced in *Escherichia coli* whose sequence is that of the nuclease A from the Foggi strain of *Staphylococcus aureus*. Mutants in this background will be indicated by the one-letter codes for the original and mutated amino acid and the residue number: for example, nuclease P117G.

[2]Nuclease H1214L denotes the recombinant nuclease variant overproduced in *E. coli* in which histidine 124 is replaced by leucine. Its sequence is identical to the naturally-occuring nuclease A isolated from the V8 strain of *Staphylococcus aureus.*

[3]The symbol "pH*" denotes the pH value of a ^2H_2O solution uncorrected for the deuterium isotope effect.

[4]Superimposition of X-ray strucutures of nuclease variants containing substitions in the loop containing Pro^{117} has shown that V^{111} and N^{118} are the N- and C-terminal residues closest to the ends of that loop whose atomic positions are essentially invariant (Hodel, *et al.*, 1994).

[5]Wild-type nuclease contains no cysteine residues.

[6]m-values are defined as the (linear) dependence of the equilibrium unfolding constant as a function of GdmCl concentration.

11

A Solid-State NMR Approach to Structure Determination of Membrane-Associated Peptides and Proteins

S.J. Opella, L.E. Chirlian, and B. Bechinger

Department of Chemistry
University Pennsylvania
Philadelphia, PA 19104 USA

Structural biology relies on detailed descriptions of the three-dimensional structures of peptides, proteins, and other biopolymers to explain the form and function of biological systems ranging in complexity from individual molecules to entire organisms. NMR spectroscopy and X-ray crystallography, in combination with several types of calculations, provide the required structural information. In recent years, the structures of several hundred proteins have been determined by one or both of these experimental methods. However, since the protein molecules must either reorient rapidly in samples for multidimensional solution NMR spectroscopy or form high quality single crystals in samples for X-ray crystallography, nearly all of the structures determined up to now have been of the soluble, globular proteins that are found in the cytoplasm and periplasm of cells and fortuitously have these favorable properties. Since only a minority of biological properties are expressed by globular proteins, and proteins, in general, have evolved in order to express specific functions rather than act as samples for experimental studies, there are other classes of proteins whose structures are currently unknown but are of keen interest in structural biology. More than half of all proteins appear to be associated with membranes, and many cellular functions are expressed by proteins in other types of supramolecular complexes with nucleic acids, carbohydrates, or other proteins. The interest in the structures of membrane proteins, structural proteins, and proteins in complexes provides many opportunities for the further development and application of NMR spectroscopy.

Our understanding of polypeptides associated with lipids in membranes, in particular, is primitive, especially compared to that for globular proteins. This is largely a consequence of the experimental difficulties encountered in their study by conventional NMR and X-ray approaches. Fortunately, the principal features of two major classes of membrane proteins have been identified from studies of several tractable examples. Bacteriorhodopsin (Henderson *et al.*, 1990), the subunits of the photosynthetic reaction center (Deisenhofer *et al.*, 1985), and filamentous bacteriophage coat proteins (Shon *et al.*, 1991; McDonnell *et al.*, 1993) have all been shown to have long transmembrane hydrophobic helices, shorter amphipathic bridging helices in the plane of the bilayers, both structured and mobile loops connecting the helices, and mobile N- and C-terminal regions. Porins, on the other hand, are predominantly β sheet (Weiss *et al.*, 1991).

Recent results from solid-state NMR experiments on several polypeptides associated with phospholipid bilayers are discussed in the context of helical membrane proteins in this chapter. Studies of amphipathic helical peptides with about 25 residues illustrate the possibilities. M2δ, which has a sequence derived from the functional pore of the acetylcholine receptor and self-assembles to form oligomeric ion channels in membranes (Montal, 1990), and magainin, an antibiotic peptide originally found in frog skin (Zasloff, 1987), are shown by NMR spectroscopy to represent the two principal arrangements of helices found in membrane proteins (Bechinger *et al.*, 1991). The membrane-bound forms of residue filamentous bacteriophage coat proteins with 45 - 50 residues are representative of a typical membrane protein with a hydrophobic membrane-spanning helix and an amphipathic bridging helix (Shon *et al.*, 1991; McDonnell *et al.*, 1993).

Solid-state NMR spectroscopy has been used since the late 1970s for the characterization of immobile and non-crystalline proteins. Solid-state NMR spectroscopy and solution NMR spectroscopy provide quite different windows on spectroscopic and molecular information because the profound effects of rapid overall molecular reorientation present in solution are absent in immobile samples examined by solid-state NMR spectroscopy. Structure and dynamics can be described in a highly integrated fashion with this approach, because of the absence of the influence of overall reorientation on the timescales of the experiments. This is of substantial value, since the dynamic properties of backbone and sidechain sites are essential in the interpretation of all results. This is especially true for membrane proteins which characteristically have mobile loops and termini juxtaposed with rigid, structured helical regions.

There are a wide range of solid-state NMR experiments that can be used to measure distances (Raleigh *et al.*, 1988; Gullion and Schaefer, 1989) or angles

(Opella *et al.,* 1987) between individual sites in proteins, and both types of parameters provide a basis for determining protein structures. Since neither rapid molecular reorientation in solution nor single crystals are required for any of the solid-state NMR experiments, the sample requirements are substantially relaxed compared to those of the other methods of structural biology. The time-average position of atomic sites can be determined from distance measurements from unoriented, immobile samples in the presence of magic angle sample spinning. Oriented samples enable angular factors to be measured, giving an independent view of protein structure that is particularly relevant to membrane proteins with their distinctive architecture in the context of phospholipid bilayers.

Multidimensional solution NMR methods are applicable to membrane peptides and proteins solubilized in detergent micelles because the reorientation of the protein-micelle complex in solution averages the nuclear spin interactions to their isotropic values, enabling local dynamics to be monitored through relaxation phenomena as well as to provide spatial information in the form of relative proximities derived from homonuclear NOE measurements. However, the methods of solution NMR spectroscopy are limited, often severely, by the broad lines and efficient spin-diffusion that accompany the relatively slow reorientation of protein-micelle complexes in solution. Optimization of the isotopic labeling and sample conditions helps to minimize these limitations, as does the use of high-field spectrometers. In principle, the structures of membrane proteins can be determined independently in micelles and bilayers, since multidimensional solution NMR and solid-state NMR experiments are based on very different spectroscopic phenomena. In micelles, the short inter-proton distances detected with homonuclear NOEs, trends in chemical shifts, and selected spin-spin couplings can readily describe secondary structure as well as the three-dimensional folding of the protein. In solid-state NMR experiments on oriented systems, it is feasible to measure a sufficient number of angles between various bonds and groups and the axis of orientation to fully characterize molecular structures. The results from NMR studies of micelles and bilayer samples are complementary, enabling membrane protein conformation to be determined with currently available instruments, and this combined approach provides a means for assessing the reliability of the results. It is possible to identify the main structural elements of membrane proteins on the basis of the solution NMR spectral parameters measured in micelles and then orient these structural domains by means of solid-state NMR spectral measurements in bilayers. The delineation of the secondary structure elements, their arrangement relative to the plane of the bilayer, and the determination of backbone dynamics constitute a substantial step toward characterizing membrane proteins. The overall architecture of a membrane protein is sufficient, in some cases, to explain essential features of protein functions, and it provides a foundation for full three-dimensional structure determination (Opella, 1994).

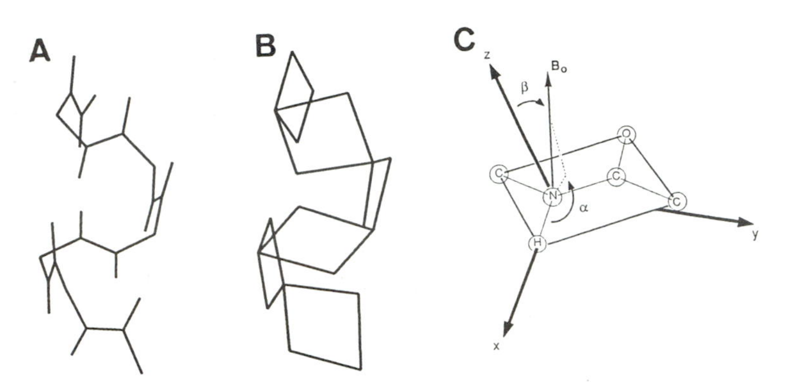

Figure 1: Representations of the structure of the polypeptide backbone with the helix axis aligned vertically. (A) Vector drawing with a line between each non-hydrogen atom; (B) planar representation with connected peptide planes; (C) definition of the axis system used to relate the measured NMR spectral parameters to the peptide plane.

Protein structure determination by solid-state NMR spectroscopy of oriented samples

The three-dimensional structure of the polypeptide backbone of a protein is described equivalently by lines representing bonds between non-hydrogen atoms, as in Figure 1A, and by the rectangular outlines of the peptide planes, as in Figure 1B. The peptide bond, together with its directly bonded atoms, forms a rigid planar unit that is a convenient focus for the determination and the presentation and analysis of protein backbone structures by solid-state NMR spectroscopy. The orientation of a peptide plane with respect to the direction of the applied magnetic field, B_0, is described by two polar angles, α and β, defined in Figure 1C. These angles can be determined from the spectral parameters measured in solid-state NMR experiments on uniaxially oriented samples. Once the orientations of all of the individual peptide planes in a protein are determined, then they can be assembled into a complete protein structure because they are all related by the common axis system defined by B_0, the direction of the applied magnetic field. The overall protocol for structure determination by solid-state NMR spectroscopy of oriented samples is outlined in Figure 2 (Opella *et al.,* 1987).

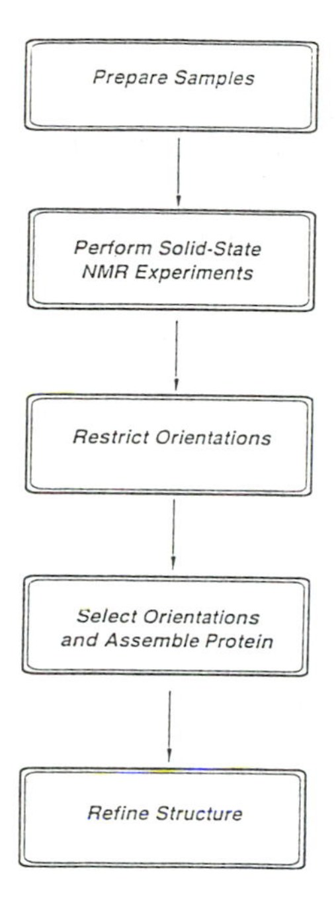

Figure 2: Protocol for structure determination by solid-state NMR spectroscopy of oriented samples.

Prepare samples

The NMR experiments take advantage of specific, selective, or uniform isotopic labeling of amino acid residues for the placement of the spin-interactions of interest. ^{15}N labeling has received the most attention, but examples with ^{2}H and ^{13}C labeling have also provided valuable information. As the spectroscopy develops, parameters from ^{1}H and ^{14}N sites will become increasingly accessible. Labeled peptides and proteins can be prepared through automated solid-phase peptide synthesis or expression in bacteria grown on defined media, depending on their size and other characteristics.

There are two basic sample requirements for this approach to structure determination: they are that the labeled sites are immobile on the timescales of the relevant nuclear spin interactions, defined by having any large-amplitude molecular motions occur less frequently than 10^4/s, and are oriented along the direction of the applied magnetic field. Residues that are part of helices are immobilized by a combination of intra- and inter-molecular interactions in phospholipid bilayers. While uniaxial sample orientation can be obtained through interactions of the sample with the magnetic field, a property which has proven to be useful for solid-state NMR studies of several proteins, most

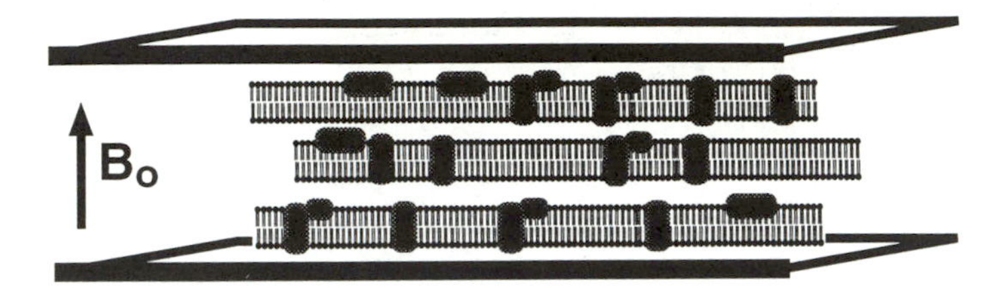

Figure 3: Diagram showing the alignment of protein and lipids relative to glass plates in an oriented membrane sample.

membrane proteins are oriented along with the lipids between glass plates, as shown schematically in Figure 3.

A very high degree of sample orientation can be achieved when small amounts of the peptide-lipid mixture are placed between glass plates and the extent of hydration adjusted under controlled conditions. Geometric considerations suggest that the use of large square glass plates results in optimal bilayer alignment, since edge effects are minimized. NMR experiments can be performed most efficiently on these samples by using a flat coil probe (Bechinger and Opella, 1991) that combines optimal sample orientation with a high filling factor, resulting in improved signal to noise ratios compared to conventional solenoidal coil arrangements.

Perform solid-state NMR experiments

Since nuclear spin interactions are anisotropic, all measurable spectral parameters available from the nuclei present in the peptide plane can be used in structural studies. Most measurements have been made on samples labeled in only one or a few sites. This has the advantage of providing assignments along with the individual spectral parameters and without concern for resolution among overlapping resonances of the expense of preparation of multiple samples. However, methods and instrumentation are being developed so that multidimensional experiments can be used to resolve among resonances and separate their spectral parameters in uniformly labeled samples (Ramamoorthy *et al.,* 1995a). This will extend the approach to almost any protein that can be expressed in bacteria or other organisms that grow on isotopically labeled media.

Restrict orientations

The measured values for the NMR spectral parameters are interpreted in the context of established structures and properties of peptides and proteins, which is similar to how known bond lengths and angles are used in X-ray crystallography. Nuclear spin interactions and their associated spectral parameters have characteristic dependencies on the orientation of the molecular axis system relative to the direction of the applied magnetic field. A single measurement generally does not yield a unique peptide plane orientation since a number of different orientations can have the same spectral features. Experimental error as well as uncertainty in the values of the underlying molecular and spectroscopic parameters must be accounted for in the analysis, expanding the number of orientations consistent with any given measurement. A graphical method has proven to be a convenient and reliable approach for data analysis, since it explicitly takes all of these factors into account (Stewart *et al.*, 1987; Chirlian and Opella, 1990).

The determination of unique orientations from measurements that have inherent ambiguities and errors is a global problem. Consider a modern explorer at any (initially unknown) location on the Earth. Identifying this location with a minimal amount of experimental data is a formidable task, since all possible locations, represented by the map of the world in the top of Figure 4, are equally likely. While a single measurement such as latitude is valuable in restricting the possible locations, it is far from sufficient to determine the actual location. The graphical analysis of a latitude measurement shows two dark bands in Figure 4A. The experimental error and any uncertainty resulting from improper calibration of the sextant used in the measurement are reflected in the breadth of each band, while the ambiguity arising from the lack of knowledge of the direction of the measurement with respect to the equator (north or south) is reflected in the presence of two symmetric bands. A separate measurement of longitude gives the bands on Figure 4B. The longitude and latitude measurements have basically the same drawbacks. However, these two parameters have dramatically different orientational dependencies, and since the actual location must be consistent with all of the measurements, it has to be one of the four points of intersection of the latitude and longitude shown in Figure 4C. This demonstrates that experimental measurements with different angular dependencies are highly effective in reducing the number of possible locations. It is their different angular functions rather than high levels of precision or accuracy in the measurements that serve the dominant role in reducing the possible locations consistent with two or more experimental measurements.

A completely analogous approach can be taken for the analysis of the solid-state NMR data. Instead of geographical locations on a world map, orientations of peptide planes consistent with the experimental measurements of spectral parameters are described as sets of α, β angle pairs (defined in Figure 1C) and represented as dark areas in an α versus β plot, which we refer to as a restriction plot; since only those α, β pairs consistent with the data are marked, these plots provided restrictions for the possible orientations.

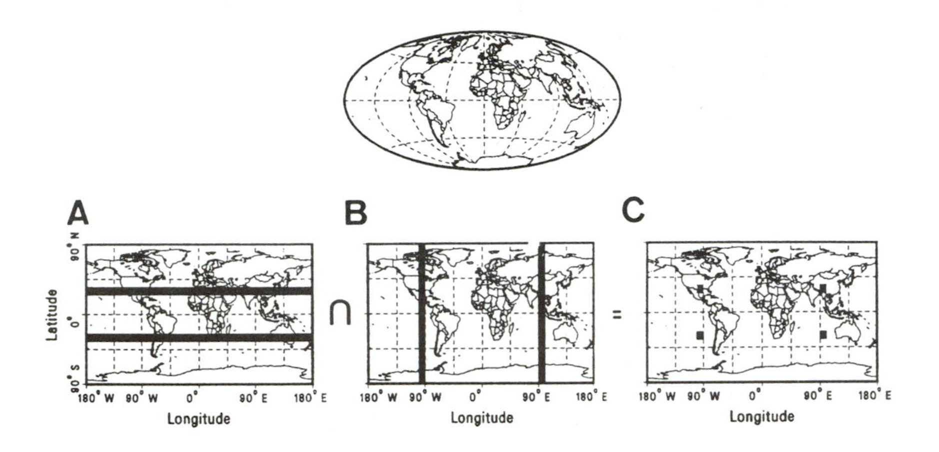

Figure 4: (Top) Global view of the world. (Bottom) Projections of the globe onto a coordinate system defined by latitude and longitude. (A) Dark bands representing the areas of the world consistent with the latitude measurement. (B) Dark bands representing the areas of the world consistent with the longitude measurement. (C) Four small squares representing the areas of the world consistent with both the latitude and longitude measurements obtained by finding the intersection of (A) and (B).

Figure 5 contains restriction plots representing actual experimental data obtained from a ^{15}N labeled residue in an oriented membrane peptide (Ramamoorthy *et al.,* 1995b). The data set consists of the ^{1}H chemical shift (9.1 ppm), ^{1}H-^{15}N heteronuclear dipolar coupling (0.6 kHz), and the ^{15}N chemical shift (73.8 ppm) frequencies. Each of these spectral parameters has its own geometrical dependence and serves as an independent source of angular information, just as shown for the latitude and longitude measurements in Figure 5. The dark areas reflect the estimates of errors in the experimental measurements and any variations in the principal elements of the chemical shift and dipolar interactions. The intersection in Figure 5D shows that there are four symmetry related peptide plane orientations consistent with the experimental measurements. These peptide plane orientations are shown in the bottom of Figure 5.

Restriction plots can also play an important role in guiding the design of experiments, especially in situations where specific labeling is required for each measurement. The values of spectral parameters can be calculated from the coordinates of the polypeptide backbone segment. For example, the restriction plots corresponding to the ^{1}H-^{15}N dipole-dipole splittings associated with the orientations of the 6 residues shown in Figures 1A and B are contained in Figure

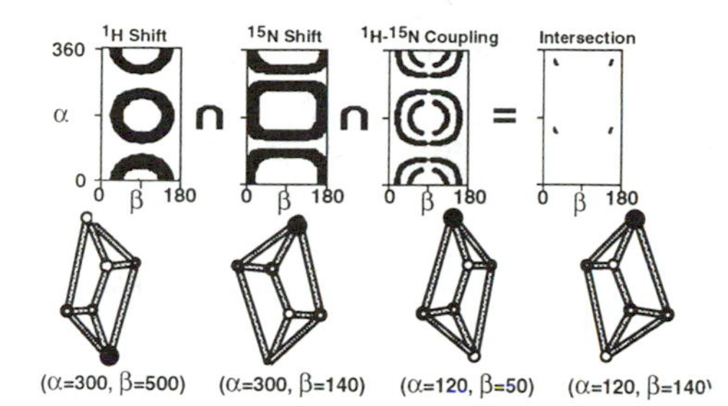

Figure 5: Restriction plots derived from the data for Phe 16 of magainin oriented in lipid bilayers. The three frequencies are measured from the spectrum in Figure 10. The four possible peptide plane orientations consistent with the intersection of the individual restriction plots are shown at the bottom. In this representation the magnetic field direction is directly out of the page toward the viewer.

6. At some orientations, a single measurement provides significant restrictions (Figure 6B, D, E, F), while at others, only a small portion of the potential peptide plane orientations is ruled out (Figure 6A, C). Additional experiments on the latter planes would provide more valuable information than on those planes whose orientations are already well defined. In this way the graphical analysis becomes part of an interactive process with the orientational restrictions obtained from one set of measurements influencing the design of subsequent experiments.

Select orientations and assemble protein

Once the restrictions on peptide plane orientations have been established for all residues, the individual planes are assembled into a complete structure. The selection of correct linkages and peptide plane orientations is a tractable

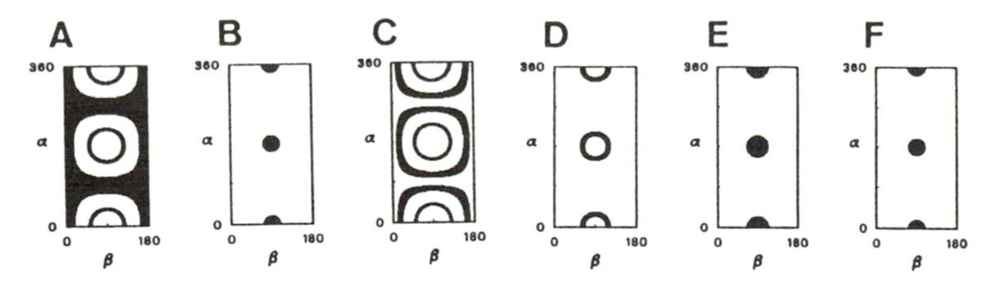

Figure 6: Restriction plots for the six plane orientations shown in Figure 1B. Each plot is for the magnitude of the ^{1}H-^{15}N heteronuclear dipolar coupling that would be observed for that orientation. The same errors and uncertainties are used for each plot. The number of orientations consistent with each measurement varies depending on the angle between the N-H bond and the applied magnetic field. Near parallel there are few possibilities.

computational task only because protein structures follow well-established, regular patterns, and backbone conformational energies have been analyzed (Opella and Stewart, 1989).

To illustrate how well-established guidelines can be used to assemble the backbone structure of a protein, consider the four possible geographical locations identified by latitude and longitude as possible destinations relative to the single starting location shown as black and gray squares, respectively, in Figure 7. Only one of these four sites can be the correct destination. The destination must be on flat, dry land and must be within 5,500 miles of the originating location, ensuring that the flight will be intentionally non-stop and the airport accessible. Only the site at 30° N, 95° W fulfills both of these criteria. The selection among possibilities at both sites is based on local criteria (dry land) and a rule (closer than 5,500 miles).

Proteins provide analogous situations for connecting the peptide plane orientations on two adjacent residues. The geometry of the peptide plane and the energy for each conformation described by a ϕ,ψ pair provide the rules used to select the correct orientation and connect the peptide planes. Linkages are ranked on the basis of Ramachandran energies and ruled out when they exceed a predefined value. Restriction plots for two consecutive planes taken from an actual membrane protein (residue 228 and 229 of the L subunit of the photosynthetic reaction center; Deisenhofer *et al.*, 1985) hypothetically oriented in membrane bilayers are shown in Figure 8. These planes may be assembled by this process using the computer program TOTLINK which tests each possible

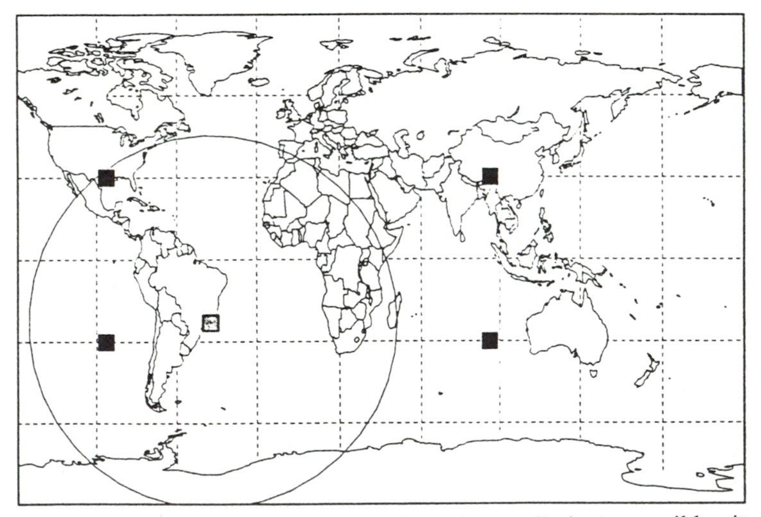

Figure 7: World map showing how rules and criteria can eliminate possible sites that are consistent with the experimental measurements of locations. This is how linkage of sites reduces the problems caused by symmetric redundancy.

orientation for the first plane with each orientation for the second. A low energy linkage for these two planes is shown in Figure 8C. This linkage $\phi = 64°$, $\psi = 30°$, is in close agreement with the crystal structure data.

When data are available for more than two consecutive planes TOTLINK traces a tree-like structure with branches representing the different plane orientations. A branch is terminated when only high energy linkages are possible. In this way, the entire structure can be constructed sequentially. Once the entire structure is assembled restrained molecular dynamics or other calculations may be used to refine the structure. This method may be extended to side chains. The reliability of the model can be assessed by back calculation of the NMR spectrum and comparison with the experimental data. This general procedure may be applied to any peptide or protein that can be oriented in the NMR spectrometer.

Examples

Amphipathic 23 residue peptides in membranes

NMR studies of peptides offer opportunities to develop and implement new technology on well-defined and relatively small polypeptides in the membrane environments of micelles and bilayers that present all of the technical difficulties observed for proteins. The functional analysis of membrane proteins using peptides is particularly direct, since many peptide sequences can be identified as

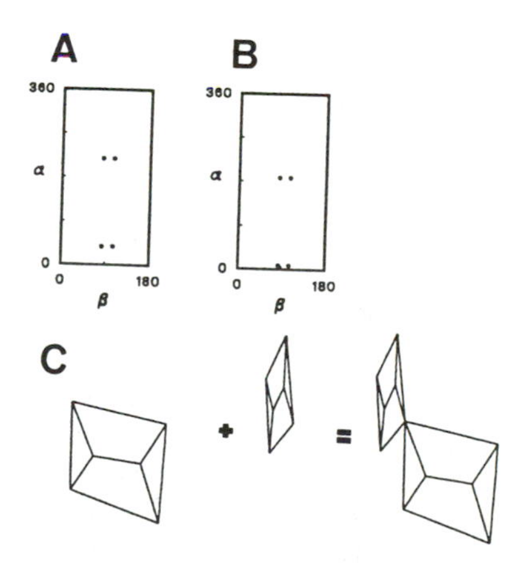

Figure 8: (A) and (B) are restriction plots showing four possible plane orientations consistent with experimental measurements. (C) shows schematically the linkage of one plane orientation from (A) and one from (B) to form a dipeptide. Other possibilities are eliminated by simple principles of protein structure.

structural and functional entities of isolated molecules, oligomers, or domains of larger proteins. Hydrophobic helical regions consisting of 20 - 25 residues are found to cross the bilayer in membrane proteins like the photosynthetic reaction center and bacteriorhodopsin as well as individual peptides. Leader peptides, which are generally classified as hydrophobic are an exception since they are found to be the in the plane of the bilayers (Bechinger *et al.,* 1995). Amphipathic helical peptides are thought to form trans-membrane ion channels by aggregating in oligomeric bundles with their hydrophilic residues on the inside, forming a pore for ions, and their hydrophobic residues on the outside, interacting with the hydrocarbon chains of the lipids (Montal, 1990). Amphipathic helical peptides also associate with the interface region of phospholipid bilayers in the plane of the bilayer, as found in membrane proteins. Multidimensional solution NMR methods have been successfully applied to peptides in solution in selected cases (Dyson and Wright, 1991), although sorting out conformational averaging of peptides in aqueous solution is typically a complex and difficult problem. The lipid environments provided by both micelles and bilayers enhance the opportunities for characterizing a stabilized peptide conformation.

Two types of amphipathic helical peptides in membrane environments provide examples for spectroscopic development; these include peptides derived from the pore-forming segments of ion channel proteins and the magainin antibiotic peptides from frog skins. The secondary structure of both types of peptides is a helix based on several physical measurements, including multidimensional solution NMR spectroscopy of representative peptides in micelles. This provides a starting point for the solid-state NMR studies. The 23-residue M2δ peptide, EKMSTAISVLLAQAVFLLLTSQR, was synthesized with Ala12 specifically labeled with ^{15}N, and the 23-residue magainin2 peptide, GIGKFLHSAKKFGKAFVGEIMNS amide, was synthesized with Ala15 specifically labeled with ^{15}N. Figure 9 compares experimental solid-state ^{15}N NMR spectra of the specifically labeled magainin2 and M2δ peptides in oriented phospholipid bilayers to simulated ^{15}N amide chemical-shift powder patterns and oriented spectra (Bechinger *et al.,* 1991). The anisotropies of the chemical shift and dipole-dipole interactions, averaged out by rapid molecular reorientation in solution, are present in these immobile samples and serve as sources of structural and dynamic information. In the spectra presented in Figure 9, the heteronuclear ($^{1}H/^{15}N$) dipole-dipole interactions are decoupled, therefore all spectral features reflect the ^{15}N chemical shift interaction of a specific ^{15}N labeled site in each peptide sample. The magnitudes and orientations in the molecular frame of the principal elements of the chemical shift tensor are prerequisites for interpretation of spectral features. The ^{15}N chemical shift tensor of a nitrogen in a peptide bond has been determined (Wu *et al.,* 1995): σ_{33} is the most downfield element, and σ_{11} and σ_{22}, of these nearly axially

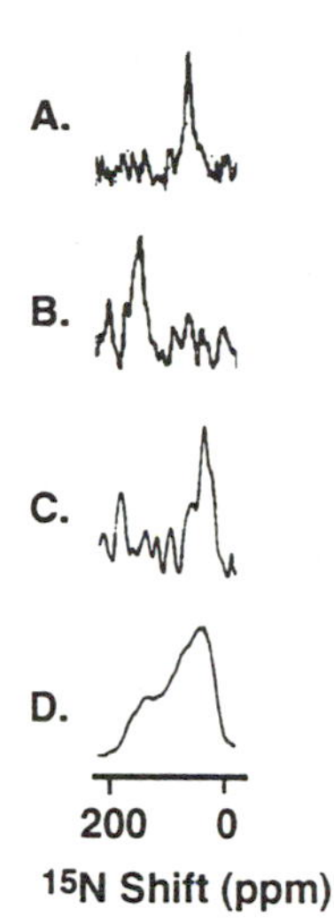

Figure 9: Experimental ^{15}N solid-state NMR spectral of labeled polypeptides in oriented phospholipid bilayers. (A) Magainin. (B) M2δ channel peptide. (C) fd coat protein labeled at Leu14 and Leu41. (D) Powder pattern from an unoriented sample of fd coat protein labeled at Leu14 and Leu41, both of which are immobile.

symmetric tensors, are the most upfield elements. Since σ_{33} is approximately parallel to the N-H bond axis, the observation of resonance frequencies near either extreme leads directly to qualitative determinations of the orientations of helices containing the labeled residue (Bechinger *et al.,* 1991).

The single resonance from the ^{15}N labeled magainin peptide is located near 80 ppm, while the chemical shift of labeled M2δ is located at the opposite end of the chemical shift range, near 200 ppm, relative to $^{15}NH_3$. The signal intensities observed at opposite ends of the available chemical shift range illustrate the role molecular orientation has on a spectral parameter, since the chemical site and the peptides themselves are very similar. The dramatic difference in chemical shift between the spectra in Figures 9A and B is due to the differences in molecular orientation. The set of α, β angles that describe the possible orientations of the peptide planes in agreement with the experimental measurements are calculated and depicted in Figures 10A and B, respectively. No common pair of α, β angles exists that could describe the orientations of the magainin and M2δ peptide planes, indicating that these amphipathic helices assume different orientations with respect to the magnetic field. Calculated pairs of all α, β angles that are consistent with helices oriented perpendicular or parallel to the bilayer normal or parallel to the bilayer surface are shown in Figures 10C and D, respectively. Comparison of the restrictions in α,β angles obtained from just the ^{15}N chemical shift measurements with these calculated values indicates that the experimental data agree with an orientation of the M2δ helix axis parallel to the surface of the bilayer. This is consistent with the model of a pore formed by transmembrane helical bundles. The alignment of the magainin helices approximately parallel to the membrane surface indicates that these peptides do not cross the membrane but rather reside in the bilayer interface. The restrictions derived solely from the chemical shift are valuable in the example because the peptides are known to be helical from other experiments and the resonance frequencies are near the extremes of the ^{15}N chemical shift span.

Three- and four- dimensional solid-state NMR experiments (Ramamoorthy *et al.,* 1995a,b,c) provide a way to measure multiple spectral parameters in a single spectrum. This is illustrated with the three-dimensional solid-state NMR spectrum in Figure 11 which was obtained on an oriented sample of magainin in phospholipid bilayers where the peptides are ^{15}N labeled at Phe16 and Val17 (Ramamoorthy *et al.,* 1995b). Each peak is characterized by the frequencies from the ^{1}H chemical shift, ^{1}H-^{15}N dipolar coupling and ^{15}N chemical shift. The values used to illustrate the roles of restriction plots in determining protein structures of oriented samples by solid-state NMR spectroscopy in Figure 5 were measured from this spectrum. The resonances from the two labeled sites in Figure 11 are resolved along a l l three frequency axes, and this demonstrates the

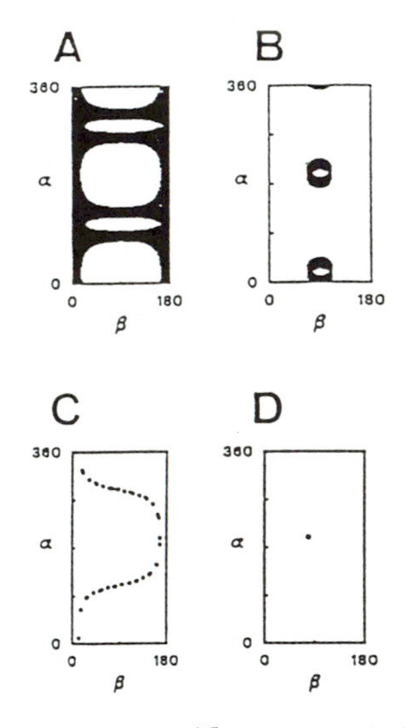

Figure 10: (A) Restriction plot for the ^{15}N chemical shift data in Figure 9A. (B) Restriction plot for the ^{15}N chemical shift data in Figure 9B. (C) Calculated possibilities for a perfectly aligned in-plane helix. (D) Calculated possibilities for a perfectly aligned trans-membrane helix.

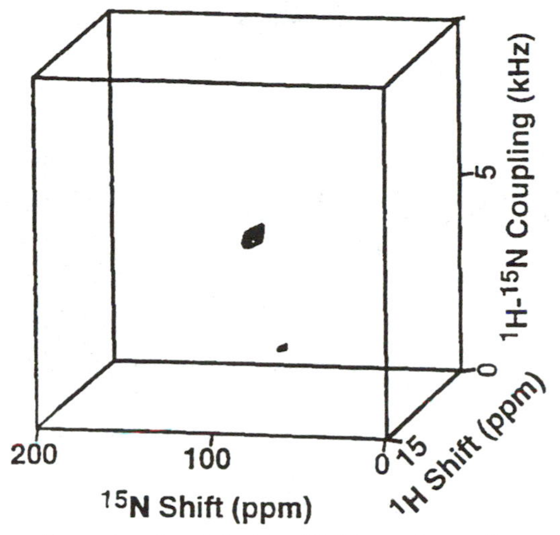

Figure 11: Three-dimensional solid-state NMR spectrum of a sample of magainin labeled at Phe16 and Val17 in oriented bilayers.

potential for resolving among the resonances present in multiply labeled samples, which would reduce the need for specifically labeling samples for structure determination. This would have a particularly strong impact on the feasibility of these studies, because each resolved resonance in these spectra enables the measurement of three different frequencies that yield angular restrictions. Thus, in a single experiment it may be possible to obtain all of the information necessary to determine the structure of a peptide in membrane bilayers. Of course, the assignment problem remains, as it does in multidimensional solution NMR spectroscopy, however, systematic assignment methods are feasible through extensions to four-dimensional spectroscopy (Ramamoorthy *et al.,* 1995).

Filamentous bacteriophage coat protein

Several thousand copies of the major coat protein are arranged symmetrically around the DNA to form a filamentous bacteriophage particle. The coat proteins are synthesized within the infected host cell and are processed through the cell membrane where they are assembled into the mature coat. Structural differences between the membrane bound and the viral forms of the protein are being determined by solid-state NMR spectroscopy. The virus particles orient spontaneously in the magnetic field of the spectrometer, and membrane bound coat proteins can be oriented along with the phospholipids between glass plates. The coat proteins of two different filamentous bacteriophages, Pf1 (46 residues) and fd (50 residues) have been studied by NMR spectroscopy (Stewart *et al.,*

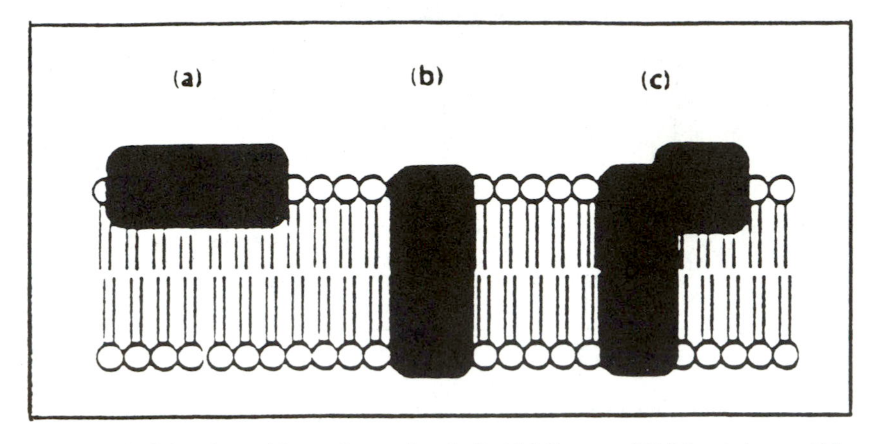

Figure 12: Models of peptides and proteins in lipid bilayers. (A) Magainin peptide. (B) M2 channel peptide. (C) fd coat protein.

1987; Shon *et al.,* 1991; McDonnell *et al.,* 1993). Despite a lack of sequence homology, both proteins exhibit similar structural features, consisting of a mobile N- terminus, a short amphipathic helix, and a long hydrophobic helix. The C- terminus is mobile in the membrane bound form and structured in the viral form. The presence of the helical segments in the membrane forms has been verified by multidimensional solution NMR measurements in micelles. Solid-state NMR experiments establish the orientations of the two helices, both with respect to each other and the lipid bilayer. Spectra of fd coat protein oriented in lipid bilayers are shown in Figure 9. The spectrum in Figure 9C of a selectively ^{15}N labeled sample sample monitors the two orientations of the two helices since the peak at 58 ppm is attributed to Leu 14 in the amphipathic helix and the peak at 190 ppm to Leu 41 in the hydrophobic helix. The powder pattern spectrum in Figure 9D from an unoriented sample shows both sites to immobile and structured. Based on the observed chemical shift frequencies in the oriented sample and the knowledge that these two residues are in different helical segments it is possible to deduce the arrangements of the helices in the bilayers with the hydrophobic helix trans-membrane and the amphipathic helix in the plane of the bilayers (McDonnell *et al.,* 1993).

Concluding remarks

The motivation for developing NMR spectroscopy for the study of membrane proteins is clear, since it is highly desirable to be able to determine the structures of non-crystalline samples of proteins in membrane environments. Progress has been made in adapting multidimensional solution NMR methods for studying membrane proteins in micelles and in developing solid-state NMR methods for studying membrane proteins in bilayers. The pace of technical improvements is increasing and should be greatly accelerated by the implementation of spectrometers with very high field magnets and higher dimensional NMR experiments in both solution and solid-state NMR protocols.

The experimental results shown in Figure 9 are summarized in Figure 12. Three different polypeptides are shown schematically in the bilayer environment. The single trans-membrane and in-plane helices, as observed for a channel M2 peptide and magainin, respectively, are compared to the longer polypeptide corresponding to a filamentous bacteriophage coat protein with two helical segments. Figure 12 accounts for many examples not mentioned here, since membrane peptide and proteins appear to be organized along the relatively simple lines deduced from the glimpses at structure obtained from the initial examples.

Solid-state NMR spectroscopy is emerging as an independent method for determining the structures of proteins. A significant advantage of solid-state NMR spectroscopy is that neither single crystals nor rapidly reorienting samples are required, both of which are difficult or impossible to obtain for membrane proteins. In addition, since there are no fundamental size limitations in solid-state NMR spectroscopy, it is possible with this approach to consider studies of large proteins, as well as those in complexes with lipids and other biomolecules.

Acknowledgements

We thank our collaborators L.M. Gierasch, M. Montal, J. Tomich, and M. Zasloff for their contributions and A. Ramamoorthy and F.M. Marassi for the many results in Figures 5 and 11. This research is supported by grants R37GM24266 and RO1GM29754 from the General Medical Sciences Institute and grant RO1AI20770 from the Allergy and Infectious Disease Institute, National Institutes of Health. It utilizes the Resource for Solid-State NMR of Proteins at the University of Pennsylvania, which is supported by grant P41RR09731 from the Biomedical Research Technology Program, Division of Research Resources, National Institutes of Health. B.B. was supported by EMBO long term fellowship ALTF 454-1989. L.E.C. was supported by postdoctoral fellowship F32GM13256 from the National Institutes of Health.

References

Bechinger, B., and Opella, S.J. (1991). *J. Magn. Reson.* **95**, 585.

Bechinger, B., Kim, Y., Chirlian, L.E., Gesell, J., Neumann, J.-M., Montal, M., Tomich, J., Zasloff, M., and Opella, S.J. (1991). *J. Biomol. NMR* **1**, 167.

Chirlian, L.E., and Opella, S.J. (1990). *New Polymeric Mater.* **2**, 279.

Deisenhofer, J., Epp, O., Miki, K., Huber, R., and Michel, H. (1985). *Nature* **318**, 618.

Dyson, H.J., and Wright, P. (1991). *Annu. Rev. Biophys. Biophys. Chem.* **20**, 519.

Gullion, T., and Schaefer, J. (1989). *J. Magn. Reson.* **81**, 196.

Henderson, R., Baldwin, J., Cesko, T., Zemlin, F., Beckmann, E., and Downing, K. (1990). *J. Mol. Biol.* **213**, 899.

McDonnell, P.A., Shon, K., Kim, Y., and Opella, S.J. (1993). *J. Mol. Biol.* **233**, 447.

Montal, M. (1990). *FASEB J.* **9**, 2623.

Opella, S.J., Stewart, P.L., and Valentine, K.G. (1987). *Q. Rev. Biophys.* **19**,7.

Opella, S.J., and Stewart, P.L. (1989). *Meth. Enzymol.* **176**, 242.

Opella, S.J. (1994). in *Membrane Protein Structure: Experimental Approaches* (S.H. White, ed) Oxford, 249.

Raleigh, D.P., Levitt, M.H., and Griffin, R.G. (1988). *Chem. Phys. Lett.* **146**, 71.

Ramamoorthy, A., Marassi, F.M., Zasloff, M., and Opella, S.J. (1995b). *J. Biomol. NMR, in press.*

Ramamoorthy, A., Wu, C.H., and Opella, S.J. (1995a). *J. Magn. Reson.* **B107**, 88.

Ramamoorthy, A., Gierasch, L.M., and Opella, S.J. (1995). *J. Magn. Reson.* **B109**, 112.

Shon, K., Kim, Y., Colnago, L.A., and Opella, S.J. (1991). *Science* **252**, 1303.

Stewart, P.L., Valentine, K.G., and Opella, S.J. (1987). *J. Magn. Reson.* **71**, 45.

Weiss, M., Abele, U., Weckesser, J., Welte, W., Schiltz, E., and Shultz, G. (1991). *Science* **254**, 1627.

Wu, C.H., Ramamoorthy, A., Gierasch, L.M., and Opella, S.J. (1995). *J. Amer. Chem. Soc.* **117**, 6148.

Zasloff, M. (1987). *Proc. Natl. Acad. Sci. U.S.A.* **84**, 5449.

12

NMR Approaches to the Heat-, Cold-, and Pressure-Induced Unfolding of Proteins

K. Akasaka, T. Yamaguchi, H. Yamada, Y. Kamatari, and T. Konno

Department of Chemistry
Kobe University
Kobe 657, Japan

Proteins are unique at least in two aspects. First, the atoms constituting a protein molecule act highly cooperatively in constructing its unique folded structure. As a result, their conformational transitions also occur in a highly cooperative fashion (often following a two-state transition). Secondly, the free energy balance between the folded (native) and unfolded (denatured) conformers are surprisingly marginal (usually less than 10 kcal/mole protein), despite the fact that interactions of thousands of atoms are involved in the folding. Such unique properties have been acquired by protein molecules through a countless number of random experiments and subsequent selections during the course of evolution of life, so that to our eyes, at present, they look as if they were carefully designed by Nature. It is important to recognize that such random experiments occurred in a dominantly aqueous environment. In order to understand the underlying principles of design as generally as possible, we need, at least, to characterize the factors that contribute to (1) the stability of protein structures, (2) the structural details of the folded and unfolded conformers, and (3) the kinetics of folding and unfolding reactions. In all these, the involvement of water has crucial importance. NMR can provide unique information not only on aspect (2) above, but on all the above aspects when used under appropriate design. In this presentation, some examples will be shown from our current research.

Proteins are easily deformed

Our daily experience in the kitchen shows that proteins are easily deformed (denatured); a boiled egg can be prepared just by heating to not more than 100°C in water. The easy deformability (which is, in fact, a global conformational transition) in aqueous environment is not merely a matter of interest in a kitchen, but is a quality of design for proteins by Nature. A global conformational transition (unfolding) of a protein molecule occurs even under physiological conditions, although infrequently, as evidenced by hydrogen exchange of peptide NH protons that are completely buried in the folded structure.

The dynamic equilibrium between the folded (N) and unfolded (D) conformers

however, seems to be designed differently for different proteins; as one typical example, the case of a proteinaceous protease inhibitor, *e.g.*, Streptomyces subtilisin inhibitor (SSI) that we have been studying for some time (Hiromi *et al.*, 1986), will be mentioned. We found by hydrogen exchange experiments that in SSI the unfolding reaction is extremely infrequent under physiological conditions (Akasaka *et al.*, 1985). However, we also found that a single amino acid substitution in the core part of this dimeric protein, such as Trp 86 into His, makes this protein easily subject to digestion by the protease with consequent loss of its function as an inhibitor (Tamura *et al.*, 1991), although two-dimensional (2D) NMR investigation showed that there were minor changes in the folded structure. We showed evidence that the loss of resistance against protease attack in the mutant is brought about by the nearly three-fold increased rate of unfolding. This and other similar experiments on SSI showed that the dynamic equilibrium between the folded and unfolded conformers depends critically on the selection of amino acids in the core part, where the atoms are heavily packed. Indeed, in all SSI-like inhibitors widely found in Streptomyces in nature, Trp is always conserved at position 86 (Taguchi *et al.*, 1992; 1993). This finding indicates that the amino acid at position 86 is carefully chosen by Nature so that the global fluctuation may be suppressed to an unusually low level in SSI-like inhibitors. It appears certain that global fluctuation is one important target of design in proteins by Nature. It may be emphasized that the study of non-native species of a protein has non-trivial importance not only for a protease inhibitor, but for most functional proteins, because non-native species should be abundant in cells in complex with chaperons (Zahn *et al.*, 1994) as well as in targeting and secretion.

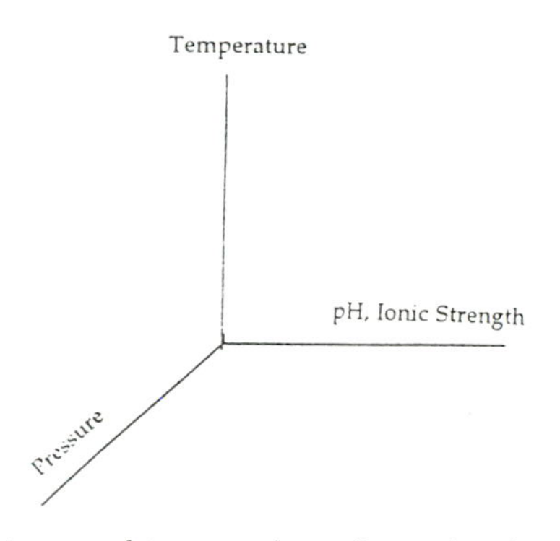

Figure 1: External parameters used to perturb conformational stability of proteins.

To study thermodynamics and structures of non-native conformers in a laboratory, we may bring the equilibrium near the transition region and produce non-native species in high concentrations by external perturbations such as temperature and pressure along with pH (Figure 1). In this report, I discuss the use of temperature and perturbations in NMR measurements to bring unexplored categories of non-native states of proteins into direct observation by NMR spectroscopy. These include the investigation of protein unfolding by cooling (cold denaturation), by high pressure (pressure denaturation), and by temperature-jump.

Cold-induced unfolding. A new method for producing intermediate conformers?

Cold denaturation of streptomyces subtilisin inhibitor

It is well known that the stability of the folded structure of a protein is maintained by a combined interplay of a number of different contributions, including hydrogen bonding, electrostatic interaction, van der Waals forces, all of which are interactions within the protein chain, and hydration free energy, which basically arises from water molecules surrounding the polypeptide chain (Privalov and Gill, 1988). Among these, electrostatic interactions can be controlled easily by changing the pH and/or ionic strength of the solution, and among acid denatured states, a compact structure "molten globule" may appear at high concentrations of the ions. Another important interaction, which greatly affects the stability of the protein structure, is the hydration energy of the protein, particularly the hydration of hydrophobic residues which increases dramatically upon unfolding. The hydration energy of a hydrophobic residue is very sensitive to temperature (Privalov and Gill, 1988). Thus the stability of a globular protein with a hydrophobic core may be decreased simply by lowering

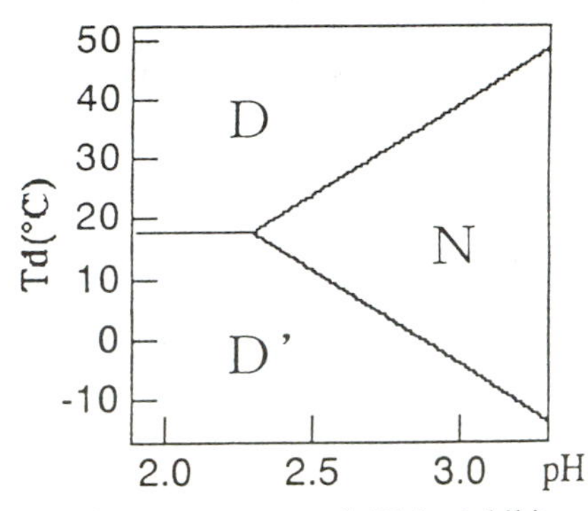

Figure 2: Phase diagram of streptomyces subtilisin inhibitor (SSI), a homo-dimeric protein with 2 X 113 amino acids, as determined by far-UV CD spectroscopy (Tamura *et al.*, 1991a).

the temperature of the solution, resulting in unfolding at low temperature (cold denaturation). So far, however, very little is known about structures of proteins in the cold denatured state, although such a study is important in view of the fact that the stability of the hydrophobic core depends critically on the hydration energy.

Streptomyces subtilisin inhibitor or SSI is a homo-dimer with 113 X 2 residues, and is an inhibitor against a considerable variety of proteases, including subtilisin BPN' (Hiromi *et al.*, 1985). Since its discovery (Murao & Sato, 1972), SSI has been intensively studied mainly by a group of scientists in Japan. Soon it became clear that SSI has a unique structure with a highly developed hydrophobic core made out of two identical subunits strongly associating by dominantly hydrophobic contact (Mitsui *et al.*, 1979). By now, it is believed that most strains of Streptomyces produce SSI-like inhibitors, *i.e.*, homo-dimers with presumably similar three-dimensional structures (Taguchi *et al.*, 1993).

Tamura *et al.* (1991a) found from circular dichroism and other methods that the protein SSI is denatured at low pH, but that the structure of the denatured protein takes different conformations at high and low temperature. It became clear that SSI can exists in at least three distinct thermodynamic states, the native (N, dimer), heat-denatured (D, monomer), and cold-denatured (D', monomer), with corresponding conformations of the polypeptide chain, depending on pH and temperature (Figure 2) (Tamura *et al.*, 1991a). The proton NMR spectrum in the cold denatured state (D', Figure 3) has peaks at the chemical shift positions of a typical unstructured polypeptide, *e.g.*, methyl protons of Val and Leu at 0.9 ppm, methyl protons of Ala at 1.4 ppm, etc., as usually expected for a denatured protein. In addition, however, the spectrum for

D' shows several characteristically shifted signals, *e.g.*, Leu and Val methyl proton signals at 0.9 ppm, methyl proton signals of Met70 and Met73, and ε-proton signals of His43 and His106. The last result clearly indicates that in D' a unique tertiary structure is present, although from the CD spectrum at least half of the secondary structure is lost. In accordance with this, a recent small angle X-ray scattering study (Konno *et al.*, submitted) indicates that the polypeptide chain has a globular shaped part, but in total the chain is substantially extended (Konno *et al.*, submitted). We conclude that in the cold denatured state of SSI the polypeptide chain forms, in part, a globular portion with hydrophobic residues and a part disordered to a comparable extent as in the heat-denatured state.

Because of the strong cooperativity among amino acid residues, most proteins are considered to exist in two thermodynamic states, the folded (N) and the unfolded (D). SSI, which it exists at least in three distinct states, N, D and D', in which the heat denatured (D) and cold denatured (D') states are thermodynamically and conformationally different, does not obey this generally accepted concept.

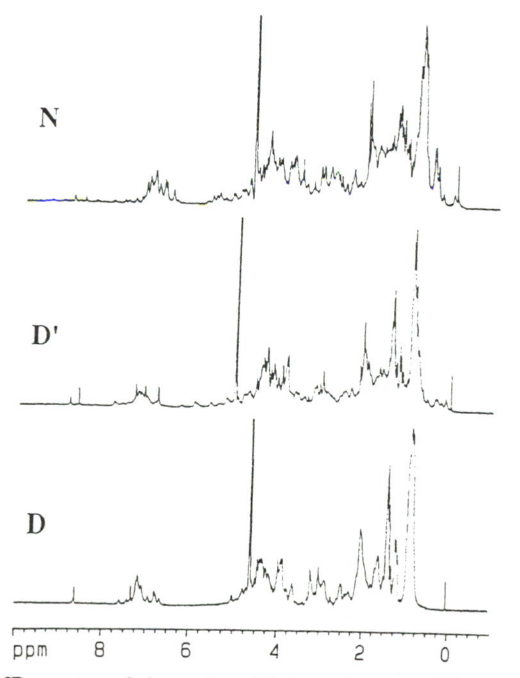

Figure 3: Proton NMR spectra of the native (N), heat-denatured (D) and cold-denatured (D') species of SSI. Experimental conditions were pH 2.6, 25 °C, pH 1.4, 35 °C, and pH 1.4, 35 °C, respectively. The spectra were measured at 600 MHz.

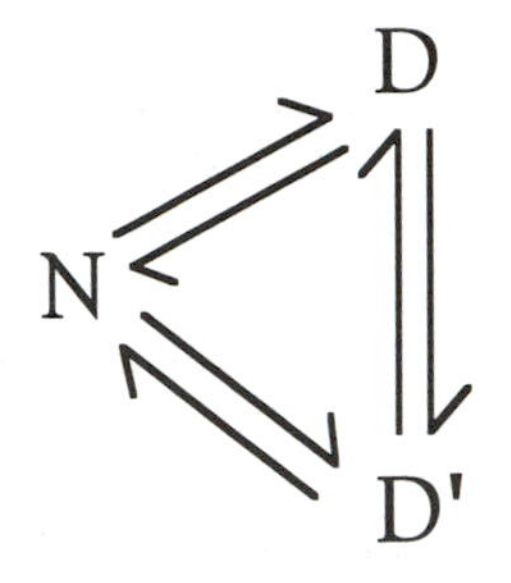

The reason why SSI showed three distinct structures instead of two may be puzzling at a first glance. However, if we consider that the factors that drive a protein into unfolding may be different between the cold denaturation and the heat denaturation, it may not be surprising that the conformation in cold-induced unfolding is different from that in heat-induced unfolding, although this does not necessarily mean that they belong to different thermodynamic states. In general, hydrophobic interactions decreases at low temperature. If subtle differences exist in the degree of hydrophobic cooperativity among different regions of a protein molecule, they may collapse at different temperatures, and maybe in steps with decreasing temperature. For example, if in a multi-subunit protein the inter-subunit cooperativity is a different unit from that within the subunit, then it could happen that the former cooperativity is lost entirely at one temperature, *e.g.*, around 0 °C, but that the latter cooperativity is not fully lost and, therefore, some structure remains within a subunit at the same temperature. In such a case, full unfolding might take place at a still lower temperature, and in this sense the observed D' species in SSI should correspond to an unfolding intermediate in a complete phase diagram. This result suggests that cold-induced unfolding can reveal a more subtle variety of intermediate stages of folding than hitherto explored at higher temperature.

Pressure-induced unfolding. Another hydration effect?

Pressure effects on unfolding of ribonuclease A studied by high-pressure NMR

In cold denaturation, free energy of water surrounding non-polar residues of the protein could be a major driving force for unfolding of the protein chain. Pressure is another factor that affects strongly the thermodynamic state of water, and, therefore, may also affect the folding-unfolding equilibrium of proteins.

We have recently succeeded in utilizing a high pressure glass cell (Yamada, 1974; 1991) to measure ^{1}H NMR spectra of ribonuclease A under high pressure (up to 2000 atm) at 400 MHz (Yamaguchi *et al.*, 1994). We could clearly distinguish His ε1 proton signals from the folded conformer and those from the unfolded one (Figure 4). Under the assumption of two-state transition,

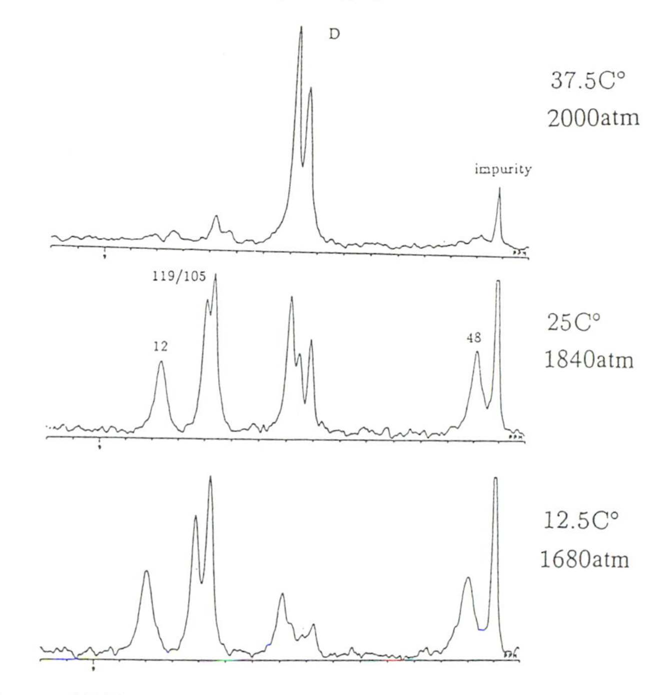

Figure 4: Proton NMR spectra (at 400 MHz) of the four His ε1 protons of ribonuclease A (6.2 mM, 0.15 M KCl, pH 1.0) under the pressure and the temperature indicated in the figure.

measurement of these signal intensities gave equilibrium constants and consequent ΔG values associated with conformational transition as functions of pressure in the temperature range between 7.5 °C and 40 °C (Yamaguchi, T. *et al.*, submitted). We found that the volume changes (ΔV) due to pressure are negative and are temperature-dependent, changing from -10 cm^3 /mol at 7.5 °C to -30 cm^3 /mol at 37 °C. Negative (ΔV) is usually observed in hydration of a non-polar substance and is attributed to the effect of hydration of non-polar groups upon unfolding of a protein chain.

Another important finding in this experiment was that ΔCp decreases with increasing pressure from 1.79 kcal/mol K at 1 atm to 1.08 kcal/mol K at 2,000 atm. Since a positive ΔCp is known to arise from hydration of non-polar groups upon unfolding, the decrease of this quantity with pressure indicates that pressure decreases the effect of this hydration; in terms of ordering of hydrated water either the order of the hydrated water or the amount of the "structured water" decreased with pressure. The strong dependence of ΔCp on pressure, along with negative ΔV and its temperature dependence, suggests strongly that pressure-induced unfolding is brought about mainly through the change in the state of water surrounding the exposed non-polar groups of ribonuclease A.

Temperature-jump NMR. A new technique for kinetic intermediates.

Direct NMR observation of folding or unfolding processes in ribonuclease A

The above study, as well as many other studies carried out on different proteins, suggests that it is necessary to investigate transient intermediate conformers as well as the denatured conformers in equilibrium, by a technique with higher spatial resolution. NMR is obviously a technique to satisfy such demands, and several different techniques have been proposed and developed recently for this purpose, including that of pulsed deuterium-labeling of the labile protons in the process of folding. We have undertaken a different approach for this purpose, and have worked to develop a temperature-jump technique useful for NMR studies.

The first version of the temperature-jump NMR utilized gas flow to change the temperature of the sample in the NMR tube (time constant = 6 seconds for

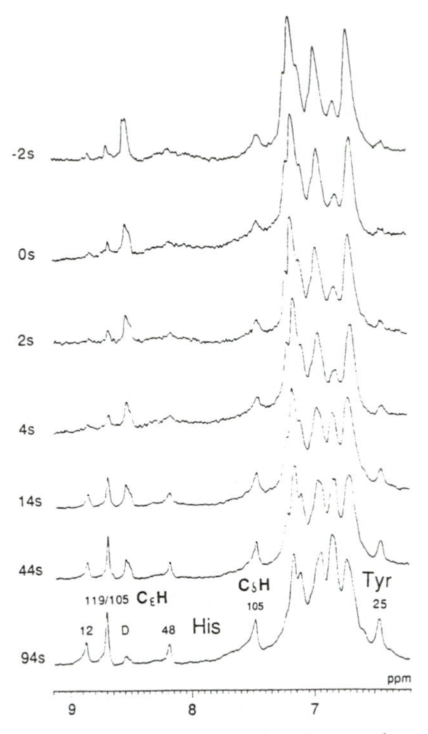

Figure 5: Proton NMR spectral changes upon temperature-jump down in the aromatic region of ribonuclease A (10% in D_2O, pH 1.2), measured at 400 MHz with a gas-flow type temperature-jump apparatus.

cooling down by 15 degrees) (Akasaka *et al.*, 1990). With this technique, one can follow a slow folding process of a protein by measuring proton 1D spectra successively after the temperature-jump. Figure 5 shows an example of the folding process in ribonuclease A upon temperature-jump from 30°C to about 45°C where time zero is taken at the time when the temperature reached 45°C (Akasaka *et al.*, 1991). We could show that, except for the small fraction of proteins that fold within the dead-time of 6 s (see the top spectrum at -2 s), the larger fraction of proteins fold very slowly with a time constant of 30 s. This time constant was common to most proton peaks observed, indicating that the folding occurs cooperatively in all parts of the polypeptide chain.

The second version of the temperature-jump NMR utilized microwave pulse from a 1.3 kW magnetron, which was quite successful for causing phase transition in a liquid crystal (nematic phase) within a time as short as several milliseconds (Naito *et al.*, 1991; Akasaka *et al.*, submitted). Similar technique was applied to a protein solution, which, in favorable cases, enabled an average of 15 degrees jump in about 200 ms in an 5mm O.D. sample tube, with rapid (30 Hz) sample spinning to average out temperature gradient. When this method was applied to an aqueous solution of RNase A, a full heat denaturation was brought about (Figure 6). It is clear that the native tertiary structure was lost very rapidly within 150 ms, although the species immediately after the microwave pulse of 150 ms should be an unfolding intermediate rather than a fully denatured protein.

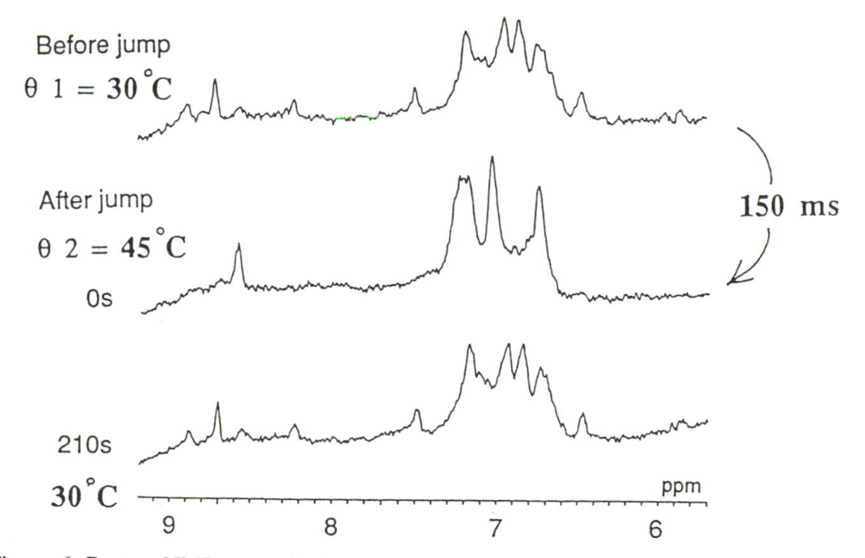

Figure 6: Proton NMR spectral changes ribonuclease A upon temperature-jump (up), measured at 400 MHz with a microwave heating.

State-correlated 2D NMR. A new area in 2D NMR spectroscopy.

Spectral correlation between the folded and unfolded ribonuclease A

The same technique can be extended into a temperature-jump 2D NMR or more generally termed "State-Correlated 2D NMR Spectroscopy" (SC-2D) (Naito *et al.*, 1990), the pulse sequence of which is shown in Figure 7. This is a 2D NMR spectroscopy, entirely different from preexisting 2D NMR methods in that this correlates NMR spectra of a single molecule between two different thermodynamic states, in this case characterized by two different temperatures, while 2D NMR spectroscopy hitherto has been limited to correlations between identical thermodynamic states.

The first successful application of SC-2D to proteins has been made on ribonuclease A (Akasaka *et al.*, 1991), the first protein studied by high resolution NMR by Oleg Jardetzky. Figure 8 is an SC-2D spectrum between the native conformer and an unfolded conformer immediately (50 ms) after the 150 ms microwave pulse in the aromatic region of RNase A. Two points should be noted. First, specific signal assignments are possible for many of the resonance lines of the unfolded conformer with this technique, and that this unfolded conformer could be one before final denaturation. Secondly, the chemical shifts of the ε1 protons of the four His residues in the unfolded conformer are dispersed, indicating substantial magnetic inequivalence among the environments of these four residues. Although this finding does not indicate immediately the existence of residual structure in the unfolded conformer, it is a likely explanation for this observation.

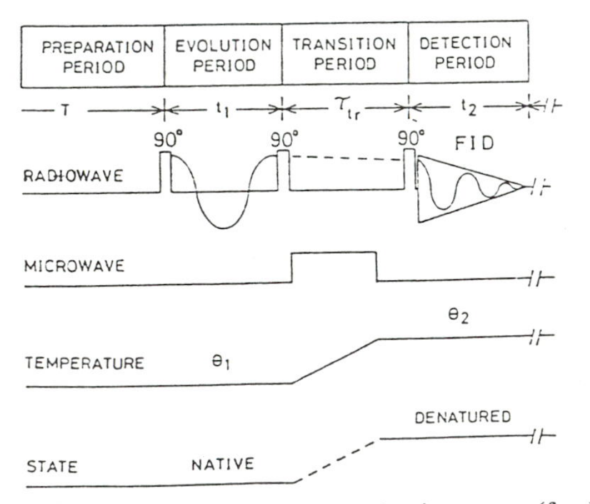

Figure 7: A combined radiowave (for NMR) and microwave (for heating) pulse sequence for measuring State-Correlated 2D NMR spectra of proteins in solution.

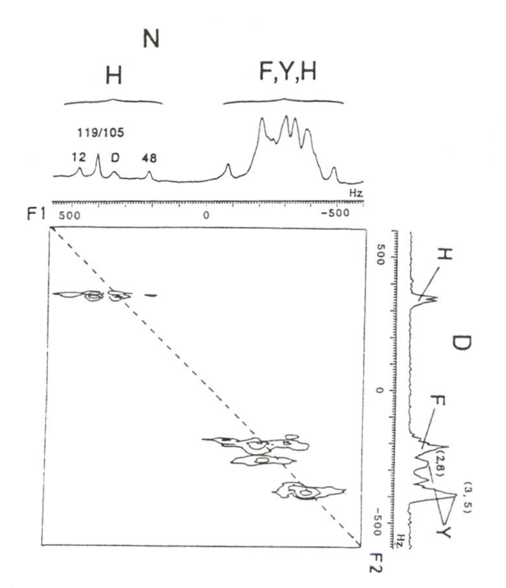

Figure 8: State-correlated two-dimensional NMR spectrum of ribonuclease A (10% in D_2O, 0.15 M KCl, pH 1.0), obtained between 30 °C (folded) and 45 °C(unfolded).

Improvement in the design of the microwave TJ probe for aqueous solutions is in progress in our laboratory.

References

Akasaka, K., Naito, A., and Imanari, M. (1991). *J. Am. Chem. Soc.* **113**, 4687.

Akasaka, K., Naito, A., Nakatani, H., and Imanari, M.(1990). *Rev. Sci. Instrum.* **61**, 66.

Akasaka, K., Naito, A., and Nakatani, H. (1991). *J. Biomol. NMR* **1**, 65.

Akasaka, K., Inoue, T., Hatano, H., and Woodward, C. K. (1985). *Biochemistry* **24**, 2973.

Hiromi, K., Akasaka, K., Mitsui, Y., and Tonomura, B. (1985). Protein Protease Inhibitor-The Case of Streptomyces Subtilisin Inhibitor (SSI), Elsevier Science Publishers, Amsterdam.

Murao, S., and Sato, S. (1972). *Agric. Biol. Chem.* **36**, 160.

Naito, A., Nakatani, H., Imanari, M., and Akasaka, K. (1991)., *J. Magn. Reson.* **87**, 429.

Naito, A., Imanari, M., and Akasaka, K. (1990)., *J. Magn. Reson.* **92**, 85.

Privalov, P. L., and Gill, S. J. (1988). *Adv. in Protein Chem.* **39**, 191.

Taguchi, S., Kojima, S., Kumagai, I., Ogawara, H., Miura, K., and Momose, H. (1992). *FEMS Microbiol. Lett.* **99**, 293.

Taguchi, S., Kikuchi, H., Kojima, S., Kumagai, I., Nakase, T., Miura, K., and Momose, H. (1993). *Biosci. Biotech. Biochem.* **57**, 522.

Tamura, A., Kimura, K., Takahara, H., and Akasaka, K. (1991). *Biochemistry* **30**, 11307.

Tamura, A., Kimura, K., and Akasaka, K. (1991). *Biochemistry* **30**, 11313.

Tamura, A., Kanaori, K., Kojima, S., Kumagai, I., Miura, K., and Akasaka, K. (1991). *Biochemistry* **30**, 5275.

Yamada, H., in NMR Basic Principles and Progress 24, Springer-Verlag, Heidelberg (1991).

Yamada, H. (1974). *Rev. Sci. Instrum.* **45**, 640.

Yamaguchi, T., Yamada, H., and Akasaka, K. (1994). *J. Mol. Biol.* **250**, 689.

Zahn, R., Spitzfaden, C., Ottiger, M., Wüthrich, K., and Plückthun, A. (1994). *Nature* **368**, 261.

13

Multidimensional NMR Investigation of the Neurotoxic Peptide Mastoparan in the Absence and Presence of Calmodulin

F. Mari, X. Xie, J.H. Simpson, and D.J. Nelson

Department of Chemistry
Florida Atlantic University
Boca Raton, FL 33431 USA

Gustaf H. Carlson School of Chemistry
Clark University
Worcester, MA 01610 USA

Calmodulin (CaM) is the major Ca^{2+} receptor in eukaryotic cells (Means and Rasmussen, 1988). This paper begins an investigation into the structural requirements for neurotoxic peptide binding to CaM. In resting cells, CaM is deficient in Ca^{2+} (the protein has the potential for binding four Ca^{2+} ions with high affinity, $pK_d > 6$ (Means and Rasmussen, 1988)). Following nerve cell excitation, intracellular levels of Ca^{2+} increase dramatically, from about 0.1 μM to about 10 μM, allowing CaM to become fully-loaded with Ca^{2+}. Ca^{2+} -loaded CaM has the ability to activate a number of neural enzymes, including cyclic nucleotide phosphodiesterase, adenylate cyclase, Ca^{2+} - CaM kinase and calcineurin (Kennedy, 1989). A tight-binding neurotoxic peptide would be expected to competitively inhibit activation of these enzymes.

The high level of intercellular coordination required by higher organisms is attained, in part, by the complex interplay of the nervous and endocrine systems. Two important second messengers are involved in information transfer processes associated with the normal operation of these two systems: cyclic AMP (cAMP) and Ca^{2+}. Cyclic AMP is involved in trans-membrane information flow following the interaction of cell surface receptors with certain hormones (*e.g.*, glucagon, epinephrine and ACTH), while Ca^{2+} is the principal information carrier in the nerve cell following stimulation of the system by membrane depolarization. CaM plays a pivotal role in second messenger function in both the nervous and endocrine systems. In the nervous system, calmodulin is the principal target for Ca^{2+}. In the endocrine system, CaM (complexed with Ca^{2+}) is responsible for activating the enzymes responsible for both cAMP synthesis (*i.e.*, adenylate cyclase) and degradation (*i.e.*, cyclic nucleotide phosphodiesterase). Additional linkage between the nervous and endocrine systems is evident from the fact that both systems are responsive to some of the same *peptide* messengers. For example, insulin, glucagon, angiotensin, and somatostatin have been found in the brain, and may function as neurotransmitters (Malencik and Anderson, 1982) perhaps through CaM mediation. Following the finding that ACTH and β-endorphin inhibit cyclic nucleotide phosphodiesterase activation by CaM by competing with the enzyme for CaM (Weiss *et al.*, 1980), Malencik and Anderson (1982) surveyed the binding of 17 different peptides by calmodulin, four of which (ACTH, β-endorphin, glucagon, and substance P) were found to bind with relatively high affinity (i.e., $K_d \sim 10^{-6}$ M). Additional insight was gained following the discovery that a 1:1 complex of high affinity (K_d = 3 nM) can be formed between CaM and melittin, a cytotoxic peptide from bee venom (Comte *et al.*, 1983). Upon binding to CaM, the α-helical content of melittin increased from 5 to 70%. It was subsequently shown that this α-helix is highly basic and amphiphilic (*i.e.*, one side hydrophobic and the other side hydrophilic). *Is it essential for CaM binding that a peptide have an amino acid sequence that has a high propensity to adopt a basic, amphiphilic α-helix?* Cox *et al.* (1985) considered this hypothesis by examining the amino acid sequences of several peptides that interact with CaM with dissociation constants of 100 nM or less. The sequences are shown in Figure 1. It is clear that the aligned sequences form a pattern of alternating hydrophobic and hydrophilic residues. (In Figure 1 aligned hydrophobic residues are shown in outlined type.) The pattern repeats every 3.6 residues, indicating that if these sequences were to fold into an α–helix, one face would be hydrophobic and the opposite face would be hydrophilic (*i.e.*, all of these peptides would form basic, amphiphilic helices).

One class of peptides presented in Figure 1 is the main focus of this paper, the mastoparans. Mastoparans are cytotoxic peptides found in the venom sacs of wasps. They function in a rather nonselective manner to stimulate exocytosis, including the release of histamine from mast cells (Hirai *et al.*, 1979). Mastoparans also show mitogenic activity for Swiss 3T3 cells (Gil *et al*, 1991). Since many of the activities of mastoparans are blocked by pretreatment of cells with pertussis toxin, the peptide has been suggested to be involved in G-protein activation. [Pertussis toxin ADP- ribosylates G_i-type G proteins, making them unresponsive to hormonal regulation.] The work of Cox *et al.* (1985) (Figure 1) also indicates that mastoparans are potent competitive inhibitors of CaM activation of CaM-regulated enzymes. Recently, Wakamatsu *et al.* (1992) determined the structure of vesicle-bound mastoparan-X in the presence of

PEPTIDE	K_d (nM)	SEQUENCE
Mastoparan (MP)	0.3	1I N L K A L A A L A K K I L
Mastoparan X (MP-X)	0.9	1I N W K G I A A M A K K L L
P. Mastoparan (P-MP)	3.5	1V D W K K I G Q H I L S V L
Mastoparan C (MP-C)	?	1L N L K A L L A V A K K I L
Mastoparan M (MP-M)	?	1I N L K A I A A L A K K L L
Melittin	3	3G A V L K V L T T G L P A L

Figure 1: Amino acid sequences of some peptides that bind tightly to calmodulin (abstracted in part from Cox *et al.*, 1985).

perdeuterated phospholipid vesicles, using two-dimensional NMR analyses, along with distance geometry and molecular dynamics calculations. The results indicated that the 12 C-terminal residues adopt an *amphiphilic, α-helical conformation* concomitant to binding to the phospholipid bilayer, a result consistent with the predictions of Cox *et al.* (1985). *Will the mastoparans adopt the same structures when bound to CaM?*

The structure of mastoparan (and other related amphiphilic peptides) when bound to biomembranes is highly relevant to peptide-receptor interactions. It has been suggested that binding of amphiphilic peptides to the membranes of target cells, induces peptide conformations that are similar to the conformations of the peptide when bound to the receptor. In the particular case of mastoparan, the membrane-bound structure of the peptide (Wakamatsu *et al.* (1992)) has been used to rationalize its G-protein activating effect. Mastoparan represents the minimal structural motif required for CaM binding. Unlike the membrane-bound structure of mastoparan, the CaM-bound structure of mastoparan is predicted to be helical throughout its sequence. We are currently in the process of synthesizing fully ^{15}N-labeled mastoparan (MP). [Other variants, mastoparan-X (MP-X), P-mastoparan (P-MP), mastoparan-C (MP-C), and mastoparan-M (MP-M), will be produced and examined later in the project.] The labeled material will allow us to use a variety of heteronuclear, multidimensional NMR experiments to determine the precise structure of this peptide when bound to CaM. [Note that mastoparan shows high affinity for other molecules, such as G-proteins. The availability of ^{15}N-labeled materials will also allow us to study any mastoparan/protein-receptor system. These studies will give a better understanding of ligand-receptor interactions.] The completed study will provide us with a better understanding of the manner in which a number of known neurotoxic peptides (from the mastoparan family of peptides) interact with an important nervous system protein, calmodulin. Calmodulin, uninhibited by neurotoxic peptides, is critical to the normal cascade of enzymatic events that lead to the inter-cellular transmission of nerve impulses. Since mastoparan serves as a useful surrogate for calmodulin-activated enzymes, insight will also be gained into the mechanism whereby CaM recognizes (and activates), target enzymes critical to neurotransmission processes.

Materials and methods

Bacterial strains and plasmids (pCaMpl/N4830-1)

The plasmid harboring the chicken calmodulin gene (Putkey *et al.*, 1985) and ampicillin resistance was used to transform *E. coli* strain N4830-1 (Clontech) carrying the temperature-sensitive repressor C1857 repressor (represses the PL promoter at 29 °C, but not at 42 °C). This temperature-inducible bacterial expression system was kindly provided by Prof. Anthony Means (Duke University).

Expression and purification of chicken calmodulin

Cells were grown on enriched media (LB broth plus ampicillin (Amp)) at 30 °C to an optical density ($OD_{650\ nm}$) > 0.6, and then induced at 42 °C for 2 h, after which the cells were harvested by centrifugation at 5,000 rpm for 10 min. at 4 °C. The cell pellet was washed in a small amount of suspension buffer (40 mM Tris-Cl, pH = 8) and in phosphate-buffered saline (pH = 8). The cell pellet (about 20 g for a 5-L culture) was resuspended in 10 mL of hypoosmosis buffer (2.4 M sucrose, 40 mM Tris-Cl, 10 mM EDTA) at room temperature, and kept on ice for 30 to 60 min. The suspension was then diluted in 40 mL of lysis buffer (50 mM MOPS, 0.1 M KCl, 1 mM EDTA, 1 mM DTT and 0.2 M PMSF) plus 10 mg lysozyme (USB chicken egg white, 23,700 units/mg). Following overnight storage at 4 °C, the cell lysate was centrifuged at 35,000 rpm at 4 °C for 1 hour. The clear, light, yellowish supernatant (about 60 mL) was brought up to 5 to 7 mM in $CaCl_2$ and applied to a phenylSepharose column (70 mL bed volume), previously equilibrated in "buffer A" (50 mM Tris-Cl, 1 mM $CaCl_2$). One hundred milliliters of buffer A and 100 mL of "buffer B" (50 mM Tris-Cl, 1 mM $CaCl_2$, 500 mM NaCl) was washed through the column consecutively before the elution buffer (50 mM Tris-Cl, 1-2 mM EDTA, 50-150 mM NaCl) was applied (Putkey *et al.*, 1985; Gopalakrishna and Anderson, 1982; Babu *et al.*, 1985). The calmodulin fractions were eluted from the column after about one bed volume of elution buffer had passed through the column. The collected fractions were monitored by UV absorbance at 280 nm. The fractions exhibiting the characteristic calmodulin-like spectra were pooled and subsequently dialyzed against 0.1 M NH_4HCO_3 overnight, with at least three changes of the dialysis buffer. Following dialysis, the calmodulin-containing solution was lyophilized, typically yielding about 25 mg protein for a 5 liter culture. SDS-polyacrylamide gel electrophoresis of the bacterially-expressed protein typically revealed a single dense band, consistent with > 95% purity for the final product.

Preparation of samples for NMR spectroscopy

Unlabeled mastoparan (INLKALAALAKKIL-NH_2) was obtained from Sigma Chemical Co. (St. Louis, MO). HPLC analysis indicated that no contaminating peptide species were present. ^{15}N-labeled mastoparan was synthesized by Michael Berne (Tufts University). The peptide samples were prepared by dissolving mastoparan in 800 μL 90% H_2O / 10% D_2O or 40% H_2O / 10% D_2O / 50% TFE-d_2 or 90% H_2O / 10% D_2O with SDS-d_{25} (Cambridge Isotopes Laboratories, Woburn, MA) micelles. (TFE-d_2 was obtained by exchange of the labile deuteron of the commercially available TFE-d_3 (Cambridge Isotopes Laboratories, Woburn, MA).) The pH of the solutions (uncorrected for isotope effects) was adjusted with the use of 0.01 M solutions of KOH and HCl. DSS was used as an internal reference in most cases.

The protein sample (1.5 - 2 mM bacterially-expressed chicken calmodulin) was made by dissolving 25 - 30 mg of CaM_{pl} in 800 - 900 μL of sample buffer (0.2 M KCl, 10 mM $CaCl_2$ in a 90% H_2O / 10% D_2O solvent system, pH = 6.5). GELMAN LC13 (0.45 μm) filters were used to remove any insoluble particles.

The protein-peptide complex sample (1:1 M/M) was prepared by adding 2.2 mg of mastoparan (mol. wt. = 1,478.9, 94% pure) slowly to the protein solution while being mixed.

NMR experiments

Proton NMR spectra were acquired on a Varian *Unity* series 500 MHz NMR spectrometer (Varian Analytical Instruments, Palo Alto, CA), with the use of the ^1H channel of a triple resonance probe (^1H / ^{13}C / ^{15}N). Spectra were processed using VNMR v3.21 on Sun 4-65 computers.

1D ^1H spectra were routinely acquired at temperatures ranging from 15 °C to 45 °C. Water suppression was accomplished by continuous wave, low power irradiation of the water resonance through the transmitter channel for a duration of 1 sec prior to the 90 ° pulse, a feature incorporated in all the pulse sequences to be used in this work. A polynomial base line correction and an exponential line broadening was used for the processing of spectra. The center of the transmitter frequency was set at the water frequency for all experiments. Broadband ^{15}N decoupling, when required, was applied during the acquisition time using WALTZ-16 heteronuclear modulated decoupling.

2D NMR spectra were acquired in the phase sensitive mode using the States-Haberkorn hypercomplex method. Water suppression was accomplished by the method described above where the presaturation period was incorporated into the 2D pulse sequence through the transmitter for a duration of 0.5 s. Further suppression of the water resonance was obtained by appropriate base line correction in the t2 dimension. 2D spectra were apodized using a gaussian window function in the t2 and t1 dimensions. 2D ^1H TOCSY experiments were acquired with 2048 data points in the t2 dimension and 2 x 256 t1 increments, with 32 scans per t1 value, a pulse delay of 0.1 sec and a MLEV17 mixing period of 120 ms. The final 2D spectra were processed with zero-filling to a final spectrum size of 2048 x 2048 data points. 2D ^1H NOESY experiments were acquired with 4096 data points in the t2 dimension and 2 x 512 t1 increments, with 64 scans per t1 value, a pulse delay of 0.1 s and with a mixing time period of 200 ms. The final 2D spectra were processed with zero-filling to a final spectrum size of 4096 x 4096 data points. A homospoil pulse subsequent to the beginning of the read pulse train followed by a 20 ms recovery delay was employed to reduce multiple quantum artifacts and to cancel the residual magnetization of the spins with long relaxation times. 2D ^1H - ^{15}N HMQC spectra were acquired with 2048 data points in the t2 dimension, and 2 x 320 t1 increments, with 64 scans per t1 value, a recycle delay of 1.5 sec, and a multiple quantum refocussing delay of 5.6 msec. The final 2D spectra were processed with zero-filling to a final spectrum size of 2048 x 2048 data points.

Results and discussion

1D ^1H NMR spectral analysis

Media composition effect.

The ^1H NMR spectra of mastoparan in H_2O, 50% TFE and in SDS showed significant differences depending on the media in which the peptide was dissolved. Since the signals from the protons attached to the peptidyl backbone nitrogens are the most sensitive to the structure of the peptide in solution, the H-N region of the ^1H spectra was used to monitor the effect of the solution on the structures of mastoparan. Figure 2 shows the H-N region of mastoparan at 15 °C and pH 3.70. In isotropic media (H_2O and 50% TFE), the resonance lines are relatively sharp (to the point that NH-αCH coupling constants can be measured easily) compared to spectra in anisotropic media (SDS-d_{25} micelles). Broader lines are expected for micelle-adsorbed mastoparan, since they tumble at a rate comparable to that of the macromolecular aggregate. However, even under these conditions, the line widths obtained for the SDS-dissolved mastoparan spectra are relatively broader than those obtained with other amphiphilic peptides of comparable size under the same conditions (Rizo *et al.*, 1993; Bairaktari *et*

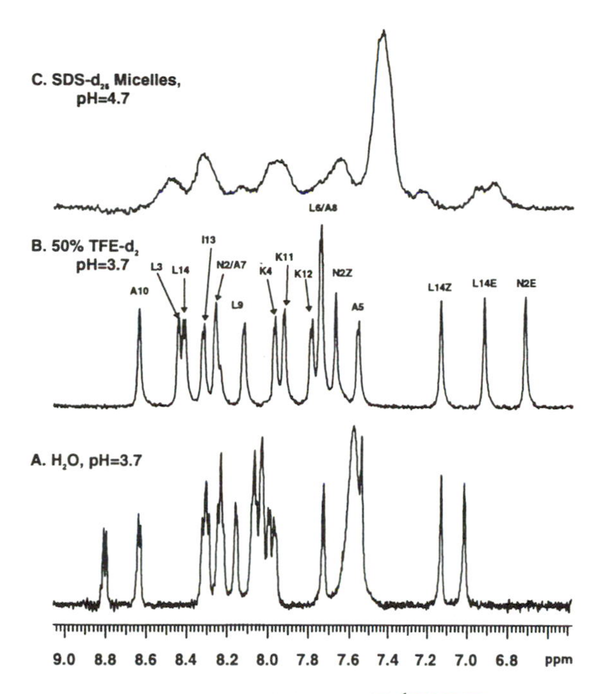

Figure 2: Amide proton regions of the 500 MHz 1D ^{1}H NMR spectra of mastoparan at 15 °C in H_2O (spectrum A), in 50% TFE (spectrum B) and in H_2O in the presence of SDS-d_{25} micelles (spectrum C).

al., 1990). This result suggests that in the presence of SDS micelles, peptide aggregation might be occurring; however, other possibilities, such as mastoparan reorganization of the lipid-water interface of the micelles, that might lead to a change in the aggregation number of SDS micelles, can also be considered.

Resonance overlap precluded sequence specific assignments for the spectra in water and in SDS micelles. However, most NH-αCH coupling constants are around 7 Hz, indicating that mastoparan is a random coil in water. Furthermore, 2D-NOESY data for mastoparan when dissolved in water (data not shown) show very few cross peaks. This seems to confirm the view that the peptide is unstructured when dissolved in water. By way of contrast, when the peptide is solubilized in SDS micelles, a large number of NOE cross correlations are observed (data not shown). The NOE patterns of the SDS solubilized peptide are very similar to the ones observed when the peptide is dissolved in a 50% TFE/H_2O mixture (Figure 3), which is a good indication that the peptide is helical when solubilized in SDS micelles (see below). However, the very broad

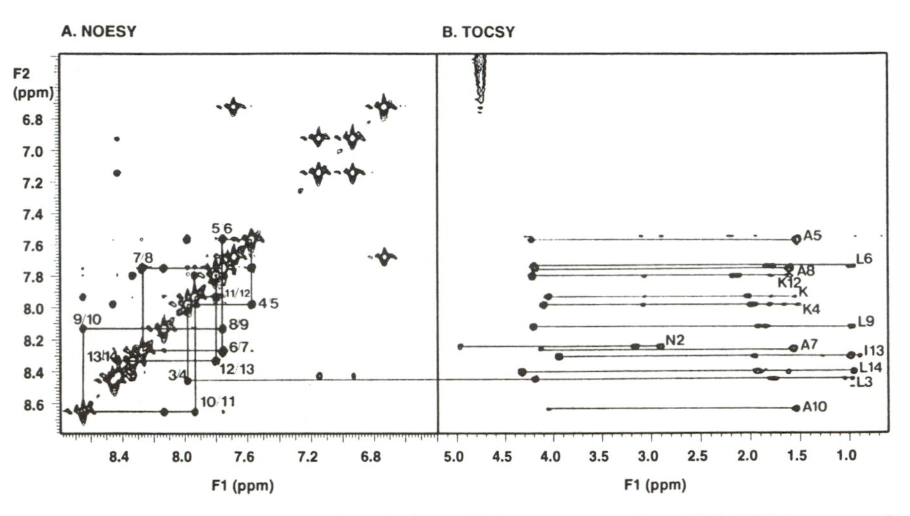

Figure 3: Portions of the 500 MHz ^{1}H NOESY (spectrum A) and TOCSY (spectrum B) NMR spectra of mastoparan at 15 °C, pH = 3.7 in 50% TFE.

lines obtained in the micelle spectrum do not allow us to establish any of the necessary details to define the precise conformation of the micelle-solubilized mastoparan.

Secondary structure of mastoparan

NOESY spectra of mastoparan in H_2O at 15 °C show very few cross peaks and no interresidue cross peaks. Possible field cancellation effects were investigated by performing NOESY at 300 MHz. The absence of cross correlations at the lower frequency indicates, in conjunction with the measured coupling constants, the lack of defined or nascent structure in water at 15 °C. At lower temperatures (-5 °C) NOESY experiments showed a few intra and interresidue cross correlations for mastoparan dissolved in water. Perhaps at such temperatures a nascent structure for this peptide can be considered; however, possible peptide aggregation might be occurring under these conditions.

NOESY spectra at 500 MHz in 50% TFE-d_2/H_2O revealed a large number of cross correlations. This is a very strong indication of the fact that, in the presence of TFE, mastoparan adopts a well defined structure. Good chemical shift dispersion is observed for the mastoparan spectra when dissolved in TFE. The use of 2D-NOESY and 2D-TOCSY data (Figure 3) allowed sequence specific assignments of all the protons for TFE-dissolved mastoparan. NOE connectivities and limiting values of

$^3J_{HN\text{-}\alpha H}$ for mastoparan in 50% TFE-d2/H_2O are shown in Figure 4. Strong NH_i-NH_{i+1} and αH_i-NH_{i+1} interactions and the weak NH_i-NH_{i+2},

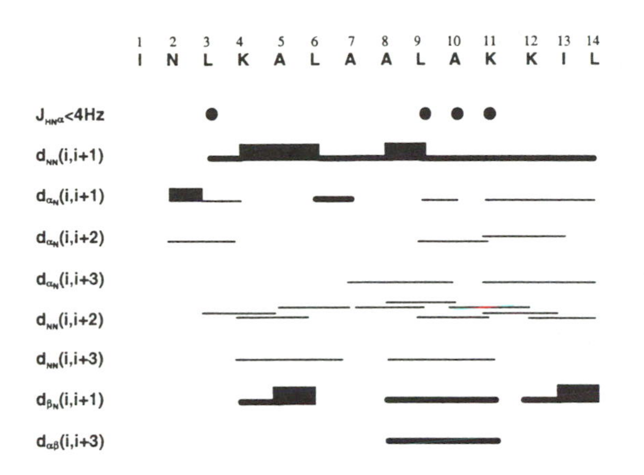

Figure 4: NOE connectivity constructed from the 500 MHz ^{1}H NOESY data on mastoparan at 15 °C, pH = 3.7 in 50% TFE.

NH_i-NH_{i+3} and αH_i-NH_{i+3} interactions are a very strong indication of the helical nature of the peptide. It has been shown by circular dichroism that mastoparan is 50% helical in TFE/H_2O mixtures (McDowell *et al.*, 1985). While CD and NOE data indicate a helical secondary structure, very few $^3J_{HN\text{-}\alpha H_3}$ show strictly helical values (< 4 Hz), although several of the measured $^3J_{HN\text{-}\alpha H}$ values were borderline, especially toward the center of the peptide.

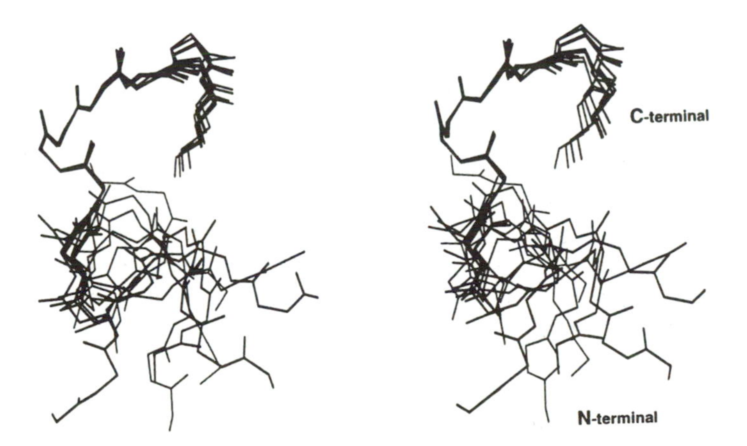

Figure 5: XPLOR structures of mastoparan processed from experimental data in the 50% TFE solvent system.

With the use of 144 NOE restrictions (*i.e.*, "off diagonal" peaks) (approximately 10 restrictions per residue) and molecular dynamics calculations, using the program XPLOR v3.0 (Brunger, Yale University, 1992), implemented on a Silicon Graphics SGI 4D/35 computer, we have resolved the structure of mastoparan in 50% TFE solution. This structure is shown to be helical from residues 6 to 11 with β-turns involving residues 2 - 5 and 11 - 14. Figure 5 shows the superposition of several of the lowest energy "stick" structures (note the "fraying" evident at the amino terminal end of the structure).

The CaM-Mas complex

In spite of the extensive conformational changes that occur in CaM upon complexation with amphiphilic peptides, only small changes in the characteristic CaM downfield protons (G25, G61, G98, I100 and N137) are observed for CaM-Mas complex with respect to the uncomplexed Ca^{2+} loaded CaM (Figure 6). These results are consistent with the findings reported for the CaM-M13 (Seeholzer and Wand, 1989; Ikura *et al.*, 1990; Roth *et al.*, 1992) and the CaM-Mas (Muchmore *et al.*, 1986; Ohki *et al.*, 1991) complexes for these particular residues. Likewise, upon mastoparan complexation, some of the CaM upfield methyl protons (I27, I100, V35, V108, V142 and L69) exhibit some upfield shifting, although the line widths for the methyl protons on I100 and V35 are much broader in the complex.

The situation for the backbone amide protons in mastoparan (A5, A7, A8 and A10) is quite different. Upon complexation with CaM, the HN of these Ala residues experience very significant chemical downfield resonance shifting. The observed chemical shifts of these protons correlate very well with chemical shifts for these residues when the peptide is dissolved in 50% TFE-d_2/H_2O (Figure 7). It was already established that mastoparan is highly helical when dissolved in the TFE/H_2O mixture, the correspondence in the HN chemical shift of the CaM-bound mastoparan with the peptide dissolved in 50% TFE is a good indication that mastoparan acquires a helical conformation upon CaM complexation. The extreme chemical shift difference found between the peptide when dissolved in water and the peptide complexed to CaM can be rationalized by the random coil to helix transition that occurs upon CaM binding.

The results described above are consistent with the results obtained for the CaM-M13 complex (Ikura *et al.*, 1992). It seems reasonable to suggest that the precise 3D structure of the CaM-Mas complex is very similar to the one reported by Ikura and coworkers for the CaM-M13 complex (Ikura *et al.*, 1992) using multidimensional NMR methods. In the CaM-M13 complex, the C-terminal and N-terminal domains of CaM remain unchanged upon complexation. This explains the very small variation in the chemical shifts of the protons indicated in Figure 6, since these residues are part of these domains and there are

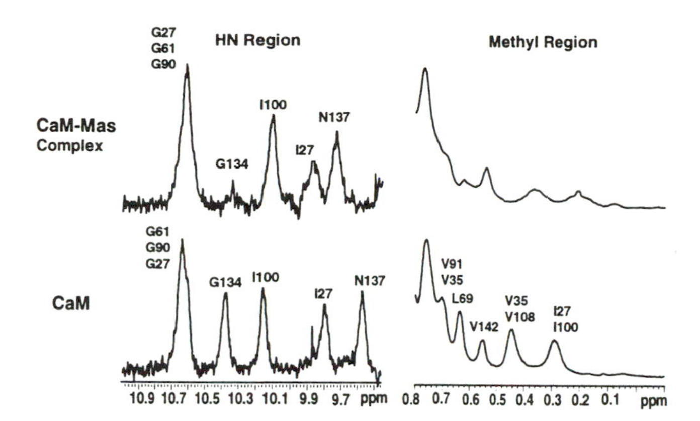

Figure 6: Downfield amide and upfield methyl regions of the 500 MHz ^1H NMR spectra of calmodulin at 35°C, pH = 6.9 in H_2O in the absence (lower spectra) and presence (upper spectra) of one equivalent of mastoparan.

no close contacts with the bound peptide. On the other hand, the long central helix revealed in the X-ray structure of CaM, which has been shown to be flexible in solution, is disrupted, upon peptide complexation, into two helices connected by a flexible loop. This loop enables the two calcium binding domains to clamp down on the peptide, which concomitantly adopts a mostly amphiphilic α-helical conformation. Most of the resonances linked to the central helix of CaM exhibit extreme chemical shift overlap. Although the overlap problem can be partially overcome by the use of 3D and 4D NMR methods, this procedure is laborious and time consuming. By labeling the mastoparan, instead of the CaM, we have been able to directly assess the conformation of the peptide in a relatively simple way.

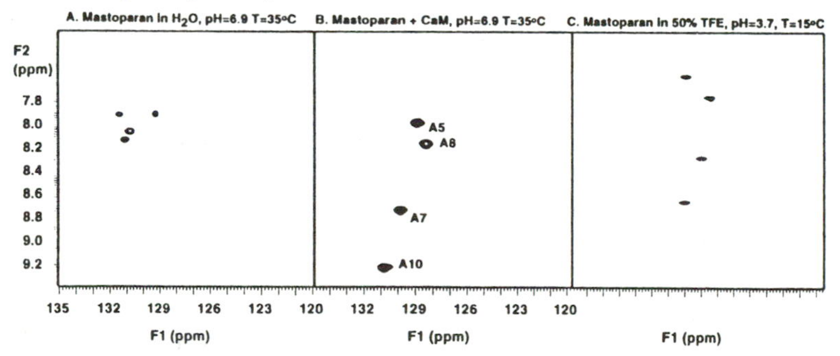

Figure 7: ^1H-^{15}N HMQC spectra of Ala-^{15}N-mastoparan at 35°C, pH=6.9 in H_2O in the absence (spectrum A) and presence (spectrum B) of one equivalent of calmodulin.

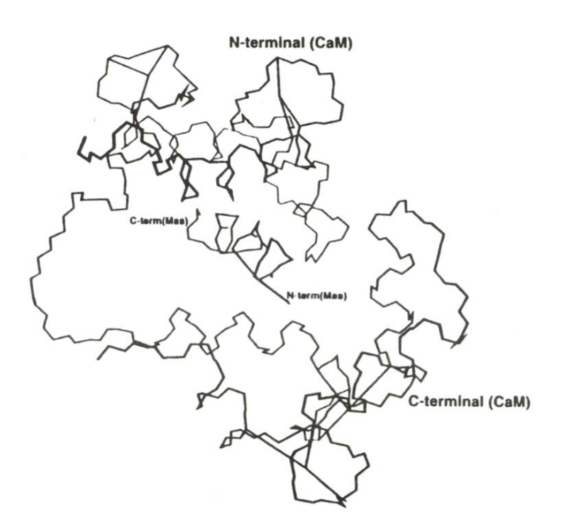

Figure 8: A proposed structure for the CaM-Mas complex.

At the current time we lack sufficient evidence to address the precise details of how Mas binds to CaM; however, a putative structure for the CaM-Mas complex can be built from the known structure of the CaM-M13 complex. On the basis of hydrophobic interactions, one can predict that residues I1, A5, A8 and L14 of Mas will occupy the same hydrophobic binding sites as in the M13 peptide. If this is the case, the putative structure of the complex can be represented as shown in Figure 8. However, Mas has additional hydrophobic residues not present in M13, namely L3, L9 and I13, which might compete for the hydrophobic binding sites proposed above. Furthermore, Mas represents the minimal binding domain of helical peptides required for CaM binding. Therefore, some of the intermolecular interactions found in the CaM-M13 complex, mainly between K18 to S21 of the M13 are not going to be present in the CaM-Mas complex. ^{113}Cd NMR evidence indicates that there are some differences between the CaM-M13 complex and the CaM-Mas complex (Ikura *et al.*, 1989) (*i.e.*, the ^{113}Cd spectrum for the CaM-M13 complex shows four distinct resonances, corresponding to the four metal ion binding sites, whereas the ^{113}Cd spectrum for the CaM-Mas complex shows only three resonances, indicative of signal overlap, with chemical shift differences up to 5.5 ppm apparent). Although it is very likely that the overall structure of the CaM-Mas complex is going to be very similar to the CaM-M13 complex, some differences can be expected. It has been suggested that other binding modes of small peptides to CaM are possible (Kretsinger, 1992). The C20W Ca^{2+} pump peptide binds to CaM in a manner that does not change the overall dumbbell-shaped structure of the protein upon binding (Kataoka *et al.*, 1991). Apparently this is not the case for the CaM/mastoparan system, since it has been shown by solution x-ray scattering that the CaM acquires a non-dumbbell shape upon complexation of the peptide to the protein, one consequence of which is a closer approach of the two Ca^{2+} binding domains (Ohki *et al.*, 1991). Other reports suggest the possibility that a second Mas molecule can bind to the N-terminal domain of CaM (Yoshino *et al.*, 1989; Matsushima *et al.*, 1989). Although this situation has not been confirmed, it will be very difficult to rationalize the binding of a second molecule of mastoparan with the present structural knowledge of CaM-peptide complexes. Precise details of the intermolecular interactions between CaM and mastoparans are needed to define the way that CaM binds this family of peptides. Further multidimensional NMR experiments are in progress to settle this matter.

Acknowledgements

This research was made possible by a shared instrument grant (in support of a Varian *Unity* 300 NMR, Grant No. RR04659 N.I.H.), and by grants from the Kresge and Keck Foundations (in support of a Varian *Unity* 500 NMR).

References

Babu, Y.S., Sack, J.S., Greenhough, T.J., Bugg, C.E., Means, A.R., and Cook, W.J. (1985). *Nature* **315**, 37.

Bairaktari, E., Mierke, D.F., Mammi, S., and Peggion, E. (1990). *Biochemistry* **29**, 10090.

Comte, M., Maulet, Y., and Cox, J.A. (1983). *Biochem. J.* **209**, 269.

Cox, J.A., Comte, M., Fitton, J.E., and DeGrado, W.F. (1985). *J. Biol. Chem.* **260**, 2527.

Gil, J., Higgins, T., and Rozengurt, E. (1991). *J. Cell Biol.* **113**, 943.

Gopalakrishna, R., and Anderson, W.B. (1982). *Biochem. Biophys. Res. Commun.* **104**, 830.

Hirai, Y., Yasuhara, T., Yoshida, H., Nakajima, T., Fugino, M., and Kitada, C. (1979). *Chem. Pharm Bull. (Tokyo)* **27**, 1942.

Ikura, M., Hasegawa, N., Aimoto, S., Yazawa, M., Yagi, K., and Hikichi, K. (1989). *Biochem. Biophys. Res. Commun.* **30**, 161.

Ikura, M., Kay, L.E., and Bax, A. (1990). *Biochemistry* **29**, 4659.

Ikura, M., Clore, G.M., Gronenborn, A.M., Zhu, G., Klee, C.B., and Bax, A. (1992). *Science* **256**, 632.

Kataoka, M., Head, J.F., Vorherr, T., Krebs, J., and Carafoli, E. (1991). *Biochemistry* **30**, 6247.

Kennedy, M.B. (1989). *Trends Neurosci.* **12**, 417.

Kretsinger, R.H. (1992) *Science* **258**, 50.

Malencik, D.A., and Anderson, S.R. (1982). *Biochemistry* **21**, 3480.

Matsushima, N., Izumi, Y., Matsuo, T., Yoshino, H., Ueki, T., and Miyake, Y. (1989). *J. Biochem. (Tokyo)* **105**, 883.

McDowell, L., Sanyal, G., and Prendergast, F.G. (1985). *Biochemistry* **24**, 2979.

Means, A.R., and Rasmussen, C.D. (1988). *Cell Calcium* **9**, 313.

Muchmore, D.C., Malencik, D.A., and Anderson, S.R. (1986). *Biochem. Biophys. Res. Commun.* **30**, 137.

Ohki, S.Y., Yazawa, M., Yagi, K., and Hikichi, K. (1991). *J. Biochem. (Tokyo)* **110**, 737.

Ohki, S., Tsuda, S., Joko, S., Yazawa, M., Yagi, K., and Hikichi, K. (1991). *J. Biochem. (Tokyo)* **109**, 234.

Putkey, J.A., Slaughter, G.R., and Means, A.R. (1985). *J. Biol. Chem.* **260**, 4704.

Rizo, J., Bianco, F.J., Kobe, B., Bruch, M.D., and Gierash, L.M. (1993). *Biochemistry* **32**, 4881.

Roth, S.M., Schneider, D.M., Strobel, L.A., VanBerkum, M.F.A., Means, A.R., and Wand, A.J. (1992). *Biochemistry* **31**, 1443.

Seeholzer, S.H., and Wand, A.J. (1989). *Biochemistry* **28**, 4011.

Wakamatsu, K., Okada, A., Miyazawa, T., Ohya, M., and Higashijima,T. (1992). *Biochemistry* **31**, 5654.

Weiss, B., Prozialeck, W., Cimino, M., Barnette, M.S., and Wallace, T.L. (1980). *Ann. N.Y. Acad. Sci.* 356.

Yoshino, H., Minari, O., Matsushima, N., Ueki, T., Miyake, Y., Matsuo, T., and Izumi, Y. (1989). *J. Biol. Chem.* **264**, 19706.

14

NMR of Larger Proteins: Approach to the Structural Analyses of Antibody

Yoji Arata

Water Research Institute
Sengen 2-1-6
Tsukuba 305, Japan

The first ^{1}H NMR spectrum of a protein, bovine pancreatic ribonuclease, reported in 1957 by Saunders *et al.* was accounted for by Jardetzky and Jardetzky (1957) in terms of the spectra of the constituent amino acids. Jardetzky and coworkers continued to report a series of important papers describing the potential usefulness of high-resolution NMR (Roberts and Jardetzky, 1970).

Modern NMR of proteins began with the classic paper published in 1968 by Markley, Putter, and Jardetzky, who beautifully demonstrated the possibility of using stable-isotope labeling for the structural analyses of proteins in solution (Markley *et al.,* 1968). Five years before the publication of this paper, Jardetzky gave an important lecture in Tokyo, stressing the importance of NMR particularly in combination with deuterium labeling as a potential solution version of X-ray crystallography for the determination of the three-dimensional structure of proteins (Jardetzky, 1965). The impact of Jardetzky's contribution was great, eventually leading to the now well-established combination of multidimensional NMR and stable-isotope labeling for the determination of the three-dimensional structure of proteins in solution.

High-resolution NMR of biological macromolecules takes advantage of the fact that ^{1}H, ^{13}C, and ^{15}N, all of which are spin 1/2 nuclei, possess long relaxation times, which primarily are due to weak dipole-dipole interactions. Thus, phase memory can be retained long enough to extract relevant information on the spin system by fully making use of multidimensional techniques. This makes high-resolution NMR special as a tool for structural analyses at atomic resolution. By contrast, relaxation times are far shorter in the case of visible, ultraviolet, infrared, and Raman spectroscopy, where much stronger interactions are involved. For this reason no structural analyses at atomic resolution are possible using these types of spectroscopy.

However, an increase in the molecular weight eventually creates difficulties in achieving sufficient spectral resolution to be able to separate and assign each of the resonances of a protein. This is due to 1) a limitation of the strength of static magnetic field available and more importantly 2) an unavoidable shortening of relaxation times originating from the slow tumbling motion of the protein molecules in solution.

Attempts to determine the entire structure of a protein in solution may be difficult to pursue, when the molecular weight of a protein exceeds 30 K. It is certainly likely that new techniques will continue to be developed to cope with the molecular weight problem. However, the relaxation time barrier eventually becomes unsurpassable and the idea of simply relying on the standard method of high-resolution NMR (Wüthrich, 1986) for the structural analyses of proteins in solution must be abandoned.

Even under these unfavorable circumstances, the immense potential of NMR is still extremely attractive in the structural study of larger proteins. X-ray crystallography determines the *entire* protein structure. Even if one is only interested in a *part* of a protein, the necessary information has to be extracted from the *entire* structure. In marked contrast, NMR inherently starts from the individual amino acid residues of a protein and therefore can be used to discuss a *part* in detail if the spectral information can be selectively extracted.

NMR is the only existing method that can give information concerning the dynamical nature of macromolecules at atomic resolution. It should in principle be possible to know the static and dynamic nature of a *part* of larger proteins even if the entire structure of the proteins cannot be solved by NMR. In order to make this approach successful, stable-isotope labeling and multidimensional techniques have to be exploited extensively. Due to the complementary nature of the information obtained from NMR and X-ray crystallography, the combination of the two methods is very useful.

The starting point of the above approach dates back to Jardetzky and coworkers, who reported NMR titration curves for the histidine resonances of bovine pancreatic ribonuclease, staphylococcal nuclease, hen egg white lysozyme, and human lysozyme. It has been noted that the titration data can be used for the determination of the pKa of histidines that exist in the active site of the enzymes (Meadows *et al.,* 1967, 1968).

Another interesting and promising possibility for the structural investigation of larger systems is the use of solid-state NMR, which has progressed tremendously in the past ten years (Opella, 1994). This should bring us an enormous amount of information on the structural aspects of supramolecular systems such as membrane proteins (Cross and Opella, 1994). Even in the case of smaller peptides and proteins, structural analyses by standard high-resolution NMR methods become difficult once they are bound to the membrane.

In the present article, attempts will be made to summarize what the author's group has achieved so far using antibody molecules with a molecular weight of 150 K.

The antibody system: Structural analyses and their relevance to the development of high-resolution NMR methods for larger proteins

Structural architecture of antibodies (Kabat, 1976; Padlan, 1994)

The simplest protein of the antibody family is immunoglobulin G (IgG). IgG is composed of 1300 amino acid residues and has a molecular weight of 150 K. As shown in Figure 1, IgG consists of two identical heavy (H) chains and two identical light (L) chains. The heavy chain comprises four homology domains, VH, CH1, CH2, and CH3, and the light chain two homology domains, VL and CL. Each of these domains has a molecular weight of 12 K.

The antigen binding fragment, Fab, consists of the entire light chain and the N terminal half of the heavy chain, and is responsible for antigen recognition. In order to cope with the tremendous diversity of antigen that antibody can encounter and has to recognize, the amino acid sequences and therefore the tertiary structure of the antigen binding site vary from one IgG to another. It is

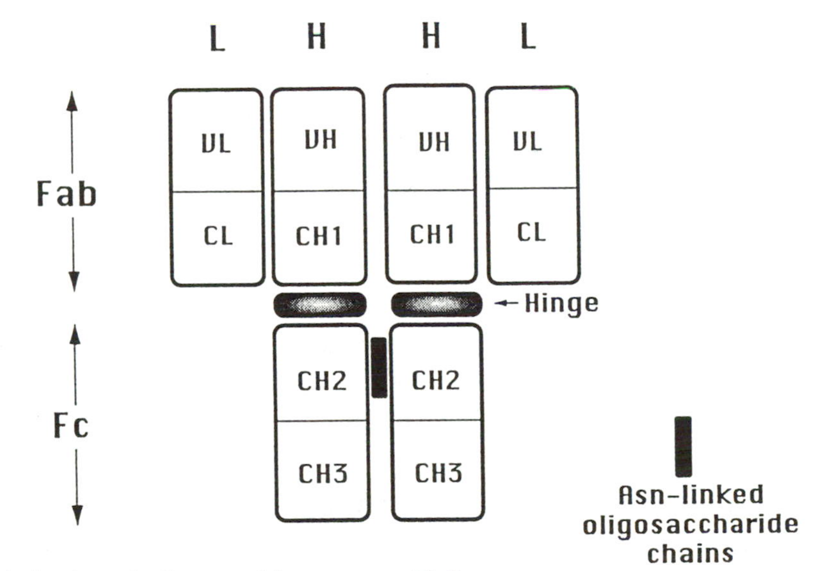

Figure 1: A schematic diagram of the structure of IgG.

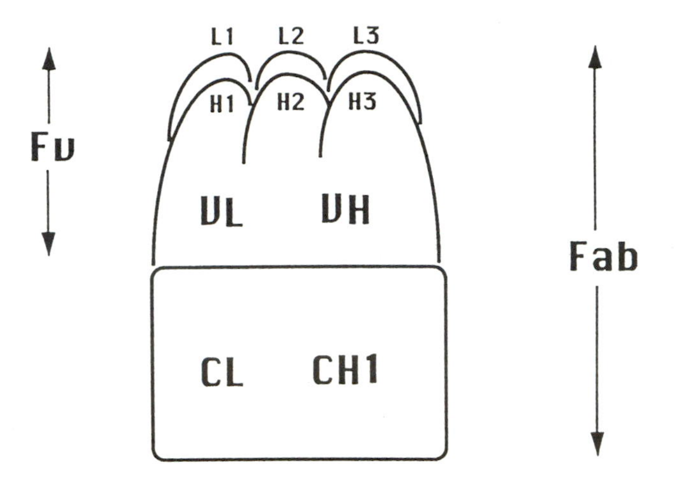

Figure 2: A schematic representation of the antigen binding site of IgG.

known that six *hypervariable* loops (i.e., H1, H2, H3, L1, L2, and L3, see Figure 2) are responsible for antigen recognition. H1, H2, and H3 originate from VH and L1, L2, and L3 from VL. X-ray crystallographic analyses have demonstrated that the six *hypervariable* loops are bundled on the surface of the IgG molecule and form the antigen binding site. The rest of the Fab region functions as a *scaffold* to support the antigen binding site constructed by the six *hypervariable* loops.

In contrast to Fab, Fc, which is formed from a set of the C terminal half of the heavy chain plays a crucial role in the expression of a variety of the effector functions that are triggered by the antigen recognition. The effector functions include the complement fixation and the binding of the Fc receptor to the Fc part of the antigen-bound antibody molecule. For this reason this part is basically the same in structure for all IgGs.

The CH1 and CH2 domains are separated by the hinge region, which is a short peptide segment comprising 5 to 15 amino acid residues. The hinge region, which is encoded by an independent exon, plays a crucial role in the expression of effector functions. It is known that two asparagine-linked oligosaccharide chains, one from each of the CH2 domains, are also essential for the expression of the effector functions.

In the following, amino acid residues originating from the heavy chain and the light chain will be identified by H and L, respectively.

Significance of the structural analyses of antibodies

X-ray crystallography has successfully elucidated the overall topology of the IgG molecule (Padlan, 1994). The application of NMR in the molecular structural analyses of IgG is definitely needed because it can address the following two important questions. How is the dynamical structure of IgG related to its biological functions and how can NMR methodology be extended to encompass structural studies on larger multifunctional proteins?

NMR and antibodies

IgG is an extremely large molecule for the application of high-resolution NMR. In view of the size of the molecule, it is obvious that one has to rely on a method that makes use of resonances with longer relaxation times. For this reason the nuclear spins under consideration have to be isolated from the heat reservoir of the lattice. In the case of ^{1}H NMR, the C2 proton of His is probably the only possibility amongst all the amino acid residues. An attempt along this line was made by our group using a 100 MHz spectrometer (Arata *et al.,* 1980). It was possible to identify the His resonance originating from the hinge region. The pH dependence of the chemical shift of the observed resonance was employed to determine the pKa of His in the hinge region. The dynamical nature of the hinge region was discussed on the basis of the result. Unfortunately most of our other attempts were less successful to use 1H NMR for the structural study of IgG. At this point it was clear that introduction of stable-isotope aided methods was necessary.

An attractive piece of work was reported by a Stanford group (Anglister, 1984). However, no unambiguous spectral assignments were provided. By that time, Kainosho and coworkers had developed a double labeling method for the unambiguous assignment of the carbonyl carbon resonances of proteins (Kainosho and Tsuji, 1982). The double labeling method was successfully introduced to our antibody project and it was possible to show using IgG selectively labeled with [1-^{13}C]Met that 1) the carbonyl carbon of IgG gave separate resonances and 2) these resonances could be assigned unambiguously by making use of the double labeling method (Kato *et al.,* 1989; Arata, 1991; Arata *et al.,* 1994). Typical examples are shown in Figure 3. The IgG labeled with [1-^{13}C]Met gave spectrum A. The IgG examined contained nine Met residues and, with the exception of two sets of resonances, all of others are clearly resolved in the ^{13}C spectrum (see Figure 3A). However, when IgG is doubly labeled with [1-^{13}C]Met and [α-^{15}N]Lys, one of the resonances is split into doublet (see Figure 3B). The observed splitting is due to the ^{13}C-^{15}N spin-coupling arising from a Met-Lys peptide in the heavy chain. On the basis of the result of this experiment, the split resonance was unambiguously assigned to Met412H.

The spectra shown in Figure 3 were the starting point of the antibody project described in the present chapter. In the following I will summarize what we have achieved so far.

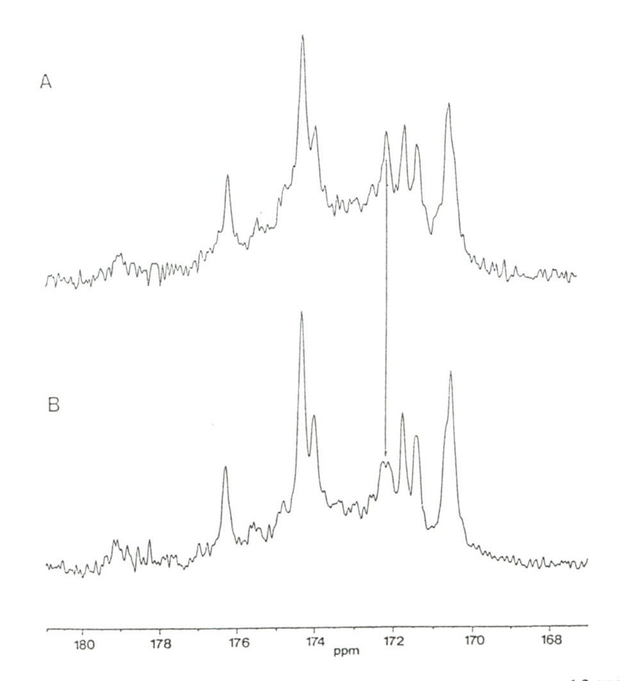

Figure 3: 100 MHz ^{13}C NMR spectra of IgG. A) IgG labeled with [1-^{13}C]Met. B) IgG doubly labeled with [1-^{13}C]Met and [a-^{15}N]Lys (Kato *et al.*, 1989).

Dynamic filtering and the structure of the hinge region

A number of physico-chemical experiments had been reported for the elucidation of the dynamical nature of the antibody molecule. All of these results suggested the existence of dynamical spot(s) in the antibody molecule. It had been concluded from protein chemical and physico-chemical data that the hinge region is actually the source of the observed flexibility. Tanford and coworkers presented an attractive model of the antibody molecule as shown in Figure 4 (Noelken *et al.,* 1965). However, no direct evidence existed to locate the flexibility of IgG to the hinge region.

On the basis of the results of a series of NMR analyses performed in the author's laboratory using the 100-MHz ^1H resonance data of His, we concluded that the hinge region is actually flexible and in addition takes an interesting dynamical *mosaic* structure (Arata *et al.,* 1980; Endo and Arata, 1985; Ito and Arata, 1985). The most difficult problem of these ^1H NMR studies was that the sidechain of only one amino acid, i.e., His, was used as an NMR probe and the important conclusion concerning the dynamical nature of the hinge region was based upon the result obtained using peptide fragments that were cleaved out of IgG by proteolyses.

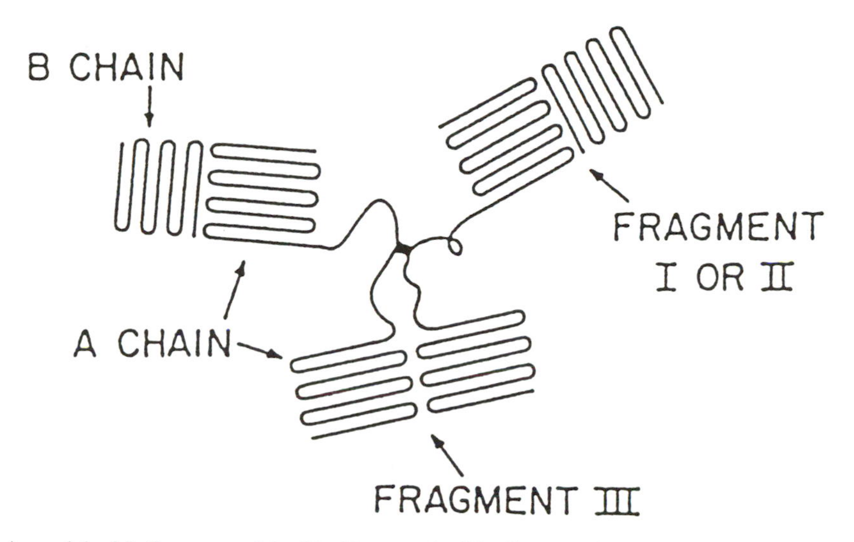

Figure 4: A model of IgG presented by Noelken et al. (Noelken *et al.,* 1965). In the present terminology, A chain, B chain, fragment II, and fragment III are called light chain, heavy chain, Fab, and Fc, respectively. The hinge region linked by the disulfide bridge (thick line) exists between Fab and Fc.

To overcome the difficulty of the sidechain ^{1}H NMR analyses and to reach a more decisive conclusion concerning the nature of the flexibility of the IgG molecule, we needed to extract information concerning the dynamics of the main chain of the molecule. Observation of the carbonyl carbon resonances of the intact IgG molecule would obviously be ideal for that purpose. Although the carbonyl carbon ^{13}C resonances are sufficiently narrow in linewidth as shown in Figure 3, we had to face the serious problem of signal overlap (Kim *et al.,* 1994). Figure 5A shows an example of the ^{13}C spectrum of IgG multiply labeled with His, Ile, Leu, Lys, Met, Tyr at the carbonyl carbons. This IgG possesses more than 150 residues of these six amino acids. The use of the CPMG pulse sequence dramatically improved the situation as shown in Figure 5B. Only five resonances had remained at the end of the CPMG sequence. The five resonances could be unambiguously assigned as indicated in the figure (Kim *et al.,* 1994). Except for Lys-447H, which is located at the C terminal of the heavy chain, Ile-221H, Lys-222H, Leu-235H, and Leu-236H all belong to the hinge region. This result clearly indicates that the hinge region of the IgG molecule is quite different from the other regions in respect to the flexibility of the main chains. By collecting more data by conducting similar experiments we were able to *map* the dynamics of the hinge region as schematically drawn in Figure 6. This indicates that the hinge region possesses a *core* region with less mobility. In the *core* region the heavy chains are linked by two disulfide

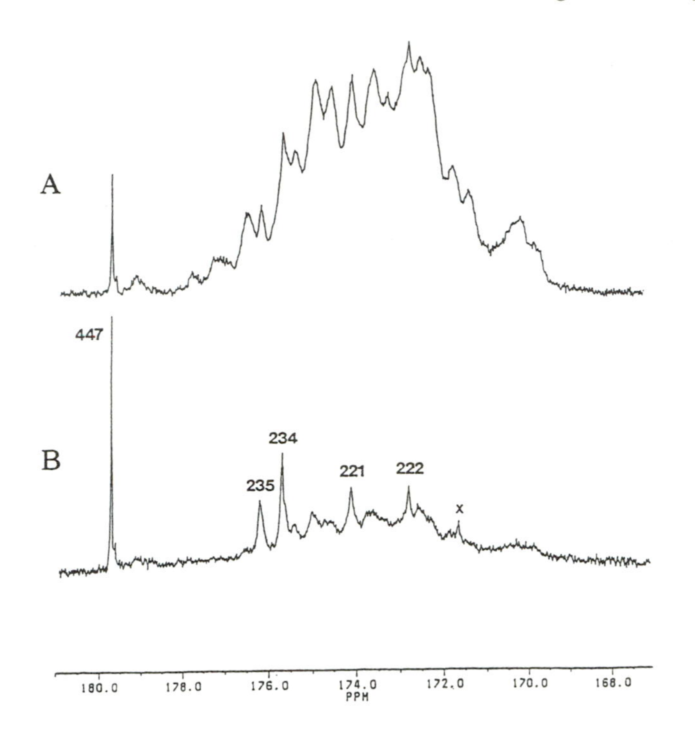

Figure 5: 100 MHz ^{13}C NMR spectrum of IgG multiply labeled with His, Ile, Leu, Lys, Met, and Tyr at the carbonyl carbons (Kim *et al.,* 1994). A) Measured using a conventional single pulse sequence. B) Measured using the CPMG pulse sequence with a delay time of 40 msec. Five signals were assigned as indicated in the figure. A small peak marked by x is due to a low molecular weight contaminant.

bridges. By contrast, a significant degree of flexibility exists in the N terminal segment (*upper* hinge) and the C terminal segment (*lower* hinge), which precedes and follows the *core* region, respectively.

The ^{13}C results confirmed our previous conclusion proposed on the basis of a limited amount of information derived using ^1H NMR data of the sidechain that the hinge region possesses a dynamical *mosaic* structure (Arata *et al.,* 1980; Endo and Arata, 1985; Ito and Arata, 1985).

Figure 6: Mapping of the dynamics of the hinge region (Kim *et al.,* 1994). Residues that give and do not give carbonyl ^{13}C resonance in the CPMG spectra are presented with open and filled circles, respectively.

It is possible that the dynamical *mosaic* structure comprising the rigid *core* flanked by the flexible segments of *upper* hinge and *lower* hinge plays a crucial role in the expression of the IgG functions. For the binding of antigens of a variety of sizes, the spatial arrangement of the two antigen binding units, Fab, in the IgG molecule has to be very flexible. The flexible *upper* hinge should contribute to provide the antibody molecule with a sufficient degree of freedom of motion for the antigen binding.

The proper quaternary structure of the effector site, which is primarily constructed by the hinge region and the CH2 domain (Burton, 1988), has to be prepared for the expression of the effector functions. It is known that Asn-linked oligosaccharide chains, one from each of the heavy chains, exist between the two CH2 domains (see Figure 1) (Deisenhofer, 1981; Sutton, 1983). The oligosaccharide chains along with the flexible *lower* hinge may play an important role in maintaining the proper spatial arrangement of the two CH2 domains in the Fc part of the molecule.

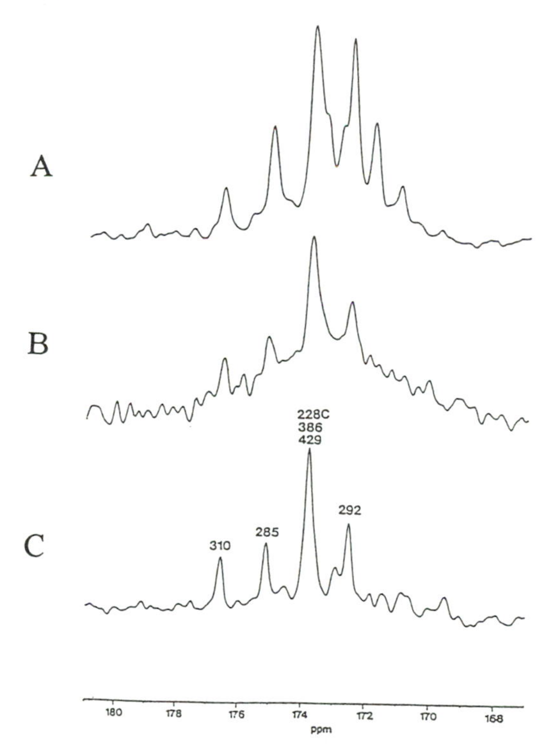

Figure 7: 100 MHz ^{13}C NMR spectrum of the immune complex (Kim 1994). An antidansyl IgG was labeled with [1-^{13}C]His. A) IgG complexed with bovine serum albumin to which an average of ten dansyl groups are linked. B) Fc cleaved from the identical IgG labeled with [1-^{13}C]His. Spectral assignments are indicated in the figure.

Toward larger systems: the ^{13}C spectrum of the immune complex with a molecular weight of 1100 K

Figure 7 shows a ^{13}C NMR spectrum of the immune complex of an antidansyl IgG with bovine serum albumin carrying approximately ten dansyl residues per molecule (Kim, 1994). The IgG was labeled with [1-^{13}C]His. The average molecular weight of the immune complex is 1100 K. Comparison of this spectrum with that of the Fc fragment cleaved out from the identical [1-^{13}C]His-labeled IgG indicates that all resonances observed in the immune complex originate from the Fc region. This strongly suggests that the existence of the flexibility in the hinge region is responsible for the observation of the Fc resonances from the immune complex with a molecular weight of 1100 K. The combination of the use of the carbonyl carbon resonances and the flexible nature of the protein structure made it possible to observe at least some of the resonances originating from the immune complex.

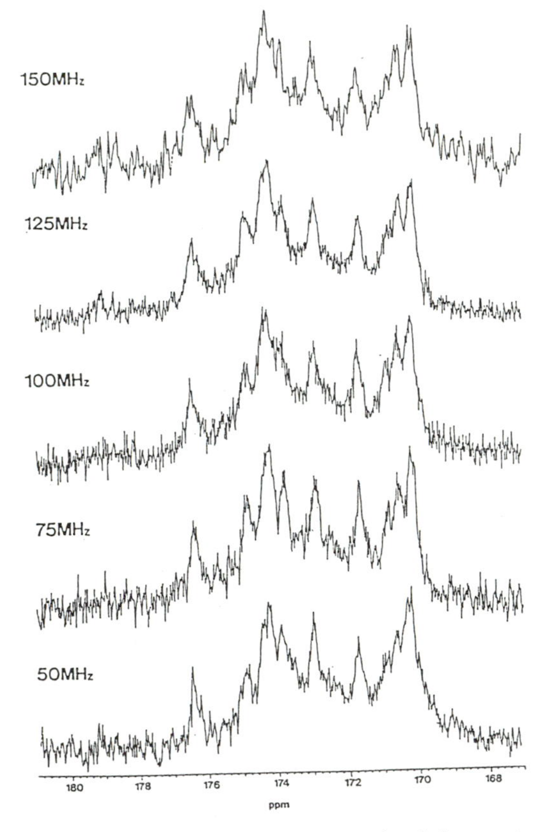

Figure 8: Dependence on the magnetic field strength of the spectra of IgG (Arata *et al.*, 1994). IgG was labeled with [1-^{13}C]Met.

Magnetic field dependence of the linewidth of the carbonyl carbon resonances

^{13}C spectra of an IgG labeled by [1-^{13}C]Met observed at different strengths of magnetic field are compared in Figure 8 (18). It can be seen that 1) the linewidths of the carbonyl carbon ^{13}C resonances increase with an increase in the strength of the magnetic field and 2) spectral resolution, which is determined by the chemical shift anisotropy of the carbonyl carbon and the magnetic field strength, becomes optimal at 50-75 MHz. It appears that, for the observation of ^{13}C spectra of larger proteins, the availability of a low-field spectrometer with maximum sensitivity is crucial.

Protein-protein interactions

The ^{13}C probe introduced at the carbonyl carbon can be used as an NMR probe in investigating protein-protein interactions in the antibody system. An example is the interaction of Fc and the B fragment of staphylococcal protein A (FB) with a molecular weight of 6 K. The chemical shift changes of the Fc resonances observed upon addition of FB are shown in Figure 9 (Kato *et al.*, 1993). In this experiment Fc was labeled with [1-^{13}C]Met. Use of more ^{13}C

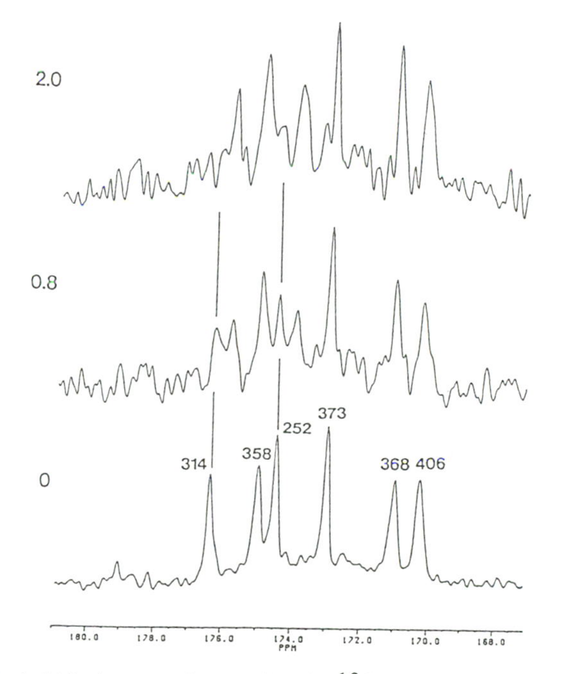

Figure 9: Chemical shift changes observed in the ^{13}C spectrum of Fc upon addition of FB (Kato *et al.*, 1993). The [FB]/[Fc] ratio is given in the figure. The spectra were observed at 100 MHz.

probes incorporated into Fc made it possible to map the binding site of FB on the Fc region. A similar approach has been employed for the analyses of the mode of protein-protein interactions of Fc with a variety of Fc-binding proteins (Kato *et al.,* 1995; Roberts *et al.,* 1995).

Antigen recognition: NMR vs. X-ray crystallography

A most appropriate system that is suited for the detailed structural analysis of the molecular mechanism of antigen recognition is Fv with a molecular weight of 25 K. As shown in Figure 2, Fv consists of VH and VL domains.

At the time that we were beginning these experiments there had only been one successful attempt at preparing a sufficient amount of Fv by proteolytic cleavage of intact IgGs and that was using a pathological IgA antibody with a rare kind of λ light chain (Inbar *et al.,* 1972). However, we were able to successfully prepare an Fv fragment of a mouse monoclonal antibody. The key to our success was the use of an IgG mutant in which the entire CH1 domain is

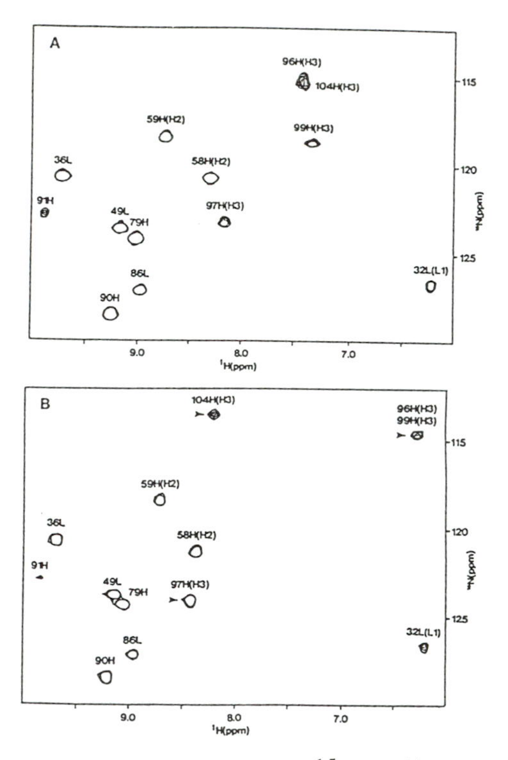

Figure 10: HSQC spectrum of Fv labeled with [α-^{15}N]Tyr (Arata *et al.,* 1994). A) In the absence of dansyllysine. B) In the presence of dansyllysine. Assignments are given in the figure.

deleted (Igarashi *et al.,* 1990; Takahashi *et al.,* 1991). Availability of the short chain mutant made it possible to produce a large amount of the Fv protein and eventually led us to start a combined use of NMR and X-ray crystallography for the structural analyses of antigen recognition.

NMR strategy for the analyses of antigen recognition

Main chain resonances were used in two ways. One was the use of ^{13}C of the carbonyl groups and ^{1}H and ^{15}N of the amide groups as NMR probes for mapping the antigen binding site of Fv. Figure 10 is an HSQC spectrum of Fv labeled with [α-^{15}N]Tyr (Arata *et al.,* 1994). Upon addition of dansyllysine, a number of peaks were shifted significantly. By accumulating this type of NMR data using ^{13}C and ^{15}N, we were able to establish that H3, one of the six *hypervariable* loops, is primarily responsible for the antigen binding. The contribution of the H1 loop was also recognized.

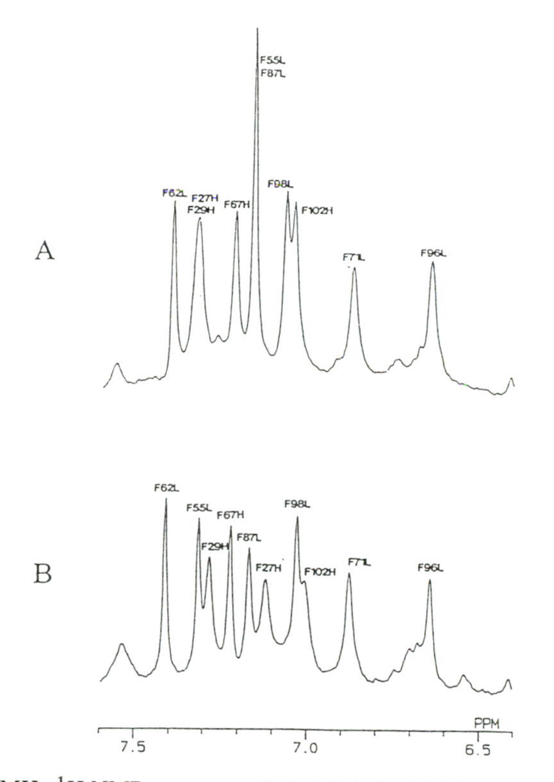

Figure 11: 400 MHz ^{1}H NMR spectrum of Fv labeled with Phe-2',4',6'-d_3 (Shimba, 1995). A) In the presence of dansyllysine. B) In the absence of dansyllysine. Assignments are indicated in the figure. Phenylalanine-2',4',6'-d_3 was kindly provided by Professor Masatsune Kainosho.

Secondly the main chain resonances were used as the starting point of the spectral assignment of sidechain resonances. So far we have been able to assign the sidechain ^{1}H resonances of all four aromatic residues (Arata, 1991; Odaka *et al.*, 1992; Takahashi *et al.*, 1991). The assignments were made by connecting the amide resonances to the 1H resonances of the aromatic sidechains by standard methods based on TOCSY and NOE connectivity. An example of the sidechain resonances observed in the presence and absence of the antigen is reproduced in Figure 11 (Shimba, 1995). The assigned resonances of the sidechains were then used for analyzing NOE data between Fv and dansyllysine.

Identification of the sidechains of the aromatic amino acid residues involved in antigen binding

Samples with different molar ratios of Fv and dansyllysine were used for the analyses of antigen-antibody interactions (Arata, 1991; Takahashi *et al.*, 1991). An Fv solution with an excess of dansyllysine was used for the identification of the NMR signals of the bound form of the dansyllysine molecule. An Fv vs. dansyllysine molar ratio of 0.5 was needed to identify the resonances originating from the Fv bound to the antigen. In this series of experiments exchange NOESY data were collected.

On the basis of these analyses it had become clear that Val2H, Phe27H, Tyr96H, and Tyr104H were involved in the antigen binding (Takahashi *et al.*, 1991; Shimba, 1995). Phe27H belongs to the H1 loop, whereas Tyr96H and Tyr104H are located in the H3 loop. This result is consistent with what was concluded on the basis of the carbonyl ^{13}C and the amide ^{1}H and ^{15}N resonance data. *vide infra*.

Structure of dansyllysine bound to Fv

The structure of dansyllysine when it is bound to Fv was determined by the conventional TRNOE method (Kim *et al.*, 1991). In this analysis, lysine was uniformly labeled with ^{13}C to extract information on the sidechain ^{1}H resonances by ^{13}C filtering. The result of the analysis indicated that the lysine part of the hapten is folded back on to the naphthalene ring.

X-ray crystallographic analysis of Fv in the presence and absence of antigen

X-ray crystallographic data became available in 1993 for Fv in the absence of dansyllysine (Nakasako *et al.*, 1993). This was followed by crystal data for the Fv-antigen complex (Nakamura *et al.*, 1995). Upon combining NMR and X-ray crystallographic data, a number of interesting features of antigen-antibody interaction have emerged as described below.

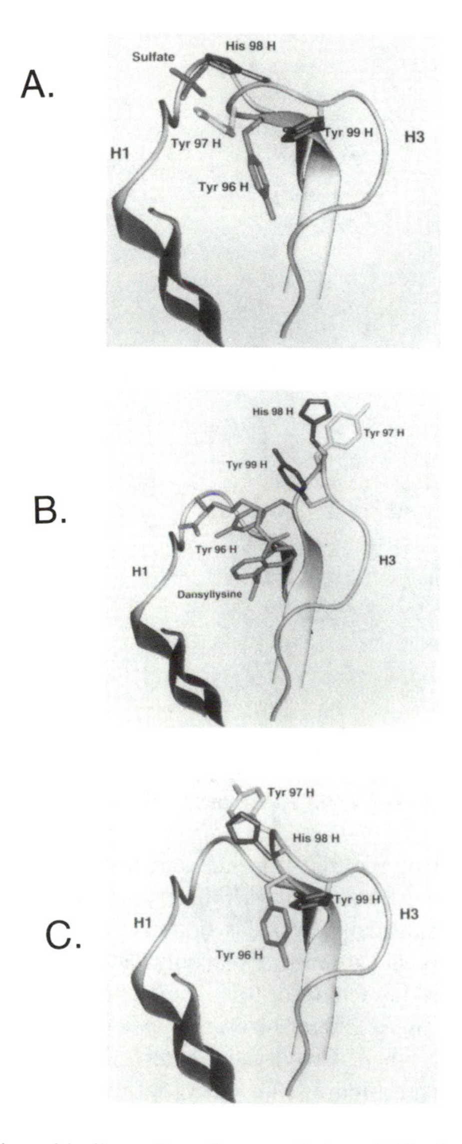

Figure 12: The antigen-binding site of an anti-dansyl Fv determined by X-ray crystallography (Nakasako *et al.*, 1993; Nakamura *et al.*, 1995). A) crystal I at 1.6 Å resolution, R factor 0.178. B) Fv-dansyllysine complex at 1.9 Å resolution, R factor 0.176. C) crystal II at 2.0 Å resolution, R factor 0.178. These figures were kindly provided by Dr. Mitsuaki Nakamura and Professor Yoshinori Satow.

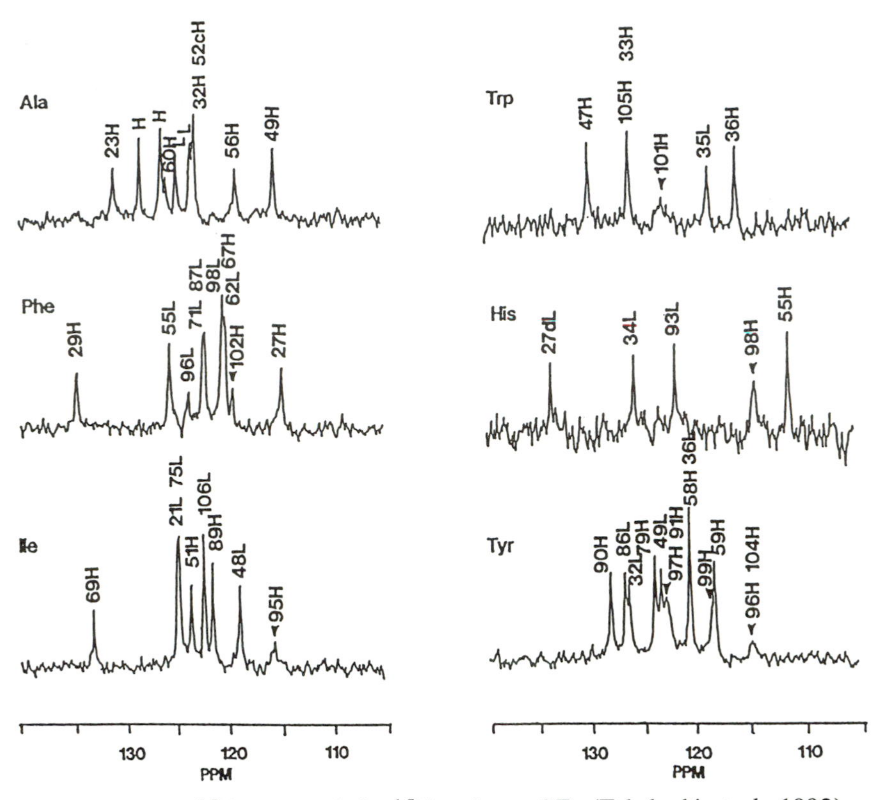

Figure 13: 60 MHz ^{15}N spectra of six ^{15}N analogs of Fv (Takahashi *et al.*, 1992). Labeled amino acids are indicated in the figure. Arrows indicate the amino acid residues that exist in the H3 loop. Assignments are also given.

The involvement of the four amino acid residues in antigen binding as elucidated by NMR was consistent with the crystal data. In addition, X-ray data gave a clear view of how dansyllysine is bound to Fv. The antigen binding site of Fv in the presence and absence of dansyllysine is compared in Figures 12A and 12B. A most striking feature of this result is that the spatial arrangement of the antigen binding site is altered to a great extent upon dansyllysine binding. This indicates that, when dansyllysine enters the binding site of Fv, the structure of the antigen binding site changes from a *closed* form to an *open* form.

In the absence of dansyllysine, the top of the antigen binding site is covered by a *lid* comprising Tyr97H and His98H. In addition, Tyr96H is located at the bottom of the binding site as a *plug*. With this structure, it would certainly be difficult for dansyllysine to enter the binding pocket and form the antigen-antibody complex.

In the complex, the *plug* is removed and the *lid* is flipped away, giving the proper space for the antigen to enter the binding pocket. In view of this interesting result obtained by X-ray crystallography, it was clear that further information is necessary concerning the dynamical nature of the antigen binding site. It was evident at this stage that NMR should be extensively used for the further analyses of antigen-antibody interactions.

Dynamic structure of the antigen binding site

^{15}N NMR data of the amide nitrogens shown in Figure 13 are of particular interest. The amide nitrogen resonances of Ile95H, Tyr96H, Tyr97H, His98H, Tyr99H, Trp101H, Phe102H, and Tyr104H are significantly broadened (Takahashi *et al.,* 1992). All these residues belong to the H3 *hypervariable* loop. The simplest interpretation of the broadening of these lines is that there is rapid chemical exchange between more than one conformation in the H3 loop. As indicated in Figure 14, these H3 resonances are significantly broadened when the observing frequency was increased from 60 MHz to 90 MHz (Takahashi *et al.,* 1992). From these results we concluded that flexibility does exist in the H3 loop.

The conclusion drawn above on the basis of the 15N data is quite consistent with the temperature factor of the X-ray structure of Fv. In the absence of dansyllysine the temperature factor is in the range 30 $Å^2$ - 40 $Å^2$ for the sidechains of Tyr97H, His98H, and Tyr99H. The rest of the molecule gives much smaller temperature factors.

The conclusion that has been reached above on the basis of the NMR and X-ray data was further reinforced by the result of another NMR experiment, which measured the temperature dependence of the HSQC spectrum of an Fv analog selectively labeled with His and Tyr at the amide nitrogen. It was clearly indicated that the ^{1}H and/or ^{15}N chemical shifts of the cross peaks originating from Tyr96H, Tyr97H, His98H, Tyr99H, and Tyr104H are shifted significantly with the change in temperature. This result again is consistent with the above conclusion that a significant degree of fluctuation exists in the H3 loop.

Two kinds of crystals were used for the structural analyses

In the present X-ray analyses the crystals of Fv were grown in a 1.7 M ammonium sulfate solution. The X-ray structural data used in the above discussion were from the crystal after soaking in a 2.8 M ammonium sulfate solution. In the following the crystal thus obtained will be referred to as crystal I. At this stage of our structural analyses, another crystal structure of Fv was solved using crystal II, which was prepared by soaking in a 2.0 M ammonium sulfate solution.

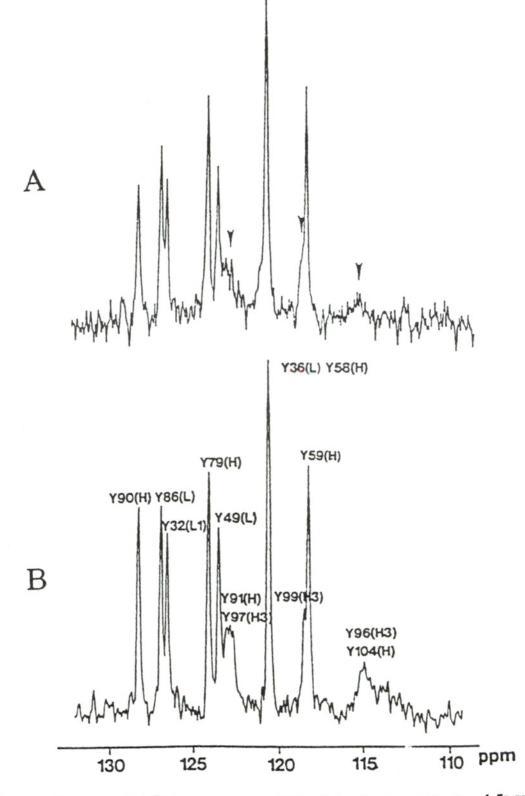

Figure 14: Field dependence of ^{15}N spectra of Fv labeled with [α-^{15}N]Tyr (Takahashi *et al.*, 1992). Observed at A) 90 MHz and B) 60 MHz. Arrows indicate the amino acid residues that exist in the H3 loop. Assignments are given in the figure.

What is most interesting is that the structure of the antigen binding site is significantly different for crystal I and crystal II. As shown in Figure 12C, in crystal II, the *lid* comprising Tyr97H and His98H is about to open, whereas Tyr96H still remains in the binding site playing the role as the *plug*. More importantly, the region with larger temperature factors is extended from the sidechains of Tyr97H, His98H, and Tyr99H to the entire part of Tyr96H, Tyr97H, His98H, and Tyr99H.

Conclusions

On the basis of the results obtained by the combined use of NMR and X-ray crystallography we conclude that 1) in the absence of dansyllysine the antigen binding site is covered by the flexible *lid* comprising Tyr97H and His98H, with Tyr96H at the bottom acting as the *plug*, 2) upon antigen binding, the *lid* is opened, the *plug* is removed, the naphthalene rings comes in replacing the *plug*, and the binding site is filled with the naphthalene ring and the lysine sidechain, and 3) the existence of the flexible *lid* is essential for the recognition of dansyllysine.

It is of interest that in crystal I the sulfate ion exists in the neighborhood of Tyr97H and His98H. By contrast, the sulfate ion disappears in crystal II, where the *lid* is about to open with the *plug* still remaining (see Figures 12A and 12C). The amino acid residues that are used to construct the inside of the antigen binding site are primarily hydrophobic in nature. This is reasonable because the Fv under investigation has to bind dansyllysine. The naphthalene ring of the hapten, which is hydrophobic, is recognized by the inner wall and bottom of the antigen binding site. However, in the absence of the hapten, the antigen binding site has to be shielded from outside which is presumably hydrophilic. It appears that it is for the reason of minimizing the hydrophobic and hydrophilic interactions that the binding site is constructed in the way which is described here.

It is quite likely that what is happening in the interaction of Fv with dansyllysine is an example of the characteristic features of antigen recognition. This is because the antibody molecule has to prepare the antigen binding site to cope with antigens of unknown structure.

In the case of the anti-dansyl IgG used in the present analyses, Fv possesses a binding constant similar to that of the corresponding Fab where the two constant domains CH1 and CL are retained. On the basis of what has been discussed above, it should be possible to discuss how the existence of the constant domains could influence the structure of the antigen binding site. The roles of the constant domains in determining the thermal stability of the antigen binding domains should also be pursued (Shimba *et al.,* 1995; Torigoe *et al.,* 1995). In conducting the analyses along this line, the close collaboration of NMR and X-ray crystallography as described in this chapter should continue to be beneficial.

Perspective

As concisely summarized by Roberts and Jardetzky in 1970, the potential of high-resolution NMR was phenomenal. Since then, NMR has continued to develop at an enormous speed and reached to the present stage where the three-dimensional structures of a number of important proteins are being solved and deposited in the *Protein Data Bank.*

It is also true that many proteins of great importance in biochemistry exceed in size the present capability of NMR. A histogram containing the number of proteins with known sequences against the number of amino acid residues is shown in Figure 15. This indicates that at least 40 % of proteins with known amino acid sequences are classified as *untouchable* as targets for three-

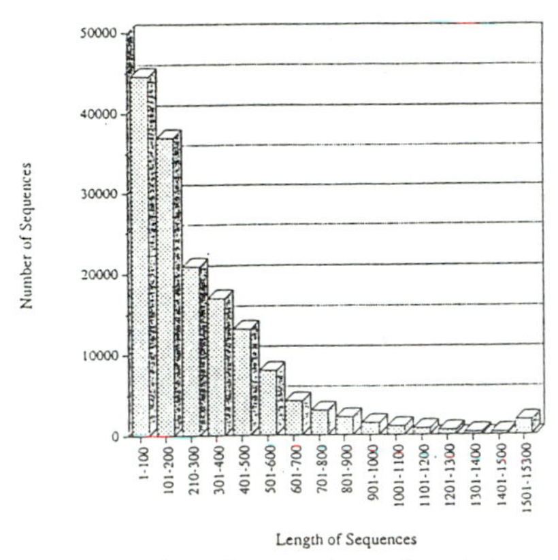

Figure 15: A histogram representing the number of proteins with known amino acid sequences vs. the length of sequences of the proteins. This data represents sequences with non-redundant amino acids. This data was kindly provided by Professor Minoru Kanehisa and Mr. Hiroyuki Ogata.

dimensional structure determination by standard high-resolution NMR methods. It is obvious that we have to develop more methods to cope with the increased size of proteins.

Our experience has clearly shown the unique and irreplaceable role of NMR in determining the structure of biological macromolecules. It is hoped that new techniques including those of solid-state high-resolution NMR (Opella, 1994; Cross and Opella, 1994) will continue to become available to extract information at atomic resolution from a wider range of biological macromolecules.

Acknowledgments

I would like to take this opportunity to express my deep gratitude to Professor Oleg Jardetzky for many interesting scientific discussions, particularly in 1971-1973 when I was visiting his laboratory at Stanford University.

Professor Kozo Hamaguchi inspired me more than twenty years ago to start my antibody project. The late Professor Tatsuo Miyazawa encouraged me all the time to advance my project. Friendship and collaboration with Professor Akira Shimizu has been invaluable in continuing NMR analyses of antibody particularly at the early stage of my career. In developing the stable-isotope aided NMR analyses described in this chapter, Professor Masatsune Kainosho's suggestions were most helpful. He has also provided us with a variety of stable-isotope labeled amino acids, which were of tremendous help to promote our NMR research. The anti-dansyl switch variant cell lines that have been used in the present studies were kindly provided by Professor Leonard A.Herzenberg and Dr. Vernon T.Oi. Analyses of the protein-protein interactions using a variety of Fc binding proteins have been performed in collaboration with Professor Gordon C.K. Roberts, Dr. Lu-Yun Lian, Professor Sture Forsén, and Dr. Mats Wikström. A most recent NMR and X-ray analyses of the antigen recognition has been made in collaboration with Professor Yoshinori Satow, Dr. Masayoshi Nakasako, and Dr. Mitsuaki Nakamura. The stable-isotope aided antibody project described in this chapter has been conducted with the help of a number of collaborators, particularly Toshiyuki Tanaka, Koichi Kato, Hideo Takahashi, and Ichio Shimada. My colleague Dr. William S. Price was kind enough to critically read the manuscript.

I deeply express my grateful thanks to all of those mentioned above.

References

Anglister, J., Frey, T., and McConnell, H.M. (1984). *Biochemistry* **23**, 1138.

Arata, Y., Honzawa, M., and Shimizu, A. (1980). *Biochemistry* **19**, 5130.

Arata, Y., Kato, K., Takahashi, H., and Shimada, I. (1994). *Methods in Enzymology* **239**, 440.

Arata, Y. (1991). in Catalytic Antibodies, Ciba Foundation Symposium 159 (Chadwick, D.J., and Marsh, J. eds.) John Wiley & Sons, p. 40.

Burton, D.R. (1988). *Biochem. Soc. Trans.* **16**, 953.

Cross, T.A., and Opella, S.J. (1994). *Current Opinion in Structural Biology* **4**, 574.

Deisenhofer, J. (1981). *Biochemistry* **20**, 2361.

Endo, S., and Arata, Y. (1985). *Biochemistry* **24**, 1561.

Igarashi, T., Sato, M., Katsube, Y., Takio, K., Tanaka, T., Nakanishi, M., and Arata, Y. (1990). *Biochemistry* **29**, 5727.

Inbar, D., Hochman, J., and Givol, D. (1972). *Proc. Natl. Acad. Sci. U.S.A.* **69**, 2659.

Ito, W., and Arata, Y. (1985). *Biochemistry* **24**, 6467.

Jardetzky, O., and Jardetzky, C.D. (1957). *J. Am. Chem. Soc.* **79**, 5322.

Jardetzky, O. (1965). An approach to the determination of the active site of an enzyme by nuclear magnetic resonance spectroscopy, International Symposium of Nuclear Magnetic Resonance, Abstract N-3-14.

Kabat, E.A. (1976). Structural concepts in immunology and immunochemistry, second edition. Holt, Rinehart and Winston.

Kainosho, M., and Tsuji, T. (1982). *Biochemistry* **21**, 6273.

Kato, K., Gouda, H., Takaha, W., Yoshino, A., Matsunaga, C., and Arata, Y. (1993). *FEBS Lett.* **328**, 49.

Kato, K., Lian, L.-Y., Barsukov, I.L., Derrick, J.P., Kim, H.-H., Tanaka, R., Yoshino, A., Shiraishi, M., Shimada, I., Arata, Y., and Roberts, G.C.K. (1995). *Structure* **3**, 79.

Kato, K., Matsunaga, C., Nishimura, Y., Wälchli, M., Kainosho, M., and Arata, Y. (1989). *J. Biochem. (Tokyo)* **105**, 867.

Kim, H., Matsunaga, C., Yoshino, A., Kato, K., and Arata, Y. (1994). *J. Mol. Biol.* **236**, 300.

Kim, H. (1994). Dynamical structure of immunoglobulin G as studied by ^{13}C nuclear magnetic resonance spectroscopy, Ph.D. Thesis, University of Tokyo, Tokyo, Japan.

Kim, J.I., Nagano, T., Higuchi, T., Hirobe, M., Shimada, I., and Arata, Y. (1991). *J.*

Am. Chem. Soc. **113**, 9392.

Markley, J.L., Putter, I., Jardetzky, O. (1968). *Science* **161**, 1249.

Meadows, D.H., Jardetzky, O., Epand, R.M., Rüterjans, H.H., and Scheraga, H.A. (1968). *Proc. Natl. Acad. Sci. U.S.A.* **60**, 766.

Meadows, D.H., Markley, J.L., Cohen, J.S., and Jardetzky, O. (1967). *Proc. Natl. Acad. Sci. U.S.A.* **58**, 1307.

Nakamura, M., Nakasako, M., Hagitani, M., Takahashi, H., Odaka, A., Shimada, I., Arata, Y, and Satow, Y. (1995). "Crystal structure of the complex of an antidansyl Fv fragment with dansyllysine". Presented at the 68th Annual Meeting of Biochemical Society of Japan in Sendai.

Nakasako, M., Noguchi, S., Satow, Y., Takahashi, H., Shimada, I., and Arata, Y. (1993). *Acta Crystallographica* A49, ps-03.03.07

Noelken, M.E., Nelson, C.A., Buckley, III, C.E., and Tanford, C. (1965). *J. Biol. Chem.* **240**, 218.

Odaka, A., Kim, J.I., Takahashi, H., Shimada, I., and Arata, Y. (1992). *Biochemistry* **31**, 10686.

Opella, S.J. (1994). *Ann. Rev. Phys. Chem.* **45**, 659.

Padlan, E.A. (1994). *Molecular Immunology* **31**, 169.

Roberts, G.C.K. and Jardetzky,O. (1970). *Advances in Protein Chemistry* **24**, 447.

Roberts, G.C.K., Lian, L.-Y., Barsukov, I.L., Derrick, J.P., Kato, K., and Arata, Y. (1995). in Techniques in Protein Chemistry VI (Crabb, J.W. ed.) Academic Press, p. 409.

Saunders, M., Wishnia, A., and Kirkwood, J.G. (1957). *J. Am. Chem. Soc.*, **79**, 3289.

Shimba, N., Torigoe, H., Takahashi, H., Masuda, K., Shimada, I., Arata, Y., and Sarai, A. (1995). *FEBS Lett.* **360**, 247.

Shimba, N. (1995). NMR and calorimetric studies of the mechanism of antigen recognition by anti-dansyl Fv, M.Sc. Thesis, University of Tokyo, Tokyo, Japan.

Sutton, B., Phillips, D. (1983). *Biochem. Soc. Trans.* **11**, 130.

Takahashi, H., Igarashi, T., Shimada, I., and Arata, Y. (1991). *Biochemistry* **30**, 2840.

Takahashi, H., Odaka, A., Kawaminami, S., Matsunaga, C., Kato, K., Shimada, I., and Arata, Y. (1991). *Biochemistry* **30**, 6611.

Takahashi, H., Suzuki, E., Shimada, I., and Arata, Y. (1992). *Biochemistry* **31**, 2464.

Torigoe, H., Nakayama, T., Imazato, M., Shimada, I., Arata, Y., and Sarai, Y. (1995). *J. Biol. Chem.* **270**, 22218.

Wüthrich, K. (1986). NMR of proteins and nucleic acids. John Wiley & Sons.

15

Selective Chemical Deuteration of Aromatic Amino Acids: A Retrospective

K.S. Matthews and R. Matthews

Department of Biochemistry & Cell Biology
Rice University
Houston, Texas 77251 USA

In 1970 when we began post-doctoral work in the laboratory of Professor Oleg Jardetzky, selective deuteration of proteins to limit the number of protons present in the system for subsequent analysis was a newly developed and effective technique for NMR exploration of protein structure (Crespi *et al.*, 1968; Markley *et al.*, 1968). This approach allowed more facile assignment of specific resonances and generated the potential to follow the spectroscopic behavior of protons for a specific amino acid sidechain over a broad range of conditions. The primary method for labeling at that time involved growth of microorganisms (generally bacteria or algae) in D_2O, followed by isolation of the deuterated amino acids from a cellular protein hydrolysate. The amino acids isolated were, therefore, completely deuterated. Selective deuteration of a target protein was achieved by growing the producing organism on a mixture of completely deuterated and selected protonated amino acids under conditions that minimized metabolic interconversion of the amino acids.

In one-dimensional spectra, aromatic amino acid resonances occur well downfield of the aliphatic resonances, and this region can therefore be examined somewhat independently by utilizing a single protonated aromatic amino acid to simplify the spectrum of the protein. However, the multiple spectral lines generated by aromatic amino acids can be complex and overlapping, precluding unequivocal interpretation. To address this complication, chemical methods were developed to both completely and selectively deuterate side chains of the aromatic amino acids, thereby avoiding the costly necessity of growing large volumes of microorganisms in D_2O and subsequent tedious isolation procedures. In addition, selective deuteration of the amino acids simplified the resonance patterns and thereby facilitated assignment and interpretation of spectra. The methods employed were based on exchange phenomena reported in the literature and generated large quantities of material for use in growth of microorganisms for subsequent isolation of selectively labeled protein (Matthews *et al.*, 1977a). The target protein for incorporation of the selectively deuterated aromatic amino acids generated by these chemical methods was the lactose repressor protein from *Escherichia coli*, and greatly simplified spectra of this 150,000 D protein were produced by this approach (Figure 1; Matthews *et al.*, 1977b). Unfortunately, even with the simplification generated by isotopic editing, the size of this protein has precluded its structure determination to this point by NMR methods. In the intervening years, however, the chemical methods developed for selective deuteration of aromatic amino acids have been utilized not only for simplifying protein spectra but also in variety of other applications. The following examples are not intended to be exhaustive but rather to illustrate the range of proteins and uses that have employed this chemical deuteration approach developed in the early 1970s.

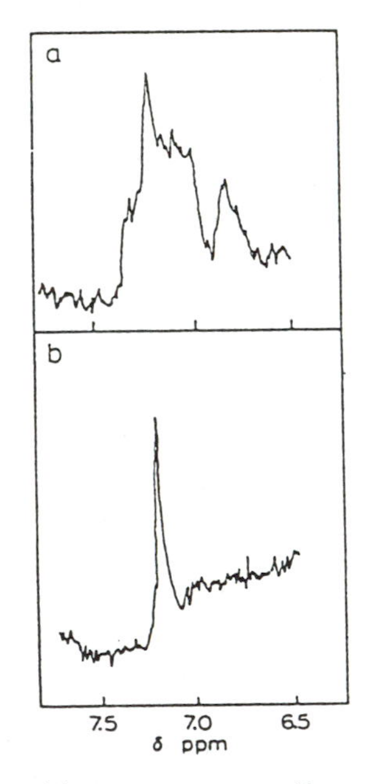

Figure 1: Spectra of denatured lactose repressor (from Matthews *et al.*, 1977b). Aromatic region spectra of fully protonated lactose repressor protein (~10 mg/ml) (panel a) and protein selectively deuterated with 3,5[$^1H_2$]tyrosine (panel b) in trifluoro[$^2H_1$]acetic acid collected at 100 MHz.

Applications to protein NMR spectroscopy

Repressors

The N-terminal DNA binding domain of the lactose repressor was examined by Arndt *et al.* (1981) using selectively labeled aromatic amino acids introduced into wild-type protein and mutant repressors that were missing a single tyrosine. This combination facilitated assignment of specific resonances and generated information that was concordant with the genetic analysis of this protein. This study also employed NOE measurements as an initial measure of the distances between aromatic residues in the N-terminal fragment of the protein. Arndt *et al.* (1983) also utilized selectively deuterated aromatic amino acids in identifying resonances in the aromatic region of the *cro* repressor of *E. coli* phage *lambda* by two dimensional NMR methods. This study was one of several demonstrations that assignments made based on the conventional methods of the time could result in significant errors in assignment of resonances. Simplification of the spectra in the latter study allowed more facile interpretation of the results of complex formation with the target DNA regulatory sequence.

Monoclonal antibodies

Studies of proteins have not been confined to bacterial systems, but have employed mammalian cells in culture, most notably hybridomas. McConnell and colleagues (*e.g.*, Anglister *et al.*, 1987) employed NMR spectroscopy and incorporation of deuterated amino acids to generate insight into the combining site of a monoclonal antibody (using Fab fragments) with a spin-labeled hapten. Using different combinations of perdeuterated aromatic amino acids in the light and heavy chains allowed determination of the cross relaxation rates between tryptophan protons and protons in the diamagnetic hapten. These studies demonstrated a single binding site for hapten based on the NMR results. Kato *et al.* (1989) used both one-dimensional NMR spectroscopy (Figure 2) and two-dimensional homonuclear Hartmann-Hahn spectroscopy to examine a selectively deuterated mouse monoclonal antibody to establish connectivity between the H^{E1} and H^{E2} protons of histidine residues in this molecule. These workers concluded that in flexible regions, even in large proteins, the labeling regime reduces the efficiency of spin relaxation thereby increasing the number of observed cross peaks. Selective labeling has also been employed to examine the Fv fragment from mouse IgG anti-dansyl monoclonal antibody (Takahashi *et al.*, 1991). Selective deuteration of this fragment (~25,000 D) in mouse hybridoma cell culture provided proton spectral information that was sufficient to deduce the mode of involvement of two tyrosine residues in antigen binding.

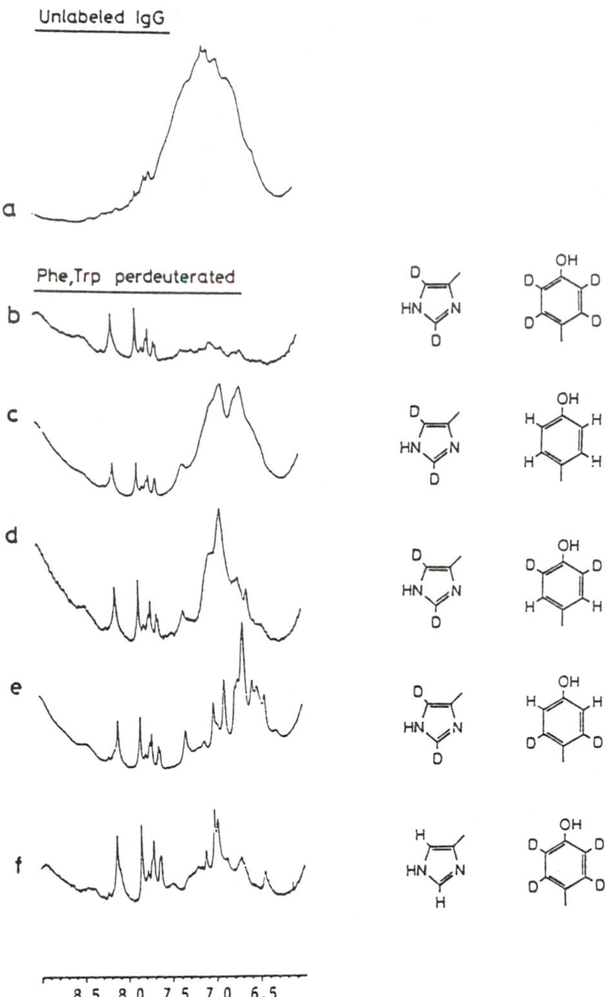

Figure 2: Spectra of mouse anti-dansyl monoclonal antibody (from Kato *et al.*, 1989). The aromatic region of the 400 MHz ^{1}H NMR spectrum for antibody unlabeled and labeled as indicated using different combinations of deuterated aromatic amino acids. Proteins (3-10 mg) were in 5 mM phosphate, 0.2 M NaCl, D_2O, pH 7.3, 30°C. A spectral width of 5000 Hz with 8000 data points was used, and ~1000 scans were accumulated. Chemical shifts are in ppm from external DSS.

Binding proteins

Periplasmic binding proteins are involved in the transport of ions into bacterial cells. Ho and colleagues have investigated the structural and dynamic properties of the binding proteins for histidine (Cedel *et al.*, 1984) and glutamine (Shen *et al.*, 1989) using selective deuteration and high field NMR studies. Changes in the binding constant for histidine as a function of pH were measured in concert with conformational changes in the protein in response to ligand binding. Selective deuteration allowed the assignment of a tyrosine and histidine to the binding site of the histidine binding protein (Cedel *et al.*, 1984). Similarly, studies on the glutamine binding protein have demonstrated conformational changes in response to ligand binding that involve hydrogen bond interactions between the protein and glutamine and within the protein structure. The data indicate that one of the residues near the intermolecular hydrogen-bonded proton between glutamine and glutamine binding protein is a phenylalanine. Interestingly, the structures surrounding the intra- and inter-molecular hydrogen bonds are stable to pH and temperature variations. The ability to monitor specific resonances and to follow the shifts with a minimum of interference from other spectral components was facilitated in these experiments by incorporation of deuterated amino acids.

Other proteins

In some cases, the availability of deuterated amino acids was found to be insufficient alone for effective simplification of the spectra. One example is the production of dihydrofolate reductase from *E. coli* (Florance and Ginther, 1981). In order to prevent metabolic interconversion, a genetic block of the *de novo* biosynthetic pathway was introduced using P1 phage into a trimethoprim resistant bacterial strain. The result of this alteration was that exogenous aromatic amino acids were required for growth, and efficient labeling of protein using deuterated aromatic amino acids was possible, resulting in significant simplification of the protein spectrum. Specific partial deuteration allowed assignment of the majority of residues in the protein thioredoxin from *E. coli* (LeMaster and Richards, 1988). By diminishing local proton density, narrower line widths were acquired with minimal reduction in sensitivity, and secondary coupling was suppressed to improve resolution. Notably, amide to amide NOE spectra yielded greater sensitivity compared to natural abundance samples, while amide to carbon-bound proton NOE spectra were minimally affected largely due to the ability to utilize longer mixing times for NOE buildup. This combination improved the capacity to make assignments in this protein.

The effects of removing the zinc ligand in gene 32 protein and subsequent reconstitution have been examined in protein with selective deuteration of aromatic amino acids (Pan *et al.*, 1989a). In particular, the effects of pH on histidine resonances were monitored to identify the specific imidazole sidechain involved in complexing with zinc, and signals from tryptophan side chains were employed to follow the conformational effects of pH and zinc removal.

Furthermore, selective deuteration of aromatic amino acids allowed identification of the aromatic protons that are shifted by the binding of gene 32 protein to oligonucleotides and polynucleotides (Pan *et al.*, 1989b). On the basis of the data obtained, it was concluded that a subset of the tyrosine, tryptophan, and phenylalanine residues in this protein appears to interact directly with bases in the oligonucleotide, while other aromatic residues are affected by the conformational changes consequent to nucleotide binding (Pan *et al.*, 1989b).

In an interesting application of the general approach of spectral editing, a synthetic diet containing selective deuterated amino acids was fed to Japanese quail, and the incorporation *in vivo* was found to exceed 80% by proton NMR studies of isolated egg white lysozyme (Brown-Mason *et al.*, 1981), resulting in a greatly simplified ^{1}H NMR spectrum (Figure 3). The authors concluded that this system has potential for production of other egg proteins and pointed out that the approach could be used for proteins found elsewhere in the organism. The feasibility of utilizing this *in vivo* labeling is predicated on the low cost of chemically perdeuterating the amino acids and on the rapid uptake of amino acids in these birds as well as the ease of isolating large quantities of proteins from the eggs. The concept was novel, but does not appear to have generated widespread interest or applications.

The use of deuterium labeling for NMR spectroscopic analysis of proteins has been reviewed in detail by LeMaster (1990a,b); these reviews provide

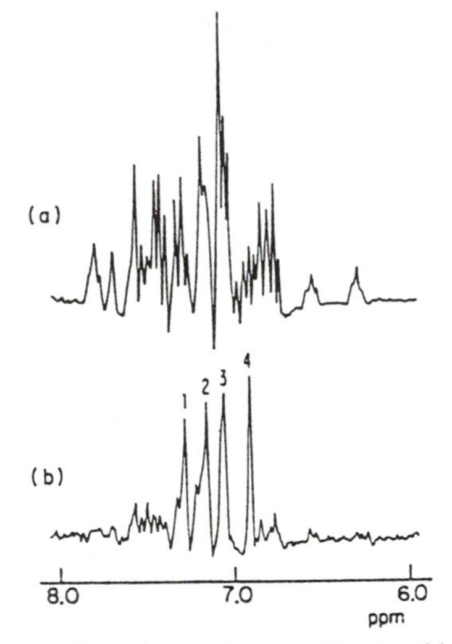

Figure 3: ^{1}H NMR Spectrum of lysozyme from quail egg white (from Brown-Mason *et al.*, 1981). Aromatic proton regions of the 270 MHz NMR spectra of (a) unlabeled and (b) lysozyme selectively deuterated *in vivo*. Protein (1 mM) was in D_2O at pH 7.5. Resonances labeled 1-4 in (b) correspond to the C(2,6) protons of the four tyrosine residues.

additional information on the more modern methods for structure determination and the mechanisms by which selective deuteration can facilitate analysis and interpretation of spectra.

Deuterium NMR spectroscopy

In addition to simplifying the proton NMR spectrum of a macromolecule, selective deuterium labeling also provides the opportunity to monitor deuterium NMR spectra directly. In particular, rotational correlation times of proteins in solution can be determined using ^{2}H nuclei under specific conditions (Schramm and Oldfield, 1983). This approach requires no assumptions about bond lengths, in contrast to utilization of ^{13}C NMR for such measurements, or relaxation mechanisms, since relaxation of ^{2}H is quadrupolar. Data using lysozyme labeled with deuterated histidine demonstrated that rotational correlation times could be monitored effectively using this approach. In addition, quadrupole coupling constants for deuterium-labeled histidine (in both imidazole and imidazolium forms) were similar measured in lysozyme in aqueous solution to those determined using solid state NMR of crystalline histidine. The only apparent limitation to this approach is that line width may preclude determinations on proteins above molecular weights of ~200,000 D.

Solid state NMR applications

Solid state NMR of ^{2}H labeled aromatic amino acids has been employed to explore the dynamics of aromatic amino acids in proteins because of the importance of these motions in the structure and characteristics of a particular protein or peptide. In particular, this approach is useful for insoluble compounds or those that cannot be crystallized, as well as for exploring dynamic properties of molecules that form crystals suitable for X-ray analysis. Labeling with ^{2}H has been employed successfully for studying the motions of aromatic rings, because the line shapes of the powder patterns from the quadrupolar interaction are influenced significantly by dynamic properties (*e.g.*, see Frey *et al.*, 1985a). Quadrupole echo ^{2}H NMR spectroscopy on crystalline [$^{2}H_5$]-phenylalanine and on this compound incorporated into the coat protein of the filamentous bacteriophage fd, Gall *et al.* (1981) demonstrated rapid flipping (180°) of the aromatic rings about the C_β-C_γ axis; these workers (Gall *et al.*, 1982) also found that tryptophan was immobile under comparable experimental conditions (Figure 4).

In similar experiments, Kinsey *et al.* (1981) examined ^{2}H labeled aromatic amino acids in a membrane protein, bacteriorhodopsin, and found that the tryptophan residues were rigid, while tyrosine and phenylalanine side chains underwent two-fold flipping similar to the motions observed in the fd coat

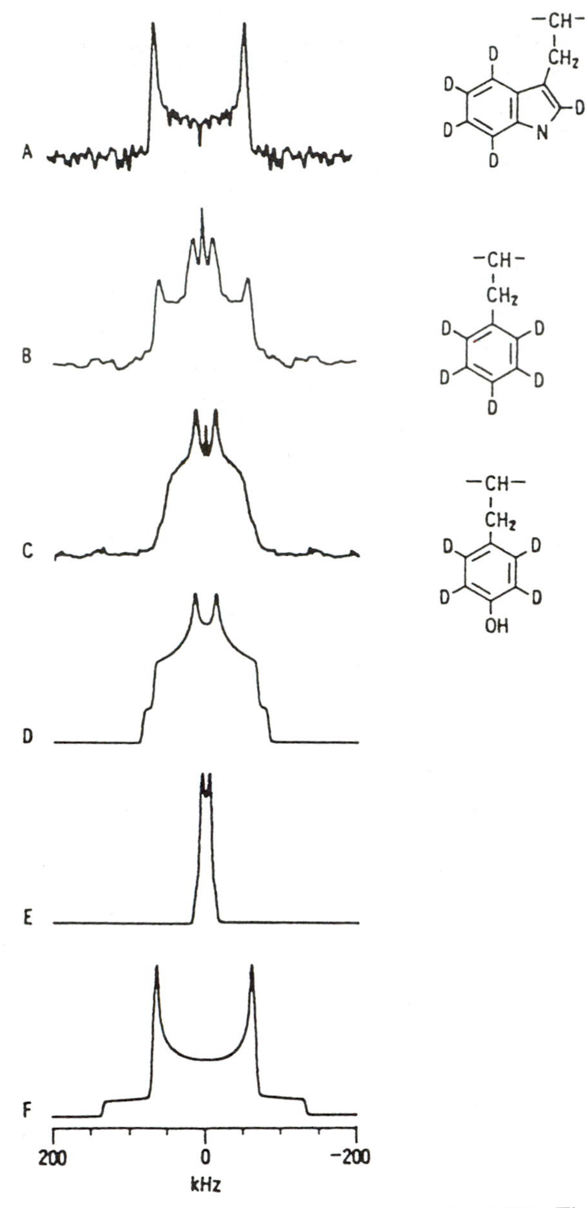

Figure 4: ^{2}H NMR Spectra of fd Protein (from Gall *et al.*, 1982). The spectra were collected at 38.4 MHz. Spectra A-C are fd protein labeled with ^{2}H tryptophan, ^{2}H phenylalanine, and ^{2}H tyrosine, respectively. Spectra D-F are calculated powder patterns for (D) rapid 180° flips about the Cb-Cg bond axis, (E) rapid rotation about the Cb-Cg bond axis, (F) for static tensor with e2qQ/h = 180.0 kHz and h = 0.05 derived from experimental spectra of polycrystalline [^{2}H]- phenylalanine and [^{2}H]-tyrosine.

protein. Upon cooling to -30°C, these motions no longer occurred, while at elevated temperature (>91°C) rapid and unrestricted motion indicated that the protein had unfolded significantly. Another example of this type of application is examination of phenyl sidechain dynamics in cyclic pentapeptides in which significant ring motion was observed (Frey *et al.*, 1985b).

Other applications of selectively deuterated aromatic amino acids

In addition to NMR investigations, selectively deuterated amino acids have been employed in a variety of other types of studies. Introduction of ^2H for ^1H alters the mass so that a deuterated amino acid and its metabolites can be detected uniquely in a mixture by mass spectrometry. Since deuterium labeling does not alter the biochemical properties of the amino acid, nor does it cause toxicity in the product, all three aromatic amino acids have been employed in tracer studies with humans using separation by capillary gas chromatography followed by detection using negative-ion chemical-ionization mass spectrometry of metabolites (Shimamura *et al.*, 1986, 1993; Hayashi *et al.*, 1986a,b; Andrew *et al.*, 1993). Furthermore, metabolic tracing has been used in examining the biosynthesis of thiamine (White, 1979). In a similar vein, the extent of incorporation of ^2H-labeled tryptophan into bacteriorhodopsin was determined by purification of the protein, hydrolysis, separation of amino acids by reversed phase chromatography and identification of the labeled amino acids by thermospray mass spectrometry (Karnaukhova *et al.*, 1989).

Partially deuterated tyrosine has been employed in Fourier transform infrared difference spectroscopy to monitor alterations in vibrational modes that differ between different conformational forms of bacteriorhodopsin that appear to be linked to the proton translocation process (Dollinger *et al.*, 1986). By different isotopic labeling of tyrosine ($^2H_2$ and $^2H_4$), unique vibrational spectra were obtained for tyrosine in the ionized and unionized forms that could be employed to interpret the small spectral changes observed in the intact protein in terms of ionization patterns. Deuterated aromatic amino acids have also been utilized to simplify assignment of peaks and interpretation of ultraviolet resonance Raman spectral changes in bacteriorhodopsin (Harada *et al.*, 1990).

Conclusion

Methods for chemical production of partially and completed deuterated aromatic amino acid side chains have found applications beyond the initial goal of simplifying the aromatic region of the one-dimensional NMR spectrum of an interesting protein. This relatively simple and inexpensive approach has provided the raw materials for a variety of studies and still serves to simplify spectral identification and interpretation. The ability to trace metabolism of compounds using ^{2}H labeling in a nontoxic and noninvasive manner is one of the more interesting and useful modes of employing these compounds. Nonetheless, the strength of the these methods lies in the ability to examine uniquely resonances in the aromatic region and to make possible facile assignment of individual aromatic amino acid resonances.

Acknowledgments

The work that led to the method of selective deuteration of aromatic amino acids was performed in the laboratory of Dr. Oleg Jardetzky at Stanford University School of Medicine. The freedom that he provided to pursue alternative methods was the crucible that led to development of these chemical approaches for selective deuteration and to the products that have enhanced a variety of efforts over the ensuing years. His continued interest in our work and efforts over the intervening years is appreciated greatly.

References

Andrew, R., Best, S.A., Watson, D.G., Midgley, J.M., Reid, J.L., and Squire, I.B. (1993). *Neurochem. Res.* **18**, 1179.

Anglister, J., Bond, M.W., Frey, T., Leahy, D., Levitt, M., McConnell, H.M., Rule, G.S., Tomasello, J., and Whittaker, M. (1987). *Biochemistry* **26**, 6058.

Arndt, K.T., Boschelli, F., Lu, P., and Miller, J.H. (1981). *Biochemistry* **20**, 6109.

Arndt, K.T., Boschelli, F., Cook, J., Takeda, Y., Tecza, E., and Lu, P. (1983). *J. Biol. Chem.* **258**, 4177.

Brown-Mason, A., Dobson, C.M., and Woodworth, R.C. (1981). *J. Biol. Chem.* **256**, 1506.

Cedel, T.E., Cottam, P.F., Meadows, M.D., and Ho, C. (1984). *Biophys. Chem.* **19**, 279.

Crespi, H.L., Rosenberg, R.M., and Katz, J.J. (1968). *Science* **161**, 795.

Dollinger, G., Eisenstein, L., Lin, S.-L., Nakanishi, K., and Termini, J. (1986). *Biochemistry* **25**, 6524.

Florance, J., and Ginther, C.L. (1981). *Biochim. Biophys. Acta* **672**, 207.

Frey, M.H., DiVerdi, J.A., and Opella, S.J. (1985a). *J. Am. Chem. Soc.* **107**, 7311.

Frey, M.H., Opella, S.J., Rockwell, A.L., and Gierasch, L.M. (1985b). *J. Am. Chem. Soc.* **107**, 1946.

Gall, C.M., DiVerdi, J.A., and Opella, S.J. (1981). *J. Am. Chem. Soc.* **103**, 5039.

Gall, C.M., Cross, T.A., DiVerdi, J.A., and Opella, S.J. (1982). *Proc. Natl. Acad. Sci. USA.* **79**, 101.

Harada, I., Yamagishi, T., Uchida, K., and Takeuchi, H. (1990). *J. Am. Chem. Soc.* **112**, 2443.

Hayashi, T., Minatogawa, Y., Kamada, S., Shimamura, M., Naruse, H., and Iida, Y. (1986a). *J. Chromatog.* **380**, 239.

Hayashi, T., Shimamura, M., Matsuda, F., Minatogawa, Y., Naruse, H., and Iida, Y. (1986a). *J. Chromatog.* **380**, 259.

Karnaukhova, E., Niessen, W.M.A., Tjaden, U.R., Raap, J., Lugtenburg, J., and van der Greef, J. (1989). *Anal. Biochem. (Tokyo)* **181**, 271.

Kato, K., Nishimura, Y., Waelchli, M., and Arata, Y. (1989). *J. Biochem.* **106**, 361.

Kinsey, R.A., Kintanar, A., and Oldfield, E. (1981). *J. Biol. Chem.* **256**, 9028.

LeMaster, D.M., and Richards, F.M. (1988). *Biochemistry* **27**, 142.

LeMaster, D.M. (1990a). *Ann. Rev. Biophys. Biophys. Chem.* **19**, 243.

LeMaster, D.M. (1990b). *Quart. Rev. Biophys.* **23**, 133.

Markley, J.L., Putter, I., and Jardetzky, O. (1968). *Science* **161**, 1249.

Matthews, H.R., Matthews, K.S., and Opella, S.J. (1977a). *Biochim. Biophys. Acta* **497**, 1.

Matthews, K.S., Wade-Jardetzky, N.G., Graber, M., Conover, W.W., and Jardetzky, O. (1977b). *Biochim. Biophys. Acta* **490**, 534.

Pan, T., Giedroc, D.P., and Coleman, J.E. (1989a). *Biochemistry* **28**, 8828.

Pan, T., King, G.C., and Coleman, J.E. (1989b). *Biochemistry* **28**, 8833.

Schramm, S., and Oldfield, E. (1983). *Biochemistry* **22**, 2908.

Shen, Q., Simplaceanu, V., Cottam, P.F., and Ho, C. (1989). *J. Mol. Biol.* **210**, 849.

Shimamura, M., Kamada, S., Hayashi, T., Naruse, H., and Iida, Y. (1986). *J. Chromatog.* **374**, 17.

Shimamura, M., Kodaka, H., Hayashi, T., and Naruse, H. (1993). *Neurochem. Res.* **18**, 727.

Takahashi, H., Igarashi, T., Shimada, I., and Arata, Y. (1991). *Biochemistry* **30**, 2840.

White, R.H. (1979). *Biochim. Biophys. Acta* **583**, 55.

16

The Interaction of Antigens and Superantigens with the Human Class II Major Histocompatibility Complex Molecule HLA-DR1

T. Jardetzky

Department of Biochemistry, Molecular Biology and Cell Biology
Northwestern University
Evanston, IL 60208 USA

The initiation and maintenance of an immune response to pathogens requires the interactions of cells and proteins that together are able to distinguish appropriate non-self targets from the myriad of self-proteins (Janeway and Bottomly, 1994). This discrimination between self and non-self is in part accomplished by three groups of proteins of the immune system that have direct and specific interactions with antigens: antibodies, T cell receptors (TcR) and major histocompatibility complex (MHC) proteins. Antibodies and TcR molecules are clonally expressed by the B and T cells of the immune system, respectively, defining each progenitor cell with a unique specificity for antigen. In these cell types both antibodies and TcR proteins undergo similar recombination events to generate a variable antigen combining site and thus produce a nearly unlimited number of proteins of different specificities. TcR molecules are further selected to recognize antigenic peptides bound to MHC proteins, during a process known as thymic selection, restricting the repertoire of T cells to the recognition of antigens presented by cells that express MHC proteins at their surface. Thymic selection of TcR and the subsequent restricted recognition of peptide-MHC complexes by peripheral T cells provides a fundamental molecular basis for the discrimination of self from non-self and the regulation of the immune response (Allen, 1994; Nossal, 1994; von Boehmer, 1994). For example, different classes of T cells are used to recognize and kill infected cells (cytotoxic T cells) and to provide lymphokines that induce the majority of soluble antibody responses of B cells (helper T cells).

In contrast to the vast combinatorial and clonal diversity of antibodies and TcRs, a small set of MHC molecules is used to recognize a potentially unlimited universe of foreign peptide antigens for antigen presentation to T cells (Germain, 1994). This poses the problem of how each MHC molecule is capable of recognizing enough peptides to insure an immune response to pathogens. In addition, the specificity of the TcR interaction with MHC-peptide complexes is clearly crucial to the problem of self:non-self discrimination, with implications for both protective immunity and auto-immune disease. Both bacteria and viruses have evolved proteins (superantigens) which subvert the specificity of the immune response, by overriding TcR affinities for distinct MHC-peptide complexes (Herman *et al.*, 1991; Kotzin *et al.*, 1993). In order to address how MHC molecules interact with both conventional peptide antigens and bacterial superantigens, X-ray crystallographic structures of the human class II MHC molecule HLA-DR1 have been solved in complexes with self-peptides (Brown *et al.*, 1993), antigenic peptide (Stern *et al.*, 1994), and two bacterial superantigens (Jardetzky *et al.*, 1994; Kim *et al.*, 1994). What follows is a description and comparison of the interactions that MHC molecules form with conventional peptide and the bacterial superantigens, based predominantly on observations made from the crystal structure of HLA-DR1 bound to the bacterial superantigen, *S. aureus* enterotoxin B.

Peptide binding to HLA-DR1

Architecture of the class II MHC peptide binding site

The structure of HLA-DR1 is shown in Figure 1. The protein is made of two chains of approximately equal length, forming a heterodimer that is anchored to the cell by two transmembrane regions. The extracellular region of each chain is divided into two domains. The N-terminal domains (α1 and β1) form the mixed α/β structure of the peptide binding site, while the lower two domains (α2 and β2) are of the immunoglobulin fold. The β2 domain has been implicated in direct interactions with the T cell accessory molecule CD4 (Cammarota *et al.*, 1992).

The peptide binding site is shown in a top view in Figure 1b. The structure is remarkably similar to the class I MHC structure (Bjorkman *et al.*, 1987; Garrett *et al.*, 1989; Madden *et al.*, 1993), given the low sequence similarity between these two classes of MHC molecules (Brown *et al.*, 1988). It is a semi-symmetric structure formed of a platform of 8-β strands supporting two long α-helices. Each domain contributes 4 strands and a long helix to form this

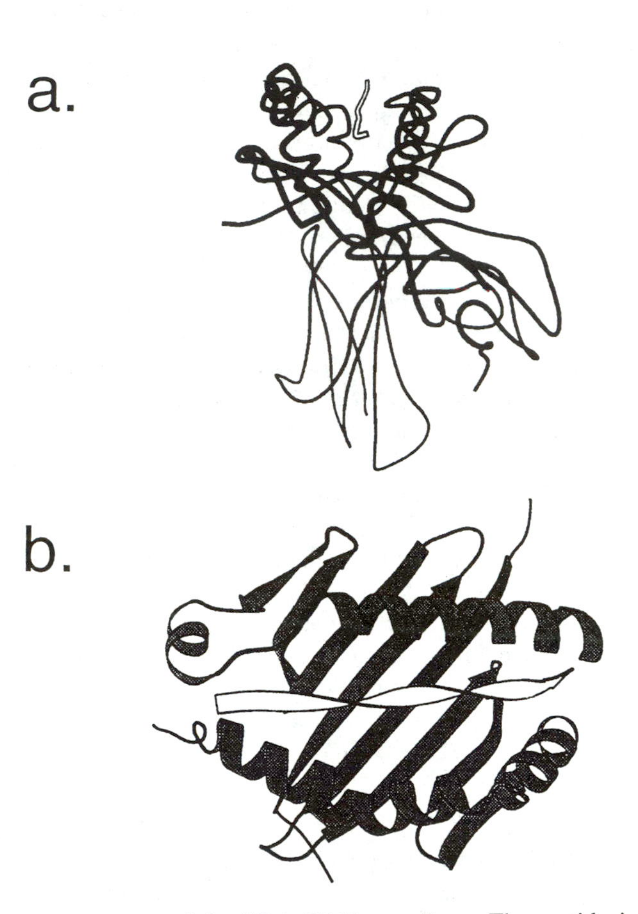

Figure 1: (a) Ribbon trace of the HLA-DR1 heterodimer. The peptide is shown in white. The α1 and β1 domains form the peptide binding site, while the lower two immunoglobulin domains (α2 and β2) are proximal to the membrane. (b) A top view of the alpha1/beta1 peptide binding site of the HLA-DR1 molecule. The lower domains are not shown. Representations were produced with the program Molscript.

structure, with side chains from both alpha and beta chains specifying the chemical nature of the binding site. The two helices and β-strands line a long groove which can accommodate extended peptide chains of approximately 15 amino acids in length. The high sequence variability associated with the many different alleles of MHC molecules maps to the peptide binding region of the molecule, providing a straightforward explanation for the sequence specificity of different MHC alleles for antigenic peptides (Bjorkman *et al.*, 1987; Brown *et al.*, 1988).

Length of peptides that bind to HLA-DR1

Peptide length and sequence variation are two important characteristics that are associated with binding to class II MHC molecules. In addition, antigenic peptides have been shown to form very stable complexes with MHC molecules with dissociation rates that typically are on the order of many days (Buus *et al.*, 1986; Jardetzky *et al.*, 1990). Any molecular description of antigenic peptide interactions with MHC molecules must try to interpret these functional attributes based on structural observations of the MHC peptide binding site.

Antigenic peptides are typically derived from exogeneously produced proteins, which can be taken up by antigen presenting cells and then processed by proteolytic enzymes into peptide fragments of variable length (Germain, 1994; Germain and Margulies, 1993). Class II molecules also bind self-peptides derived from cellularly produced proteins, in the absence of any exogenous antigens. The isolation and sequencing of these self-peptides from purified class II molecules, including HLA-DR1 (Chicz *et al.*, 1992; Chicz *et al.*, 1993) demonstrates a length variation from 13 to 25 amino acids. This variability is in contrast to the peptides found associated with class I MHC molecules, which are generally 9 to 11 amino acids long (Germain, 1994; Van Bleek and Nathenson, 1990; Falk *et al.*, 1991). This difference in the average peptide length that is

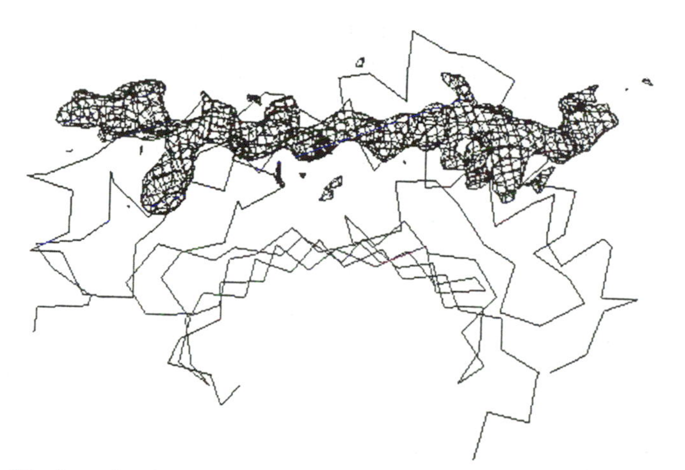

Figure 2: Electron density corresponding to the mixture of self-peptides bound to HLA-DR1. The view is from the side, with the beta1 domain in the foreground and the beta2 domain in the background. The map is a |2Fo-Fc| map calculated with HLA-DR1:SEB model phases, prior to the building and refinement of any peptide model. Note the large sidechain density projecting down into the MHC binding site to the left side (N-terminus) of the peptide binding site. The electron density clearly extends out both ends of the binding site, consistent with the binding of longer peptides.

difference in the specific antigen presentation pathways that deliver peptides to recognized by class I and class II MHC molecules reflects not only a biological these molecules, but also a distinctly different structural strategy for peptide recognition that these two classes of proteins have evolved.

Figure 2 presents a side view of the HLA-DR1 peptide-binding site, with the electron density observed for a mixture of endogenous peptides that remain bound after immunoaffinity purification from the lymphoblastoid cell line LG-2 and cocrystallization with the bacterial superantigen SEB. A significant proportion of HLA-DR1 molecules (80-90%) are occupied by a mixture of cell-derived peptides, which remain tightly associated with the class II MHC molecules (Chicz *et al.*, 1992; Chicz *et al.*, 1993). Although there are potentially hundreds of different peptides bound, two peptides occur repeatedly in a nested set of different lengths and amounts, as isolated from LG-2 HLA-DR1. Peptides from the HLA-A2 molecule and the invariant chain occur in a range of lengths (as 14-mers to 24-mers), with each peptide core sequence corresponding to an estimated 13% of the isolated peptide mass (Chicz *et al.*, 1992). However, the density in Figure 1 does not simply fit the sequences of either of these two peptides.

The electron density of this endogenous peptide mixture is continuous and readily interpretable for a segment of 13 amino acids, showing features similar to that observed for the DR1:HA peptide complex (Stern *et al.*, 1994). The density extends out both ends of the DR1 peptide-binding site (Brown *et al.*, 1993), and additional density at the C-terminus of the peptide (to the right in Figure 1) is observed at lower contour levels, consistent with the observation that 24 amino acid long endogenous-peptides are found bound to DR1 (Chicz *et al.*, 1992). The continuity of the endogenous-peptide density in DR1 is in contrast to the electron density observed for the mixtures of self peptides bound to the class I MHC molecules HLA-A2 and HLA-Aw68 (Bjorkman *et al.*, 1987; Garrett *et al.*, 1989; Saper *et al.*, 1991), where the central regions of the peptide density are less well resolved than the N- and C- termini. Although the length variation of self peptides bound to HLA-DR1 is great (13-25 amino acids), the electron density of the mixture in this crystal suggests a common conformation for a central core of 13 amino acids, with longer peptides accommodated by extension out of the peptide-binding site.

The peptide main chain conformation is relatively regular as it extends throughout the MHC peptide-binding site, with correspondingly restricted values for the peptide ϕ and ψ angles. The mean peptide ϕ and ψ angles are near -80° and 130°, respectively, for a peptide built into the endogenous peptide density, which are similar to the values observed for a polyproline type II helices (Adzhubei and Sternberg, 1993). This regular main chain conformation was also observed in the DR1:HA peptide complex (Stern *et al.*, 1994), and may be a general structural feature of peptides bound to HLA-DR1.

Sequence specificity of peptide binding to HLA-DR1

Clearly, peptide sequence specificity is a determining factor in the ability of class II MHC molecules to successfully provide a protective immune response. Rather than generating the diversity of antigen binding sites associated with TcR and antibodies, MHC molecules are designed to recognize many different peptide sequences, by having few sequence specific interactions that are absolutely required for high affinity peptide binding. In the case of HLA-DR1, this could be demonstrated by studying the interaction of an antigenic peptide from the influenza hemagglutinin (HA 306-318) (Jardetzky *et al.*, 1990). As shown in Figure 3, when forty single amino acid substitutions were made in the core sequence of such an antigenic peptide, very few of them have significant effects on the peptide binding affinity, although radical substitutions of size and charge have been tested. In this study, only one amino acid position could be shown to have a dominant effect on the binding of these peptides, the tyrosine at position 308. In fact, the relevance of all the other amino acids in determining high affinity peptide binding was tested by a global substitution of all but two amino acids of the HA peptide to alanines (Jardetzky *et al.*, 1990). A lysine was retained near the C-terminus for the sake of peptide solubility. This reduced peptide (HA-YAK) could be shown to have a comparable if not improved peptide binding affinity for HLA-DR1. In addition, HA-YAK demonstrated the characteristic slow off-rates associated with antigenic peptide binding to class II molecules, firmly demonstrating that a minimal peptide could form complexes that are relevant to normal peptide interactions. Since peptides ranging in length from 2-6 amino acids with tyrosine in their sequence did not bind HLA-DR1, a minimal peptide length was also implicated in high affinity peptide binding.

More recent experiments have defined a peptide binding motif for HLA-DR1 (Figure 3), by using the method of M13 phage-display to select peptides which bind with high affinity to purified, recombinant HLA-DR1 (Hammer *et al.*, 1992; Hammer *et al.*, 1993). These methods have detected four positions that are important in defining an allele specific motif for peptides (Hammer *et al.*, 1992; Hammer *et al.*, 1993; Sette *et al.*, 1993). Each of the four positions show preferences, but not absolute selection for particular amino acids, consistent with the broad specificity of DR1 binding. Position 1, which is the most important HLA-DR1 anchor residue position, is preferentially tyrosine or phenylalanine (Jardetzky *et al.*, 1990; Hammer *et al.*, 1993), as found previously. Two other positions show some preference for the hydrophobic amino acids methionine or leucine, while the fourth shows a preference for the smaller side amino acids alanine, serine, or glycine. None of these amino acids is absolutely required for high affinity peptide binding, but these clearly demonstrate sidechain interactions throughout the length of the peptide which influence the peptide specificity of the DR1 molecule. The limited sequence

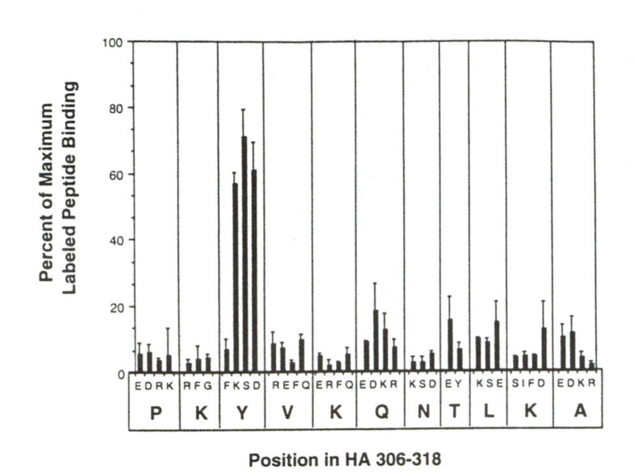

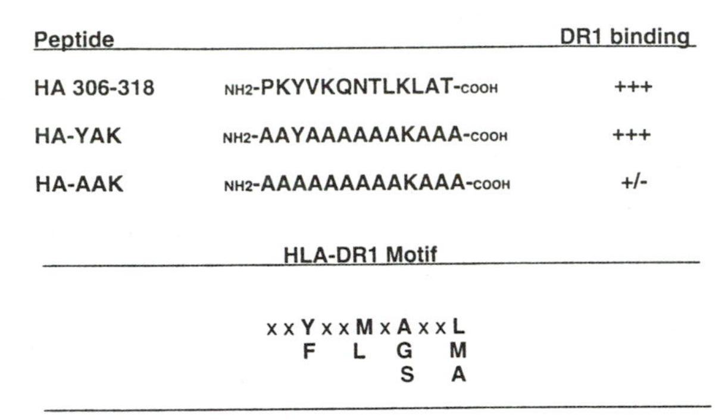

Peptide		DR1 binding
HA 306-318	NH2-PKYVKQNTLKLAT-COOH	+++
HA-YAK	NH2-AAYAAAAAAKAAA-COOH	+++
HA-AAK	NH2-AAAAAAAAAKAAA-COOH	+/-

HLA-DR1 Motif

x x Y x x M x A x x L
F L G M
S A

Figure 3: Specificity of peptide binding to HLA-DR1. The first panel shows the effects on MHC binding of a set of single substitutions introduced into the hemagglutinin 306-318 peptide. Peptides substituted at tyrosine 308 show significantly reduced binding to HLA-DR1, as measured by their ability to compete in a radioactive peptide binding assay. Panel two shows the alanine peptide (HA-YAK), which retains MHC binding with similar affinity to the original HA peptide, although the majority of side chains are absent in this peptide. The final panel shows the HLA-DR1 peptide binding motif as determined by peptide phage display experiments, with selection for high affinity binding to HLA-DR1.

specificity of class II MHC molecules for any given peptide clearly reflects the ability of these molecules to accommodate peptide sequences of great diversity.

The endogenous peptide density has sidechain characteristics that are consistent with the sequence motif defined for peptides that bind to HLA-DR1 (Figure 2). We have therefore built a model peptide containing a minimal HLA-

DR1 peptide-binding motif (NH_2-AA**Y**AA**M**AAA**L**AA-COOH) into the endogenous peptide electron density (Figure 4). Other side chains from the endogenous peptide density are equally interpretable and may reflect secondary weaker selection of the binding site for particular amino acids or conformations. Without knowledge of the true complexity of the peptides that contribute to the endogenous peptide density, further interpretation of the sidechain density remains difficult.

Figure 4 also shows a surface representation of the MHC molecule that has been colored to indicate those regions of the peptide-binding site which are polymorphic between different class II MHC molecules (Brown *et al.*, 1988). A light gray surface indicates relatively conserved residues, whereas dark gray represents a polymorphic region of the molecule. These polymorphic regions of the binding site include the sidechain binding pockets that accommodate each of the motif determining residues of the peptide (white in Figure 4). This distribution of polymorphism indicates how peptide specificity is modulated and selected between different class II alleles. Changes associated with a peptide sidechain binding pocket would clearly affect the sequence specificity of peptide recognition. The correlation between polymorphism and pockets indicates a powerful structural selection for changes that determine sidechain binding specificity.

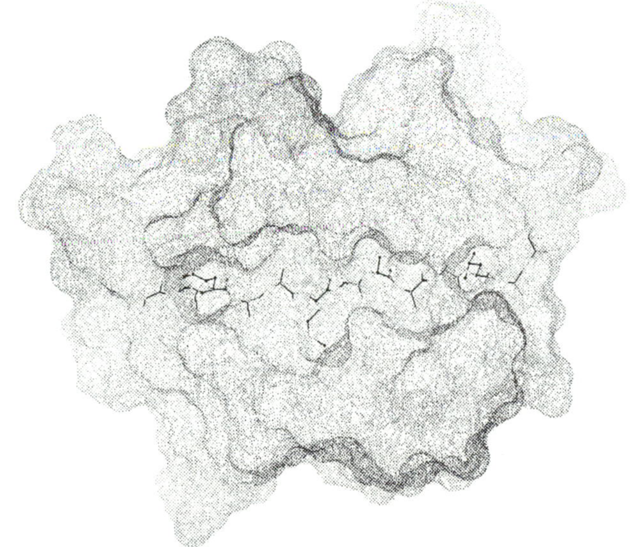

Figure 4: Surface representation of the HLA-DR1 molecule with peptide built into the endogenous peptide density shown in Figure 2. The surface is color-coded to indicate the location of MHC polymorphisms. Light gray indicates conserved regions of the MHC surface, while dark gray indicates regions of the binding site that are variable in mouse and human class II MHC sequences. The motif-determining residues of the peptide are shown in black.

Peptide sequence-independent interactions of the MHC binding site

Although peptide sidechain interactions with the MHC antigen binding site clearly have a role in determining the sequence specificity of peptide binding, sequence-independent interactions may be more important in understanding a generalizable mechanism by which all class II MHC molecules can interact with peptides. The ability of different class II molecules to bind peptides with characteristic slow dissociation rates, combined with the observation that the majority of peptide side chains are dispensable for binding, suggests that conserved MHC-peptide interactions are important. Such interactions can be found between MHC residues which are conserved in different MHC alleles and peptide main chain atoms.

Fifteen potential hydrogen bonds are observed in the complex with endogenous peptides (Figure 5), and twelve of these involve MHC atoms or residues that are conserved in the majority of class II sequences. Six of these hydrogen bonds are bidentate hydrogen bonds, from asparagines (α62, α69, and β81), which form 9- (β81) and 11-membered (α62, α69) ring structures between the peptide main-chain atoms and the amide sidechain atoms (Le Questel *et al.*, 1993). The conservation of this hydrogen bonding pattern in both the DR1:HA peptide structure (Stern *et al.*, 1994) and the DR1:endogenous peptide structure further supports the idea that these interactions will be a consistent feature of different peptides that bind to class II MHC molecules.

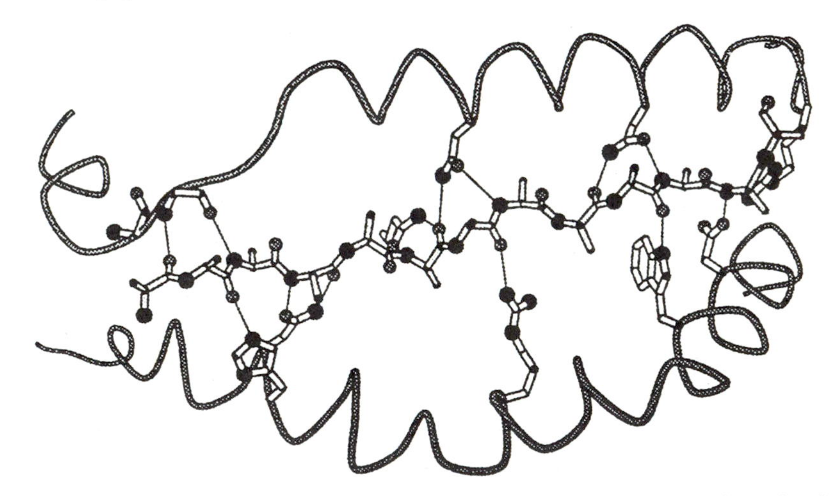

Figure 5: Hydrogen-bonding pattern between MHC residues and peptide main-chain atoms. The α1 helix is above and the β1 helix is below the peptide. Note the three conserved asparagine residues (α62, α69 and β82), which form bidentate hydrogen bonds to the peptide.

Conclusions

The mechanism by which class II MHC molecules form tightly bound complexes to peptides of diverse sequence is suggested to have two components. One component is the MHC pocket interactions with specific side chains of peptides. These pockets vary with the allotype of the class II molecules, correlating with the selection of different peptide sequence motifs. Most of the MHC pockets are able to accommodate a number of different side chains, reducing the restrictions placed on the binding of different peptide sequences. A second component to peptide binding involves conserved MHC residue interactions with the peptide main chain, which would provide a peptide sequence and MHC allotype independent mechanism for the interaction with peptide. A regular distribution of peptide phi/psi angles is observed for both the DR1:HA structure and a mixture of endogenous peptides, suggesting that there is a common (and generalizable) conformational constraint placed on the peptide by the DR1 molecule.

Superantigen Binding To HLA-DR1

Superantigens (SAGs) are a recently-identified group of T-cell antigens with significantly different properties from conventional peptide antigens. Superantigens still bind to class II major histocompatibility complex (MHC) molecules and stimulate T cells, but this binding is less allele-specific than conventional peptide binding and occurs outside of the conventional peptide binding site. The stimulation of T cells is also altered, in that a much larger percentage of T cells are stimulated by superantigens, in a manner that overrides the antigen combining site specificity of the TcR and depends most prominently on sequences in the variable β-chain domain (Vβ) of the TcR (Herman *et al.*, 1991; Kotzin *et al.*, 1993). Viral and bacterial superantigens have been identified, that include soluble enterotoxins from *S. aureus* and related proteins from other bacteria, and the mouse mammary tumor virus (MMTV) superantigen, which is a membrane bound protein (Acha *et al.*, 1993). The bacterial toxins are associated with the induction of a number of diseases, most notably food poisoning and toxic shock syndrome for the *S. aureus* toxins. The MMTV superantigen has evolved to play an important role in the life cycle of the virus, being important for viral transmission from mother to offspring by providing a pool of stimulated lymphocytes for viral infection. The powerful T cell activation properties of superantigens is thought to directly lead to respiratory distress and shock, due to the overproduction of lymphokines such as tumor necrosis factor (Miethke *et al.*, 1992). In addition, superantigens are implicated a number of autoimmune disorders (Kotzin *et al.*, 1993; Acha, 1993), as superantigen activation of self-specific T cells may provide a mechanism for triggering disease.

The interaction of superantigens with both MHC molecules and T cell receptors (TcR) is distinct from that of conventional peptide antigens. As discussed in the previous section, conventional antigens are short peptides ranging in length from approximately 12-30 amino acids that bind in an extended conformation into a defined groove formed by the $\alpha1$ and $\beta1$ domains of the class II MHC molecule (Figure 1). Antigen-presenting cells endocytose intact protein antigens and use endogenous proteases to produce peptide fragments that can associate with class II MHC molecules. In contrast, superantigens bind as intact proteins to MHC molecules, and proteolytic degradation of these proteins typically leads to loss of the superantigenic properties. The binding of bacterial superantigens to MHC molecules occurs outside of the conventional antigen binding site, requiring structural features of the folded superantigen protein to provide the interaction site.

The ability of these toxins to stimulate disproportionately large numbers of T cells also relies on a different set of interactions with the TcR than occurs with conventional antigen. Three TcR hypervariable loops from both the α– and the β–chain form a binding site for a specific peptide:MHC combination (Jorgensen, Reay, Ehrich and Davis, 1992), analogous to an antibody antigen-combining site. The site with which the superantigens interact is distinct from the antigen combining site, closely associated with the fourth hypervariable region (HV4) of the TcR Vβ domain (Kotzin *et al.*, 1993). Thus, existing sequence differences found in the Vβ gene family determines the ability of a particular superantigen to stimulate a T cell. For example, *S. aureus* enterotoxin B is able to stimulate a subset of T cells that carry Vβ3, Vβ12, Vβ14, Vβ15, Vβ17, or Vβ20 domains, while *S. aureus* toxic shock syndrome toxin stimulates T cells with Vβ2 domains. Superantigens interact with both MHC and TcR molecules outside their normal antigen specific sites, leading to the stimulation of many more T cells than observed with conventional peptide antigens.

X-ray crystallographic studies of bacterial toxins bound to the human class II MHC molecule HLA-DR1 have provided a clearer picture of the action of superantigens. To date, two such complexes have been solved. The first is a complex of the *S. aureus* enterotoxin B (SEB) with HLA-DR1 (Jardetzky *et al.*, 1994), and the second is the related *S. aureus* toxic shock syndrome toxin (TSST-1) with HLA-DR1 (Kim *et al.*, 1994). A comparison of these two structures shows some of the variability in superantigen:MHC interaction sites that can be expected from this family of toxins. In both cases, the toxins bind outside the MHC peptide-binding site using residues from their N-terminal domain to interact with the HLA-DR1 molecule. Gross conformational changes are not evident in either the MHC molecule (Brown *et al.*, 1993; Jardetzky *et al.*, 1994; Kim, *et al.*, 1994) or the toxins upon complex formation (Swaminathan, *et al.*, 1992; Acharya *et al.*, 1994; Prasad *et al.*, 1993).

Although the two toxins bind to overlapping sites on the DR1 molecule using a similar region of their N-terminal domains, the specific interactions are very different, providing two complexes with distinct orientations of the superantigen proteins relative to the MHC peptide binding site. The observed binding of these two superantigens to MHC molecules at least partially explain previous mutational studies, which indicate that SEB and TSST-1 interact with a common subset of MHC residues. The complexes of *S. aureus* toxins bound to HLA-DR1 indicate that conventional MHC:TcR interactions are blocked by superantigens and that there is significant variability in the formation of MHC:SAg:TcR ternary complexes.

Interaction of SEB with HLA-DR1

The structure of SEB bound to HLA-DR1 is shown in Figure 6, SEB is observed to bind to a region of the MHC molecule that neighbors the conventional peptide binding site. SEB itself is a 28 kD protein that is composed of two domains shown in yellow and red in Figure 6 and these two domains are observed in the other bacterial toxins whose structures have been solved. The N-terminal domain, which belongs to a family of protein folds termed the oligomer-binding fold (Murzin, 1992; Murzin and Chothia, 1992), forms the majority of contacts to the DR1 molecule, using one face of the β–sheet to interact with the DR1 α1 α-helix and two loops of the lower α1 β–sheet. This region of the MHC molecule shows a pronounced cleft that is lined by these secondary structures, providing a relatively hydrophobic surface for superantigen binding. Since SEB only interacts with residues of the DR1 α-chain, this provides a straightforward explanation for the ability of this toxin to bind to many different HLA-DR molecules. The α-chain, and therefore the entire SEB binding site, is conserved in heterodimers of different HLA-DR alleles. In contrast, HLA-DP and HLA-DQ proteins have variable α-chains and therefore the conservation of SEB binding depends on the variation of residues at these positions.

SEB binding to HLA-DR1 involves one face of the N-terminal β-barrel domain and this positions the remainder of the SEB molecule off to one side of the MHC peptide binding site. Three strands of the β-sheet (strands 1, 2 and 3) form the face of the MHC binding site on SEB, with the loop between strands 1 and 2 providing critical hydrophobic interactions with the DR1 molecule. The interface is further defined by the fifth α-helix in the SEB molecule, which lies to the left of the complex interface shown in Figure 6, and by an SEB loop which crosses the top of the interface, interacting with the HLA-DR1 α1 α–helix. While the calculated buried surface area of the complex is comparable to antigen:antibody interfaces (approximately 750Å^2 is buried on each protein upon complex formation), the affinity of HLA-DR1 and SEB is approximately

1 μM. Recent data indicates that the affinity of SEB for DR1 may be dependent on the composition of bound peptides (Thibodeau *et al.*, 1994), as revealed by the binding of SEB to DR1 expressed in different cell types, although this remains to be firmly establish with peptide:DR1 complexes in vitro. As no direct contacts are apparent between SEB and the peptide density discussed in the previous section, there is not a simple explanation for this peptide-dependent binding.

The SEB interface with HLA-DR1 can be easily divided into two subsites with distinct characteristics. One of these subsites involves a set of hydrophobic interactions, while the other allows the formation of a buried salt bridge between

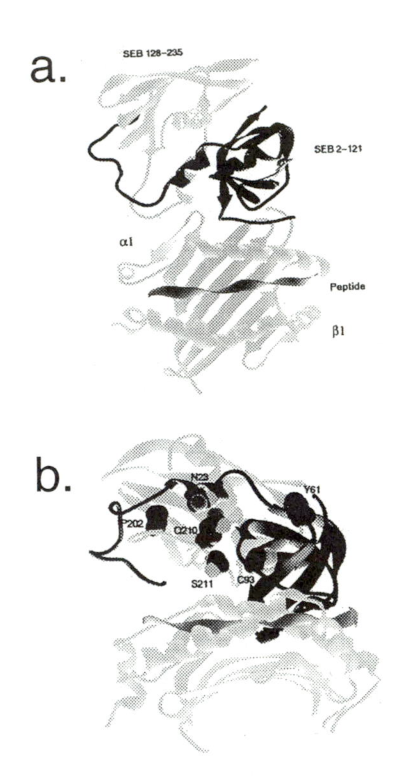

Figure 6: (a) Top view of the HLA-DR1:SEB complex. The binding of the SEB superantigen occurs exclusively to residues of the α1 domain of HLA-DR1, which lie outside the conventional antigen binding site. (b) Side view of the HLA-DR1:SEB complex. In addition, CPK models of residues are shown that affect the specificity and potency of T cell stimulation, but that have no effect on the superantigen affinity for MHC molecules. These residues line a face of the superantigen molecule above and to one side of the MHC peptide binding site, defining a potential interaction site for TcR.

the two proteins. SEB residues F44, L45, and F47 extend outwards from the loop between strands 1 and 2, providing a hydrophobic ridge which packs between the two HLA-DR1 loops at the interface. This hydrophobic ridge fills the shallow groove on DR1, which is formed on the sides by loops 1 and 3 of the α1 domain of DR1 and from below by the α1 α-helix. Mutations of residues at this interface, in either the SEB molecule or the DR1 molecule, have been found to disrupt the binding interactions. For HLA-DR1, prominent mutations which affect binding are found at Met 36 and Ile 63 which disrupt the hydrophobic interactions, and Lys 39, which disrupts the salt bridge with SEB (Thibodeau *et al.*, 1994; Thibodeau, *et al.*, 1994b).

Residues of SEB also interact with a region on top of the DR1 α1 α-helix, as shown in Figure 6. Amino acids 92-96 form an extended chain which lies across HLA-DR1 residues α60, α61, α64, α67, and α68. These interactions serve to cap off the more central and extensive surfaces described above and form a potential barrier to direct TcR interactions with the DR1 molecule. During conventional peptide antigen presentation by MHC molecules, this region of the α1 α-helix is thought to provide important interactions with the TcR, based on mutational analysis of peptide binding and T cell stimulation (Dellabona *et al.*, 1990; Ehrich *et al.*, 1993).

The structure of the toxic shock syndrome toxin (TSST-1) bound to HLA-DR1 has also been solved by X-ray crystallographic methods (Kim *et al.*, 1994). A comparison of the interaction of TSST-1 and SEB with HLA-DR1 shows that these two toxins bind to overlapping regions of the HLA-DR1 surface, but with a distinct set of atomic interactions. Although the same face of the N-terminal β– domain of TSST-1 is used to bind the HLA-DR1 molecule, this superantigen binds across the MHC peptide binding site, interacting with both the DR1 α– and β-chains. The TSST-1 toxin does maintain an analogous interaction with the hydrophobic DR1 pocket on the α-chain that forms half of the SEB binding site. In the case of TSST-1, a leucine residue in a structurally homologous position to SEB Leu 45, forms a similar set of interactions with hydrophobic DR1 amino acids, explaining the sensitivity of TSST-1 binding to mutations at DR1 Met α36. The orientation of this toxin across the peptide binding site also supports the idea the direct TcR:MHC interacts are blocked during superantigen stimulation of T cells.

Both the functional and structural data have indicated distinct binding geometries for different superantigen:MHC complexes. The structures of DR1:SEB and DR1:TSST1 directly demonstrate such differences in complex formation, although the same structural domain of the superantigen forms the MHC binding site. Mutational analyses of the related toxin SEA, indicate that it may form an additional, unique complex with MHC molecules, using a Zn binding site in its C-terminal domain to interact with a region of the MHC β1 α–helix (Herman *et al.*, 1991; Karp and Long, 1992; Fraser *et al.*, 1992;

Hudson *et al.*, 1993). This C-terminal domain interaction is probably structurally incompatible with the formation of a one-to-one complex of SEA and DR1 that simultaneously forms SEB-like interactions, assuming the conservation of the toxin fold in SEA. However, a comparison of the SEA residues which correspond to the SEB residues that form the HLA-DR1 binding site, indicates significant conservation of the α–chain binding site in SEA (Jardetzky *et al.*, 1994). Recent mutational data (R.P. Sékaly, personal communication) indicates that this binding site for the DR1 α-chain is functional in SEA, in addition to the C-terminal domain binding site for the MHC β-chain. Both N-terminal and C-terminal domains of SEA may, therefore, interact with MHC in a functionally relevant manner. It remains to be determined if SEA can thereby crosslink adjacent MHC molecules, or whether the two modes of interaction occur with independent MHC molecules. The former possibility may trigger the presenting cell to provide signals important for lymphocyte activation.

Interaction of MHC: superantigen complexes with TcR

The combination of the structure of the DR1:SEB complex with a set of functional data mapping the superantigen interaction with the TcR, allows some insight into the probable form of the ternary complexes of TcR:MHC:SAg. This involves the identification of both superantigen residues and TcR residues which determine the specificity and potency of superantigen stimulation of T cells, but that have no effect on the direct binding of superantigens to MHC molecules. The variability in toxin:MHC complexes that is discussed in the previous section suggests additional variability in ternary complex formation between TcR:MHC:SAg. In addition the TcR binding site on all superantigens may also not be conserved, as residues in TSST-1 implicated in TcR stimulation may lie along a different surface of the superantigen (Kim *et al.*, 1994).

For SEB, a series of mutations have been investigated that affect T cell activation while MHC binding remains unperturbed (Kappler *et al.*, 1992). The residues that are implicated are shown in Figure 6b. These include Asn 23, Asn 60, Tyr 61 and the Cys residues 93 and 113, which form a buried disulfide bond near the MHC binding site. These residues have been used to identify a potential TcR binding site on the SEB molecule, which lies at the interface between the N and C terminal domains of the superantigen fold. It is interesting to consider an additional set of residues which have been shown to be important in determining the specificity of SEA and SEE for different Vβ domains. These residues, Gly 200, Ser 206 and Asn 207, are found in the same region of the superantigen structure as the SEB residues (Figure 6b), suggesting a common or overlapping binding site for TcR (Hudson *et al.*, 1993; Mollick *et al.*, 1993).

Although SEA has a distinct mode of binding to the β–chain of MHC molecules, recent evidence supports the idea that it also interacts with the α–chain in a manner similar to that observed for SEB bound to HLA-DR1. Therefore, the expectation that SEA and SEB may posses a common site for TcR interactions is not unreasonable, in spite of the observed differences in MHC interactions. In the DR1:SEB complex, this potential TcR binding site is found pointing toward the MHC peptide binding site, forming an interface that is above the MHC α–helices (Figure 6b).

This superantigen binding site for TcR most likely recognizes an outer face of the Vβ domain. Initial studies of superantigen stimulate demonstrated an overwhelming correlation of superantigen activation of T cells with the specific allele of Vβ domain. Direct binding of soluble, recombinant β-chain has also been shown for the bacterial superantigens (Gascoigne and Ames, 1991; Irwin *et al.*, 1993). Mapping of Vβ residues which determine the specificity of this interaction has implicated three β-strands of the immunoglobulin domain, including a segment known as the hypervariable region 4 (HV4) that varies between Vβ sequences (Figure 7) (Irwin *et al.*, 1993). The three β-strands, B, D, and E, lie adjacent to each other in the immunoglobulin fold, forming an outer surface of the TcR α/β heterodimer, based on our understanding of antibody structures. This surface of the Vβ domain is consistent with the placement of a hypothetical model of the TcR directly above the MHC peptide binding site, as shown in Figure 7.

The development of a model of the ternary complex formed by TcR:MHC:SAg as shown in Figure 7 should be of predictive value in order to establish the validity of its overall features. In fact, this model helps to reconcile a number of experiments regarding the contributions of both MHC polymorphisms and TcR α-chain variations to superantigen stimulation (Blackman *et al.*, 1993; Smith *et al.*, 1992). In this model, direct contacts between polymorphic MHC residues, for example along the surface of the β–chain α-helix, could interact directly with residues of the TcR. In addition, the positioning of the TcR over the MHC peptide binding site is consistent with potential interactions of the TcR α-chain with the MHC molecule. Perhaps the most compelling evidence for this model comes from a mutational study which has identified DR1 residues that can influence the stimulation of Vβ6 (but not Vβ8.2) carrying T cells and that have no influence on SEB binding to DR1 (Labrecque *et al.*, 1994). These mutations are found along the top surface of the HLA-DR1 β1 α-helix (shown in Figure 7), consistent with the gross features of the ternary complex.

The ternary complex TcR:MHC:SAg is most likely quite distinct from that formed by TcR:MHC:peptide, although some overall low-resolution similarity may be maintained in these complexes. The observations of both the DR1:SEB and the DR1:TSST-1 crystal structures indicate a variability in the binding of superantigens to MHC molecules in ways that must block typical TcR:MHC interactions. However, the prevalence of bacterial superantigens that bind to MHC molecules suggests that these proteins provide an important functional aspect of bacterial evasion of immune mechanisms, in addition to their ability to induce disease. Further structural studies of the interactions of superantigenic proteins with both MHC molecules and TcR may aid in our fundamental understanding of the molecular mechanisms that trigger T cell immune responses.

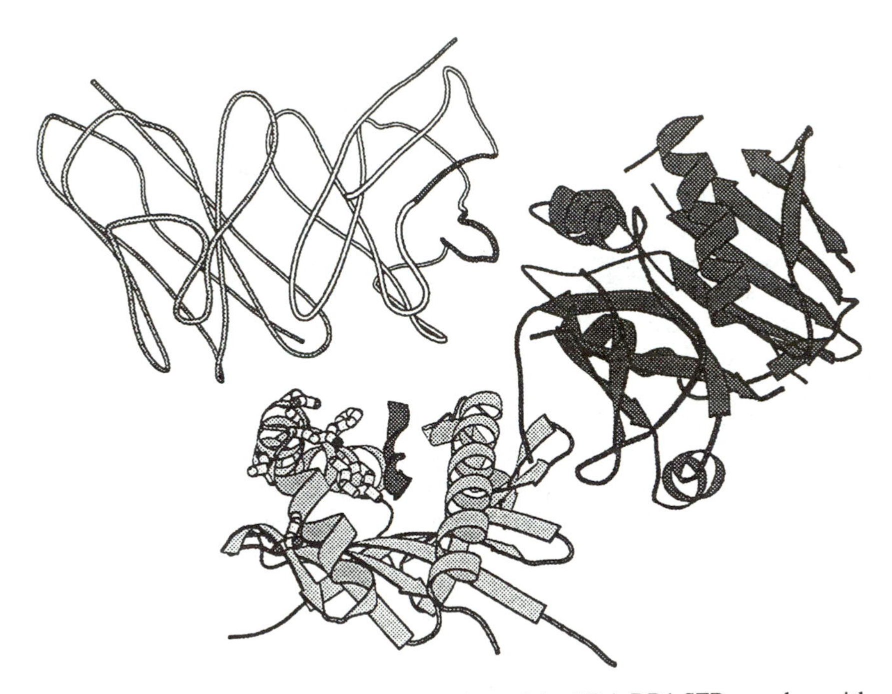

Figure 7: Hypothetical model of the interaction of the HLA-DR1:SEB complex with the variable domains of a TcR. The DR1 $\alpha 1/\beta 1$ peptide binding site is shown in medium gray at the bottom, the superantigen SEB and the peptide are shown in dark gray. MHC residues shown with sidechains influence the SEB directed stimulation of a subset of T cells without changing the SEB:MHC affinity;they may be involved in potential direct interactions with the TcR. The model of the variable α (light gray) and variable β (white) domains of the TcR are based on antibody structures. TcR residues that are implicated in direct interactions with the superantigen are located in the B, D, and E strands of the variable β-chain immunoglobulin fold, and these strands are colored dark gray to highlight their expected position in the TcR heterodimer. Juxtaposition of these Vb strands near the TcR binding site shown in Figure 6b positions the TcR above the MHC binding site, in such a way that direct contacts between the MHC molecule and the TcR could occur, in the region identified by mutational analysis.

References

Acha, O.H. (1993). *Ann. Rheum. Dis.* **52**, S6.

Acha, O.H., Held, W., Waanders, G.A., Shakhov, A.N., Scarpellino, L., Lees, R.K., and MacDonald, H.R. (1993). *Immunol. Rev.*

Acharya, K.R., Passalacqua, E.F., Jones, E.Y., Harlos, K., Stuart, D.I., Brehm, R.D., and Tranter, H.S. (1994). *Nature* **367**, 94.

Adzhubei, A.A., and Sternberg, M.J.E. (1993). *J. Mol. Biol.* **229**, 472.

Allen, P.M. (1994). *Cell* **76**, 593.

Bjorkman, P.J., Saper, M.A., Samraoui, B., and Bennett, W.S. (1987). *Nature* **329**, 512.

Blackman, M.A., Smith, H.P., Le, P., and Woodland, D.L. (1993). *J. Immunol.* **151**, 556.

Brown, J.H., Jardetzky, T., Saper, M.A., and Samraoui. (1988). *Nature* **332**, 845.

Brown, J.H., Jardetzky, T.S., Gorga, J.C., Stern, L.J., Urban, R.G., Strominger, J.L., and Wiley, D.C. (1993). *Nature* **364**, 33.

Buus, S., Sette, A., Colon, S., Jenis, D.M., and Grey, H.M. (1986). *Cell* **47**, 1071.

Cammarota, G., Scheirle, A., Takacs, B., and Doran, D. (1992). *Nature* **356**, 799.

Chicz, R.M., Urban, R.G., Laue, W.S., and Gorga, J.C. (1992). *Nature* **358**, 764.

Chicz, R.M., Urban, R.G., Gorga, J.C., and Vignali, D.A. (1993). *J. Exp. Med.* **178**, 27.

Dellabona, P., Peccoud, J., Kappler, J., Marrack, P., Benoist, C., and Mathis, D. (1990). *Cell* **62**, 1115.

Ehrich, E.W., Devaux, B., Rock, E.P., Jorgensen, J.L., Davis, M.M., and Chien, Y.H. (1993). *J. Exp. Med.* **178**, 713.

Falk, K., Rotzschke, O., Stevanovic, S., Jung, G., and Rammensee, H.G. (1991). *Nature* **351**, 290.

Fraser, J.D., Urban, R.G., Strominger, J.L., and Robinson, H. (1992). *Proc. Natl. Acad. Sci. U.S.A.* **89**, 5507.

Garrett, T.P.J., Saper, M.A., Bjorkman, P.J., and Strominger, J.L.(1989). *Nature* **342**, 692.

Gascoigne, N.R.J., and Ames, K.T. (1991). *Proc. Natl. Acad. Sci. U.S.A.* **88**, 613.

Germain, R.M., and Margulies, D.H. (1993). *Annu. Rev. Immunol.* **11**, 403.

Germain, R.N. (1994). *Cell* **76**, 287.

Hammer, J., Takacs, B., and Sinigaglia, F.(1992). *J. Exp. Med.* **176**, 1007.

Hammer, J., Valsasnini, P., Tolba, K., and Bolin, D. (1993). *Cell* **74**, 197.

Herman, A., Kappler, J.W., Marrack, P., and Pullen, A.M. (1991). *Annu. Rev. Immunol.* **9**, 745.

Herman, A., Labrecque, N., Thibodeau, J., Marrack, P., Kappler, J.W., and Sekaly, R.P. (1991). *Proc. Natl. Acad. Sci. U.S.A.* **88**, 9954.

Hudson, K.R., Robinson, H., and Fraser, J.D. (1993). *J. Exp. Med.* **177**, 175.

Irwin, M.J., Hudson, K.R., Ames, K.T., Fraser, J.D., and Gascoigne, N.R.J. (1993). *Immunol. Rev.*

Janeway, C., Jr., and Bottomly, K. (1994). *Cell* **76**, 275.

Jardetzky, T.S., Gorga, J.C., Busch, R., Rothbard, J., and Strominger. (1990). *Embo. J.* **9**, 1797.

Jardetzky, T.S., Lane, W.S., Robinson, R.A., and Madden, D.R. (1991). *Nature* **353**, 326.

Jardetzky, T.S., Brown, J.H., Gorga, J.C., Stern, L.J., Urban, R.G., Chi, Y.I., Stauffacher, C., Strominger, J.L., and Wiley, D.C. (1994). *Nature* **368**, 711.

Jorgensen, J.L., Reay, P.A., Ehrich, E.W., and Davis, M.M. (1992). *Annu. Rev. Immunol.* **10**, 835.

Kappler, J.W., Herman, A., Clements, J., and Marrack, P. (1992). *J. Exp. Med.* **175**, 387.

Karp, D.R., and Long, E.O. (1992). *J. Exp. Med.* **175**, 415.

Kim, J., Urban, R.G., Strominger, J.L., and Wiley, D.C. (1994). *Science* **266**, 1870.

Kotzin, B.L., Leung, D.M., Kappler, J., and Marrack, P. (1993). *Adv. Immunol.* **54**, 99.

Labrecque, N., Thibodeau, J., Mourad, W., and Sékaly, R.-P. (1994). *J. Exp. Med.* **180**, 1921.

Le Questel, J.Y., Morris, D.G., Maccallum, P.H., Poet, R., and Milner-White, E.J. (1993). *J. Mol. Biol.* **231**, 888.

Madden, D.R., Garboczi, D.N., and Wiley, D.C. (1993). *Cell* **75**, 693.

Miethke, T., Wahl, C., Heeg, K., Echtenacher, B., Krammer, P.H., and Wagner, H. (1992). *J. Exp. Med.* **175**, 91.

Mollick, J.A., McMasters, R.L., Grossman, D., and Rich, R.R. (1993). *J. Exp. Med.* **177**, 283.

Murzin, A.G. (1992). *Nature* **360**, 635.

Murzin, A.G., and Chothia, C. (1992). *Curr. Opin. Struct. Biol.* **2**, 895.

Nossal, G.J. (1994). *Cell* **76**, 229.

Prasad, G.S., Earhart, C.A., Murray, D.L., Novick, R.P., Schlievert, P.M., and Ohlendorf, D., H. (1993). *Biochemistry* 13761.

Saper, M.A., Bjorkman, P.J., and Wiley, D.C. (1991). *J. Mol. Biol.* **219**, 277.

Sette, A., Sidney, J., Oseroff, C., del Guercio, M.F., Southwood, S., Arrhenius, T., Powell, M.F., Colon, S.M., Gaeta, F.C.A., and Grey, H.M. (1993). *J. Immunol.* **151**, 3163.

Smith, H.P., Le, P., Woodland, D.L., and Blackman, M.A. (1992). *J. Immunol.* **149**, 887.

Stern, L.J., Brown, J.H., Jardetzky, T.S., Gorga, J.C., Urban, R.G., Strominger, J.L., and Wiley, D.C. (1994). *Nature* **368**, 215.

Swaminathan, S., Furey, W., Pletcher, J., and Sax, M. (1992). *Nature* **359**, 801.

Thibodeau, J., Cloutier, I., Labreque, N., Mourad, W., Jardetzky, T., and Sékaly, R.-P. (1994). *Science* **266**, 1874.

Thibodeau, J., Labreque, N., Denis, F., Huber, B., and Sékaly, R.-P. (1994). *J. Exp. Med.* **179**, 1029.

Van Bleek, G.M., and Nathenson, S.G. (1990). *Nature* **348**, 213.

von Boehmer, H. (1994). *Cell* **76**, 219.

17

Chorismate Mutase, Essentially a Template Enzyme

W.N. Lipscomb, Y.M. Chook and H. Ke

Department of Chemistry
Harvard University
Cambridge, MA 02138 USA

It was a particular pleasure for the senior author to open the scientific sessions which celebrated Oleg Jardetzky's 65th birthday anniversary. Oleg was a student in my Physical Chemistry course at the University of Minnesota in 1950, and received an M.D. from the Medical School in 1954. At that time he came to my office to ask for a Ph.D. research problem in statistical mechanics of membrane processes, and I suggested that he study instead the NMR quadrupole line broadening of Na^+ in solutions of small biologically interesting molecules as they interact with Na^+. This research, with John Wertz, is surely an early study of biologically interesting problems using NMR, and I consider it a privilege to have helped to start Oleg on his outstanding career in this area of science.

Chorismate mutase

The enzyme chorismate mutase from *Bacillus subtilis* forms the topic of this chapter, and the method is single crystal X-ray diffraction. It is a very recent study and highlights some questions. Only 127 amino acids are present in the polypeptide chain, and the structure would be a candidate for NMR pulse methods, except that the molecule in the solution and in the crystal is trimeric. (We were told by the biochemists that it was dimeric!) Non-crystallographic symmetry (based on vector distances) was helpful in solving the structure, and I

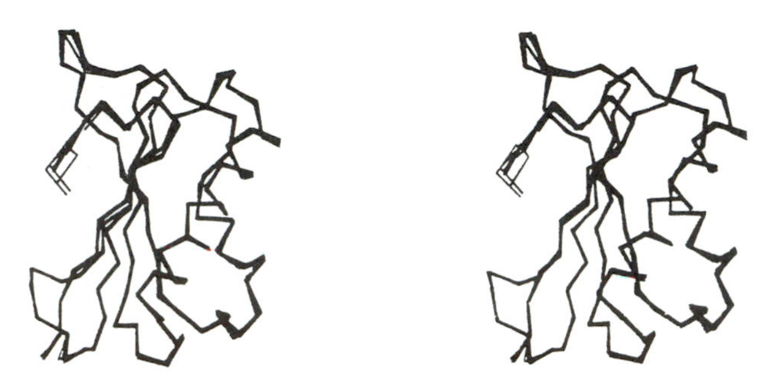

Figure 1: Superposition of the 12 monomers of chorismate mutase from *Bacillus subtilis.*

wonder if there is some partly equivalent use of molecular symmetry (based on scalar distances) that would simplify the analysis of the NMR spectrum of the trimer in solution.

The second question is how nearly the same are the monomers in the crystallographic unit. In the crystal there are actually 12 monomers (*i.e.*, 4 trimers) in the asymmetric unit of the crystal. When all 12 are superimposed the polypeptide chains are very similar indeed (Figure 1). No doubt the trimerization reduces the distortions of structure below those expected for isolated monomers.

The third question relates to the chemical mechanism by which chorismate mutase isomerizes chorismate to prephrenate in a pericyclic reaction, the only pericyclic reaction that is known to be catalyzed by an enzyme. Four plausible mechanisms are shown in Figure 2, and we shall see that this structural study indicates that Mechanisms 3 and 4 are less probable, while Mechanisms 1 and 2 remain likely.

Chorismate mutase initiates the phenylalanine, tyrosine branch of the biosynthetic pathway for aromatic amino acids (Weiss and Edwards, 1980) (Figure 3). The rate enhancement by a factor of 2×10^6 over the rate of the uncatalyzed reaction (Andrews *et al.*, 1973) has been a model for catalysis of a unimolecular isomerization in which no covalent intermediates occur during the catalyzed reaction. The activation enthalpy ΔH^* is 21 kcal/mole and ΔS^* is -13 eu for the uncatalyzed reaction (Andrews *et al.*, 1973). For a rate enhancement of 2×10^6 the corresponding change in ΔG^* value is 9 kcal/mole. If ΔS^* is the same for both the catalyzed and uncatalyzed reactions, ΔH^* is reduced by the enzyme to 12 kcal/mole, whereas if ΔS^* is zero for the catalyzed reaction ΔH^* is reduced by 5 kcal/mole.

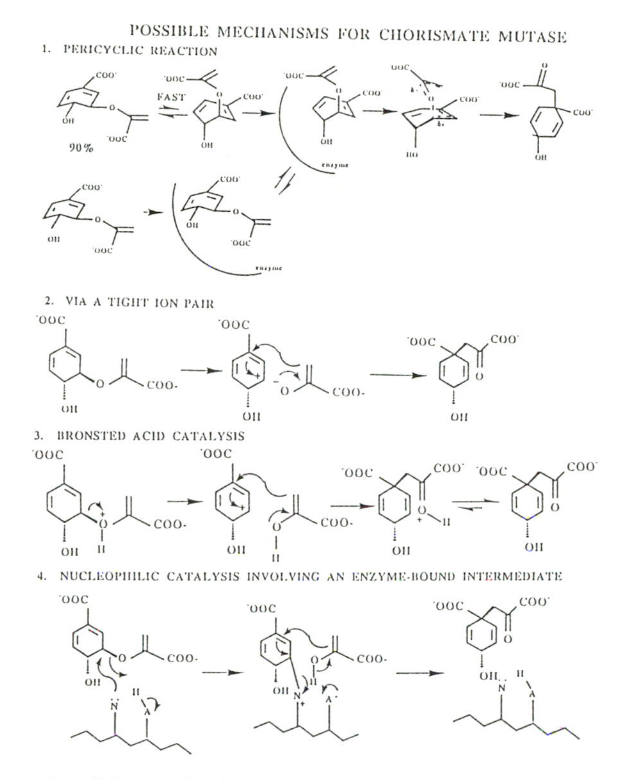

Figure 2: Four plausible mechanisms for enzymatic conversion of chorismate to prephrenate.

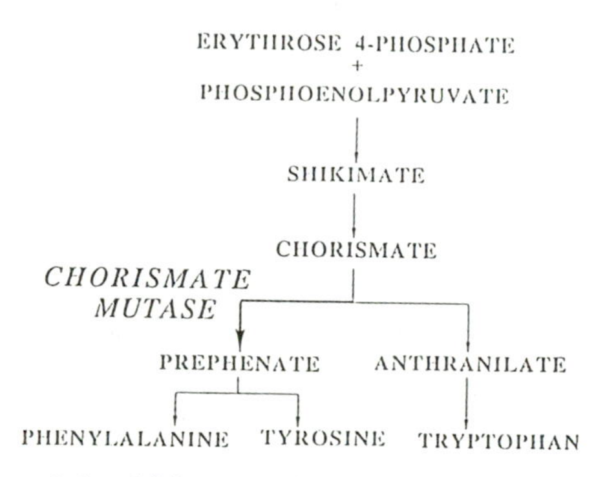

Figure 3: Diagram of the shikimate pathway.

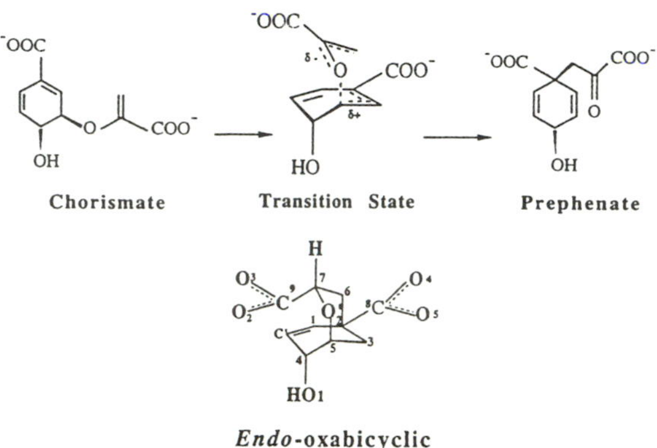

Figure 4: The reaction pathway showing the putative transition state and the endo-oxabicyclic transition state analogue.

An endo-oxabicyclic inhibitor (Bartlett and Johnson, 1985) (K_d = 3μM) which resembles the putative transition state (Bartlett and Johnson, 1985; Sogo *et al.*, 1984; Copley and Knowles, 1985) (Figure 4) was supplied to us by P. A. Bartlett (Bartlett and Johnson, 1985). The structure of the complex of this inhibitor with the enzyme is described below.

The non-enzymatic transformation of chorismate to prephrenate is concerted and asynchronous, and probably occurs via a distorted chair-like transition state (Copley and Knowles, 1985). Most probably the enzyme-catalyzed rearrangement proceeds through a similar transition state (Sogo *et al.*, 1984; Copley and Knowles, 1985; Gray *et al.*, 1990).

Antibodies raised against Bartlett's endo-oxabicyclic inhibitor have been examined for catalytic activity. The antibody 11F1–2E11 (Jackson *et al.*, 1988) shows a ratio k_{cat}/k_{uncat} of 2 x 10^4, whereas the 1F7 (Bowdish, *et al.*, 1991; Hilvert *et al.*, 1988; Haynes *et al.*, 1994) antibody shows a ratio of 190 at 25°C. These values are to be compared to the ratio of 2 x 10^6 shown by the *B. subtilis* chorismate mutase (Gray *et al.*, 1990).

Our X-ray diffraction study was made on the chorismate mutase from the Marburg strain of Bacillus subtilis, which has 127 amino acid residues per polypeptide chain (Lorence and Nester, 1967; Gray *et al.*, 1990). Unlike other chorismate mutases (Koch *et al.*, 1971; Schmidheini *et al.*, 1990), this enzyme is monofunctional, non-allosteric, unaffected by amino acids in the pathway, and characterized by Michaelis-Menten kinetics.

The structure

Crystals are monoclinic in the space group $P2_1$. The unit cell has parameters a = 102.4 Å, b = 68.3 Å, c = 102.8 Å and β = 105.6°. Attempts to solve the structure using heavy atom derivatives gave poor results owing to the unusual number of 12 monomers in the asymmetric unit and due to changes in cell dimensions upon treatment of the crystals with heavy atom reagents. The poor phases that were initially obtained were then used to solve the 72 selenium sites in a selenomethionine substituted biosynthetic derivative of the enzyme. The final model of the native enzyme consisted of 1380 residues and 522 water molecules, and yielded a crystallographic R value of 0.19 for the 71,847 structure factors in the range of 8.0 Å to 1.9 Å. In addition, the inhibitor-enzyme and the prephrenate-enzyme complexes were solved to 2.2 Å resolution. Further details are given in a preliminary note (Chook *et al.*, 1993), a full paper (Chook *et al.*, 1994), and an earlier communication to remain unpublished in *Nature*.

The structure of the monomer shows a single domain containing five β-strands, an 18 residue α-helix and a 6 residue 3_{10} helix (Figure 5). Owing to lack of electron density, the N-terminal Met and the C-terminal residues 118–127 are omitted. Originally proposed as a homodimer (Gray *et al.*, 1990), the electron density reveals that the molecule is actually a homotrimer (Figure 6) consistent with a recent gel filtration study (Rajagopalan *et al.*, 1993). The trimer resembles a pseudo-αβ-barrel which has an approximate non-crystallographic threefold axis. Near the surface of the trimer, the interfaces

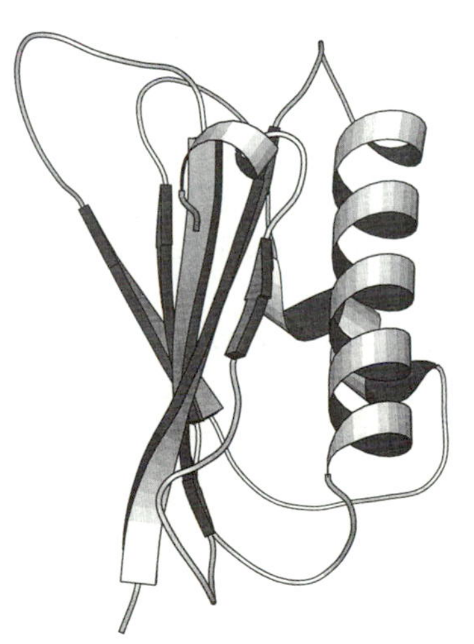

Figure 5: The monomer of *Bacillus subtilis* chorismate mutase.

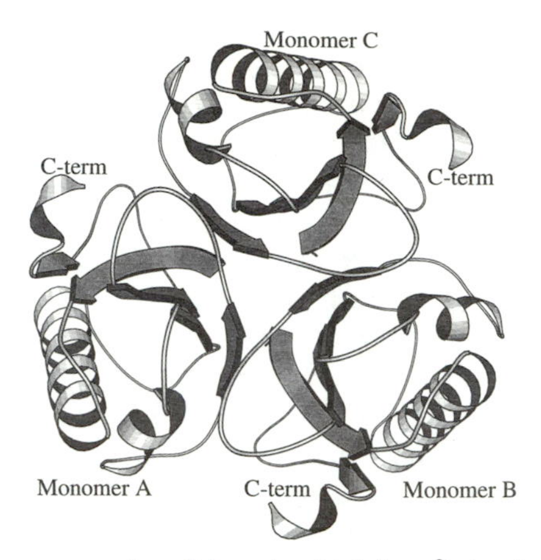

Figure 6: The trimer as a pseudo-αβ-barrel, which has β structure inside and helical structures on the outside.

between monomers form three equivalent cone-like clefts. The walls of the cleft contain residues from the 3_{10} helix (59, 60, 62 and 63), a loop (65–72) and β-strand III (73–77) from one subunit; and, from the adjacent subunit, β-strand V (107–109), β-strand I (2–11), β-strand IV (88–97) and the C-terminal tail (111–115). The cleft is capped by a loop (78–87) from this adjacent subunit.

This cleft contains the active site, which is shared by adjacent monomers. Binding occurs at this site for either the endo-oxabicyclic inhibitor (Figures 7 and 8), which has a K_i of 3 μM (Haynes *et al.*, 1994), or prephrenate (Figure 9). Very little change is observed in the conformation of the enzyme as either of these two ligands bind.

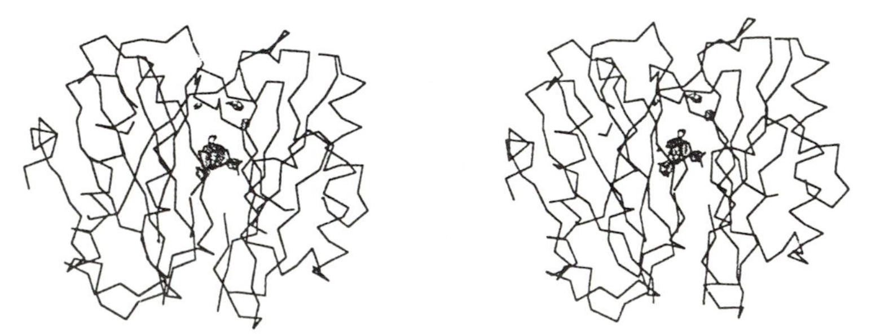

Figure 7: Stereoview of the electron density of the inhibitor in the cleft between adjacent monomers of the trimer.

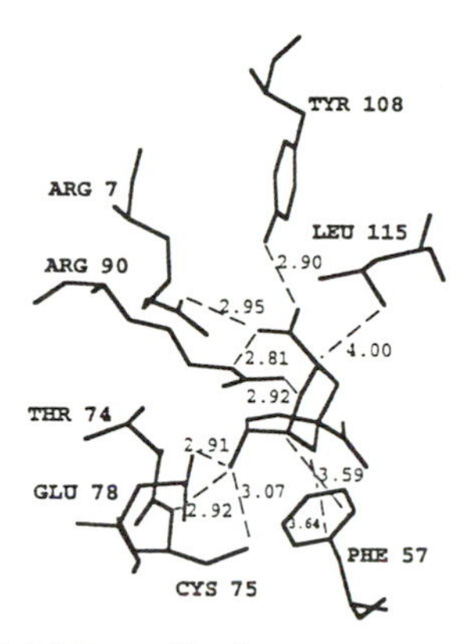

Figure 8: Interactions of the inhibitor with the enzyme.

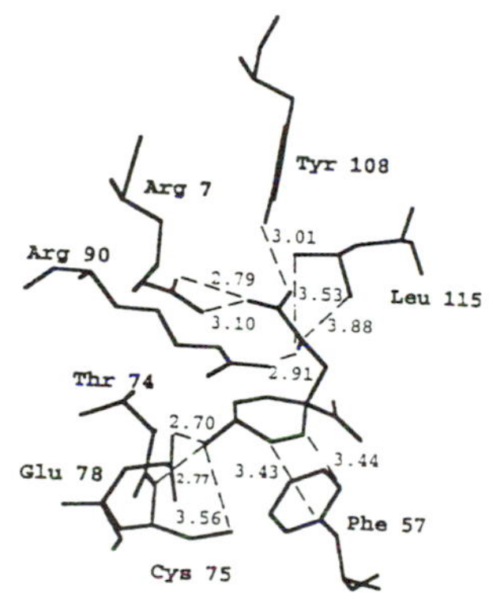

Figure 9: Selected interactions of prephrenate with the enzyme.

Relation of structure to mechanism

There are no functional groups on the enzyme that can transfer a proton to the ether oxygen O_2' of the inhibitor (Figure 4). This result is consistent with the flat pH profile of k_{cat}/K_m between pH 4 and pH 9, and the lack of solvent isotope effect for this *B. subtilis* enzyme (Gray *et al.*, 1990). Neither are there any nucleophiles near the C_5 atom of the C_5–O_2' scissile bond. Thus, the structures of the enzyme and these two complexes support a pericyclic process similar to that of the uncatalyzed reaction. However, the transition state of the uncatalyzed reaction is polar, as indicated by the dependence of rate on solvent polarity (Copley and Knowles, 1987). The active site shows favorable interactions that may promote this polar transition state. Thus, the guanidium group of Arg 90, 2.9 Å from the bridging oxygen O_2' (Figure 8) may stabilize a developing negative charge on O_2'. To a lesser extent the developing positive

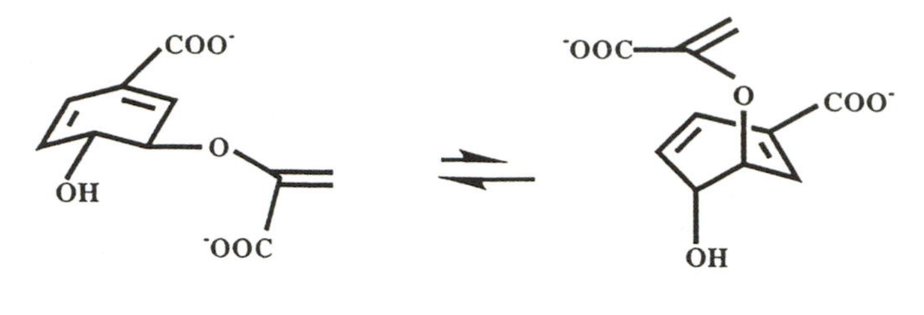

pseudo-diequatorial pseudo-diaxial

Figure 10: Extremes of the flexible conformation of chorismate in solution where 90% is in the pseudo-equatorial conformer, and 10% is in the pseudo-diaxial conformer which is preferred by the enzyme.

charge on C_5 could be stabilized by the π-electron system of Phe 57 (at 3.6 Å) and by a carboxylate oxygen of Glu 78 (at 4.2Å). Indeed, these influences would also stabilize the extreme polar state of a tight ion pair, O_2–...C_5+ (Figure 2a). Thus, distinction between Mechanisms 1 and 2 of Figure 2a is difficult to prove at present.

In all, there are some 25 contacts less than 4 Å, including seven hydrogen bonds, between the inhibitor and the enzyme. About 95% of the surface area of the inhibitor or product (prephrenate) is buried in these two complexes. One aspect of this rather complete interaction is that the enzyme selects the pseudo-diaxial conformation from the 10:90% equilibrium in solution of the pseudo-diaxial and pseudo-equatorial isomers (Figure 10). A second aspect to this near-complete coverage of ligands is the restriction of the leaving of prephrenate by the side chains of Phe 57 and particularly of Leu 115 which is disordered in the unligated enzyme, whereas it is ordered near O_2 and C_7 of prephrenate and O_2, C_7 and C_9 of the inhibitor in the enzyme complexes. This "cap" over the active site is strikingly revealed by the temperature (disorder) factors in the refinement of the X-ray data.

Thus, these results provide a structural basis for stabilization of developing charges in the putative transition state of the reaction, and for the biochemical results on rate limiting steps: at low concentration of substrate the rate limiting transition state occurs before the transition state of the chemical steps (Addadi *et al.*, 1983), whereas under saturating conditions the dissociation of prephrenate is rate limiting (Gray *et al.*, 1990).

Acknowledgments

We thank J. R. Knowles for his initiation of this study, P. A. Bartlett for a gift of the inhibitor, N. h. Xuong for the data collection facility, the Pittsburgh Supercomputing Center (Grant CMB900069P), M. K. Rosen and N. S. Sampson for help in the Se–Met protein preparation, and the National Institutes of Health GM06920 for support.

References

Addadi, L., Jaffe, E.K., and Knowles, J.R. (1983). *Biochemistry* **22**, 4494.

Andrews, P.R., Smith, G.D., and Young, I.G. (1973). *Biochemistry* **18**, 3492.

Bartlett, P.A., and Johnson, C.R. (1985). *J. Am. Chem. Soc.* **107**, 7792.

Bowdish, K., Tang, Y., Hicks, J.B., and Hilvert, D.J. (1991). *Biol. Chem.* **266**, 11901.

Chook, Y.M., Ke, H., and Lipscomb, W.N. (1993). *Proc. Natl. Acad. Sci. U.S.A.* **90**, 8600.

Chook, Y.M., Gray, J.V., Ke, H., and Lipscomb, W.N. (1994). *J. Mol. Biol.* **240**, 475.

Copley, S.D., and Knowles, J.R. (1985). *J. Am. Chem. Soc.* **107**, 5306.

Copley, S.D., and Knowles, J.R. (1987). *J. Am. Chem. Soc.* **109**, 5008.

Gray, J.V. , Eren, D., and Knowles, J.R. (1990). *Biochemistry* **29**, 8872.

Gray, J.V., Golinelli-Pimpaneau, B., and Knowles, J.R. (1990). *Biochemistry* **29**, 376.

Haynes, M.R., Stura, E.A., Hilvert, D., and Wilson, I.A. (1994). *Science* **263**, 646.

Hilvert, D., Carpenter, S.H., Nared, K.D., and Auditor, M.T.M. (1988). *Proc. Natl. Acad. Sci. U.S.A.* **85**, 4953.

Jackson, D.Y., Jacobs, J.W., Sugaswara, R., Reich S.H., Bartlett, P.A., and Schultz, P.G. (1988). *J. Am. Chem. Soc.* **110**, 4841.

Koch, G.L.E., Shaw, D.C., and Gibson, F. (1971). *Biochim. Biophys. Acta* **229**, 795.

Lorence, J.H., and Nester, E.W. (1967). *Biochemistry* **6**, 1541.

Rajagopalan, J.S., Taylor, K.M., and Jaffe, E.K. (1993). *Biochemistry* **32**, 3965.

Schmidheini, T., Mosch, H.-U., Evans, J.N.S., and Braus, G. (1990). *Biochemistry* **29**, 3660.

Sogo, S.G., Widlanski, T.S., Hoare, J.H., Grimshaw, C.E., Berchtold, G.A., and Knowles, J.R. (1984). *J. Am. Chem. Soc.* **106**, 2701.

Weiss, U., and Edwards, J.M. (1980). *The Biosynthesis of Aromatic Amino Compounds* (Wiley, New York) 134.

Section 3: Nucleic Acids

18

Computing the Structure of Large Complexes: Modeling the 16S Ribosomal RNA

R. Chen, D. Fink, and R.B. Altman

Section on Medical Informatics
Stanford University
Stanford, CA 94305 USA

Ribosomes are the sites of messenger RNA (mRNA) translation to protein, and thus are crucial to the normal functioning of all cells. These ribonucleoprotein particles are composed of a small (30S) subunit and a large (50S) subunit. The 30S subunit, in turn, is composed of a strand of RNA (16S rRNA) and 21 proteins ranging in molecular weight from 9 kD to 61 kD. Studies have demonstrated that ribosomal RNA is necessary for normal ribosome function and protein production (Dahlberg, 1989; Noller, 1991). In particular, 16S rRNA is essential for normal assembly and function of the 30S subunit, which is responsible for translation initiation (Hardesty and Kramer, 1985). Elucidating the structure of 16S rRNA could greatly aid our understanding of the molecular mechanisms for protein translation, and such basic structural information could ultimately have wide-ranging importance in fields such as pharmacology and drug design.

Because of the difficulties associated with X-ray analysis of large complexes such as the ribosome (Eisenstein *et al.*, 1991), high-resolution structural data for the 16S rRNA remain sparse. However, neutron diffraction studies have determined the relative positions of the 30S proteins (Capel *et al.*, 1988), which, along with the reported 16S rRNA–protein interactions (Noller, 1991, Noller *et al.*, unpublished; Brimacombe, 1991), enable low-resolution structural models—showing how the RNA associates with the protein components—to be built. Several studies have sought to take advantage of these structural data for the 30S subunit. Stern *et al.* have used interactive model building to produce a three-dimensional 16S rRNA structure (Stern *et al.*, 1988). This method can produce viable models, but is hindered somewhat by subjectivity intrinsic to the process and by the nonexhaustive nature of its conformation search. Hubbard and Hearst have used distance geometry techniques to model the RNA structure, but did not incorporate neutron diffraction data on the protein positions (Hubbard and Hearst, 1991). Malhotra and Harvey have used an energy minimization technique to produce a set of possible conformations for 16S rRNA; their study, however, depends on electron microscopic studies on the molecule to provide initial information on surface topology (Malhotra, 1994). Our goal in this work is to use a new set of labeling data, and report on the family of conformations compatible with this data.

Constraint satisfaction — PROTEAN

In our study, we use a constraint satisfaction program called PROTEAN, a program originally created to calculate protein structure from constraints derived from nuclear magnetic resonance (NMR) data, and developed by the groups of Oleg Jardetzky and Bruce Buchanan at Stanford (Altman and Jardetzky, 1989; Brinkley *et al.*, 1988). PROTEAN uses a simplified representation of the 16S rRNA secondary structure, in which helices are modeled as cylindrical objects and proteins as spherical objects (Altman and Jardetzky, 1989). From the perspective of constraint satisfaction, these objects represent nodes in a network. The labels for these nodes are the set of discrete, valid positions for the objects (Mackworth, 1977). Arcs between nodes represent constraints between objects within the 16S rRNA model. These arcs can be either tether constraints between helices representing the range of distances compatible with the single-stranded RNA connecting them, or proximity constraints between objects representing the range of distances compatible with the experimental protection data detailing specific RNA-protein interactions.

An initially large number of labels (list of possible locations and orientations) for each node in the constraint network is reduced by an iterative process involving satisfaction of these arcs. First-order consistency is achieved when all constraints involving one node are satisfied, thus reducing the number of legal positions for the objects involved in these constraints. Similarly, second-order consistency involves satisfaction of all constraints involving two nodes. More generally, nth-order consistency refers to the network when all constraints involving n nodes or less are satisfied. When n node consistency is achieved, the number of locations for each object has been pared down to a number many orders of magnitude smaller.

Achieving first-order consistency

To start, an object (or set of objects with fixed relative positions) in the network is designated the anchor. The positions of all other objects, anchorees, are defined in the anchor's coordinate system. To achieve first-order consistency, all constraints between the anchor and the anchorees are satisfied. PROTEAN exhaustively samples (in a systematic grid search) all locations and orientations for each object, to find ones that satisfy the constraints. This list of valid positions and orientations for each object represents a starting point for further "pruning" using higher-order consistency checks, which will be discussed in the next section.

In the case that an object has no direct constraints to an anchor, the object is first anchored to an anchoree that does have direct constraints to the anchor, and then appended to the original anchor. This is done by taking the cross-product between all locations of the object to the anchoree and of the anchoree to the anchor.

Achieving higher order consistency

High-order consistency checking involves further reducing location lists by simultaneously satisfying all constraints between n nodes, where n ranges from two to the total number of objects in the structure. This is called an n-yoke operation. To do a two-yoke operation, for instance, all possible pairs of locations in the location lists for two objects are considered at a time. If a location for one object cannot satisfy all constraints between the two objects (for any of the possible locations of the second object), then that location is eliminated. Iterative application of this two-yoke operation to all pairs of objects eventually results in second-order consistency.

Similarly, higher-order consistency can be achieved by taking groups of three or more objects, and eliminating locations that are not compatible with at least one location for all other objects. The process results in minimal location lists for each object, where every location in each of the lists is guaranteed to be part of at least one structure that satisfies all the constraints simultaneously. Such a structure is called a coherent instance.

One major problem, however, is that the computational complexity of the yoking operation grows exponentially with the number of objects being yoked simultaneously. The operation is O(LN), where L is the average size of the location lists and N is the number of objects being yoked simultaneously.

To make the higher-order yoking computationally practical, we employ several techniques. These involve the careful ordering of operations performed, the definition of relatively independent subnetworks in the constraint graph, and the intelligent sampling of location lists. The goal is to reduce the size of the location lists while still preserving their representative nature, and with minimal loss of information.

We have applied the PROTEAN methodology in collaboration with Harry Noller and Bryn Weiser of U.C. Santa Cruz, to interpret their data on the structure of the 16S rRNA. Specifically, we are testing the hypothesis that certain key structural features of the rRNA are defined entirely by the hydroxyl radical protection data.

Methods

The calculation of rRNA structure involves three tasks: representing the structural components and defining the distance constraints between them, determining a strategy for assembling the complex, and finding ways to reduce the computational complexity of the problem.

Representing structural components

In calculating the structure of the 16S ribosomal RNA, we drew upon three sources of information: the 16S rRNA secondary structure, the relative locations of the protein components, and the hydroxyl radical protection data that provides information about protein-RNA proximities.

16S rRNA secondary structure

The 16S rRNA secondary structure has been reliably predicted in several studies (Woese *et al.*, 1983; Gutell and Woese, 1990). It consists of 52 helical regions and single-stranded regions interconnecting the helical regions (Figure 1). Predictions of helix formation are based on standard RNA base-pairing (A-U, C-G) and also include wobble base-pairs (G-U) (Gutell and Woese, 1990). We assume that all base-paired regions organize to produce A-form helices.

In modeling the secondary structure, helices are represented as rigid, cylindrical objects of 7 Å radius (for the purposes of volume overlap checking). Helices with small internal bulges are considered one rigid cylindrical object. Single-stranded loop regions at the ends of helices are not modeled explicitly. Single-stranded regions of RNA that connect two helices are represented as tether constraints between two objects, with the upper bound on this tether constraint being the length of this region when the single-stranded RNA is fully extended.

Protein locations within the 30S subunit

There are 21 protein components within the 30S subunit, named S1, S2, ..., S21. Neutron diffraction experiments provide the three-dimensional location of

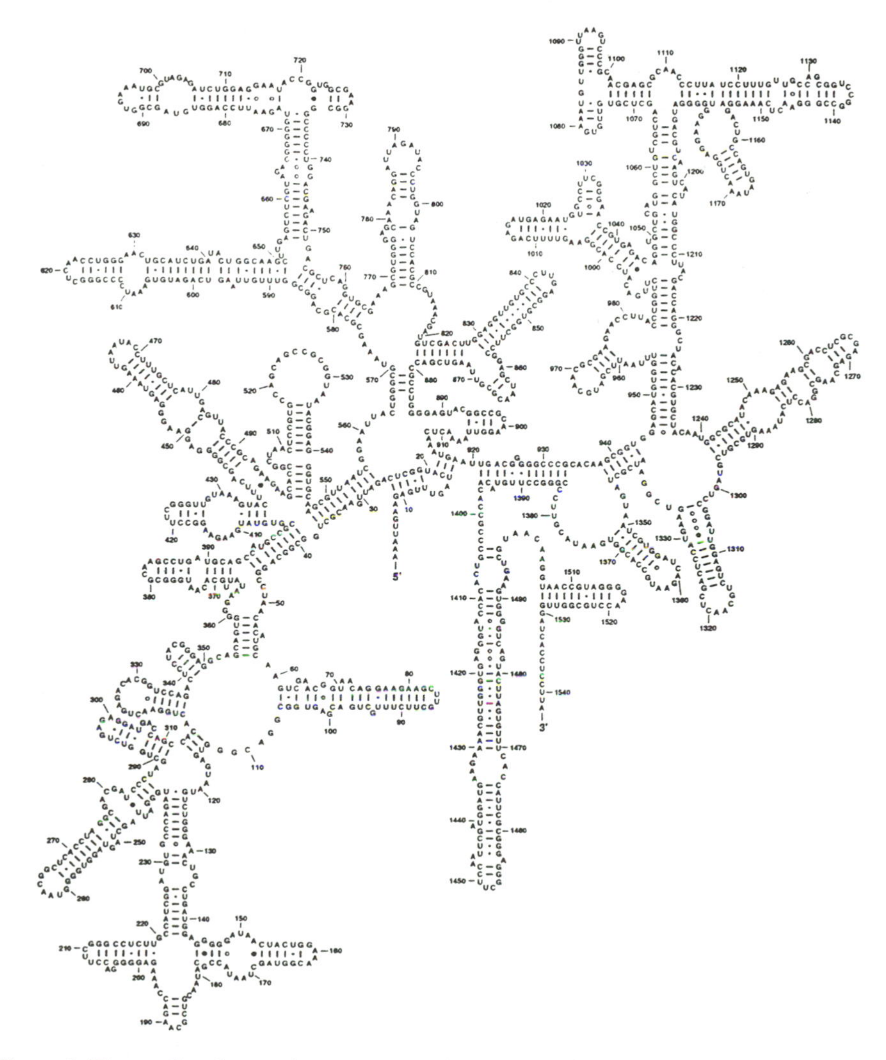

Figure 1: The predicted secondary structure of *E. Coli* 16 S ribosomal RNA (Gutell and Woese, 1990). The secondary structure is preserved across hundreds of procaryotic rRNA sequences.

these protein components. Specifically, the center of mass for each protein and the uncertainty associated with each of these measurements has been reported

(Capel *et al.*, 1987, 1988; Moore, *et al.*, 1985). All of these proteins have been shown to interact with the RNA except for S1 (Noller *et al.*, unpublished.). The exact structures of the protein components have not been determined, but based on their amino acid sequence, their minimal anhydrous radii can be calculated to range from 13.8 to 16.6 Å (Altman, *et al.*, 1994). In modeling, we use spheres based on the anhydrous radii to represent the 21 proteins. To represent the uncertainty in the positions of these proteins, we define error ellipsoids that are centered at the mean positions of the proteins and whose axes represent the error in location associated with each coordinate direction, as reported in the literature (Altman *et al.*, 1994). The standard deviation of the mean location ranges from 6Å to 17 Å (Altman, *et al.*, 1994). The volume within the error ellipsoid thus represents all possible positions for the protein center of mass (Figure 2). These proteins provide a starting framework upon which we can overlay the 16S rRNA structure. Specifically, proteins define a coordinate space within which helical objects can subsequently be positioned (anchored).

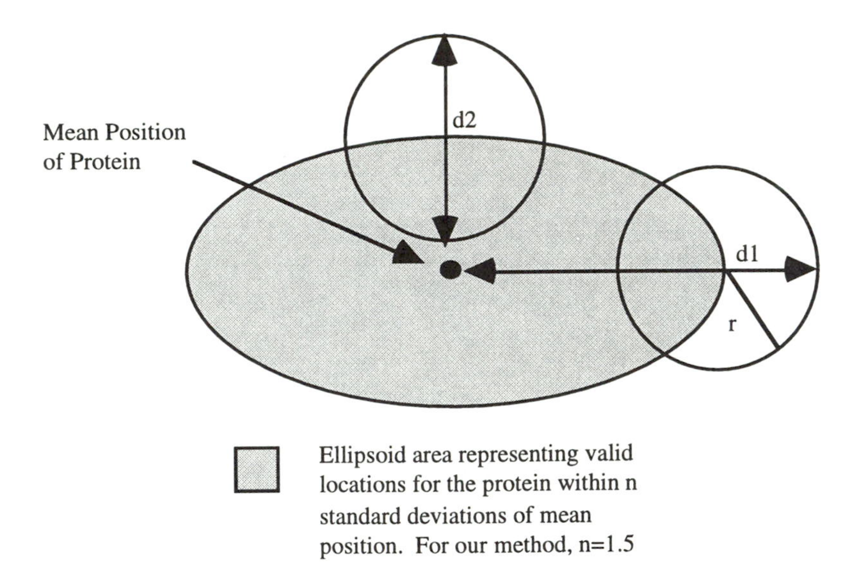

Figure 2: Checking constraints between a base and the center of mass of the protecting protein. Shown are the mean position of the protein and its ellipsoid of error for its location. The protein's anhydrous radius (r) is also shown. (d1) and (d2) refer the to largest possible distance away from the mean position that the protein edge can be. Notice that (d2) and (d1) are different because they represent deviations in mean position in different directions. This is referred to as a directional constraint.

16S rRNA-protein interactions

Hydroxyl radical footprinting experiments provide data detailing RNA-protein interactions in 16S rRNA (Noller *et al.*, unpubl.). The data set we have used in our computations is significantly different from those published previously (Stern *et al.*, 1989), and will be published separately (Noller *et al.*, unpubl.). The footprinting experiments involve binding specific proteins to the RNA and subsequently exposing the complex to hydroxyl radicals. Bases that are bound to the protein are protected from hydrolysis, thus detailing proximity relationships between RNA and protein. These interactions are graded weak, medium, or strong. Bases can be protected because of direct contact with the protein, or because of conformation-dependent cooperative protection. In order to ensure that only direct contacts are used, we use only strong protections as distance constraints.

In modeling protection data, representing the position of a protected base relative to a protein brings up several issues. First, to represent the uncertainty associated with the exact center of mass of the protein (as determined by neutron diffraction experiments), we have defined an error ellipsoid whose dimensions are 1.5 times the standard deviations in each of the three axial directions. Second, because the exact structure of each of the proteins is unknown, we have represented the proteins as spheres with their minimal anhydrous radius. Since proteins of known structure can range from 0.7 to 1.5 times their predicted spherical anhydrous radius, our use of this radius is a compromise (Altman *et al.*, 1994). To calculate the maximum possible distance of a protected base to the protecting protein (as determined from the labeling experiments), we use the equation (Altman *et al.*, 1994):

$$\mathrm{Dmax}(z,N,F) = (N \times SD) + (F \times R_{anhydrous})$$

where

z = vector representing direction from center of mass to base
N = number of standard deviations from mean center of mass allowed as valid protein positions
F = multiplier of the anhydrous radius that captures the eccentricity of the protein.
$R_{anhydrous}$ = anhydrous radius

In our calculations, we used N = 1.5 and F = 1.0 for most helices (Figure 2). We consider this to be conservative, since only if a protein is very eccentric, and is at the edge of its legal excursion, would legal positions be missed. For a few helices, we increase the value of the parameter N up to 2.5, in order to ensure that each helix starts with a reasonably large set of initial locations.

Assembling the complex

The constraint network for the 16S rRNA is given in Figure 3. Using the PROTEAN constraint satisfaction program, we processed the constraints in this network to produce a model of the tertiary structure 16S rRNA in a multi-step process.

First, we partitioned the 16S rRNA secondary structure into six roughly equal domains—the central domain and domains 1 through 5 (Figure 3). These domains were modeled individually first, and then reunited (with the exception of domain 5 which was excluded because it was severely underconstrained). There is enormous computational advantage in this segmental approach to computing the 16S rRNA structure. By satisfying the constraint network for each domain first, we can significantly reduce the number of valid locations in the search space for each object, making higher-order constraint satisfaction (involving all objects) computationally feasible. High-order consistency checks are exponential in the number of objects included. By checking high-order consistency in five domains with 10 objects each, as opposed to one domain with 50 objects, the exponent is reduced. Because the domains group objects according to physical proximity, there is little need to check for consistency between two objects on opposite sides of the structure, and so it suffices to check high order consistency in local neighborhoods, and not globally.

Second, we anchored all the helices in each domain to the fixed proteins with a sample interval of 7 Å in position and 45° in orientation. Helices H2, H14, H32, H34, and H35 required finer sampling intervals to yield valid locations. This produced an initial list of valid locations for each helical object in the search space with, on average, 10^3 to 10^4 locations per object. For objects with no direct constraints to proteins, we appended those objects to the anchor's coordinate system by the process described in the introduction. This first step achieved first-order consistency among objects in each domain.

Next, the two-yoke operation was performed iteratively until the locations lists were maximally reduced. The two-yoke achieved second-order consistency among objects in the domain by considering pairs of objects that have constraints between them and removing locations from the lists of both objects that fail to satisfy the constraint.

The fourth step involved performing higher-order consistency checks between objects in the domain until n-order consistency was achieved (n is the total number of objects in the domain). This step produced location lists for each object whose members are guaranteed to be part of at least one structure that satisfies all the constraints. Figure 4 shows how higher-order consistency checks can greatly reduced the number of valid locations for an object. Underconstrained objects, which characteristically have large location lists ($>10^3$), add considerably to the computational complexity of these higher-order

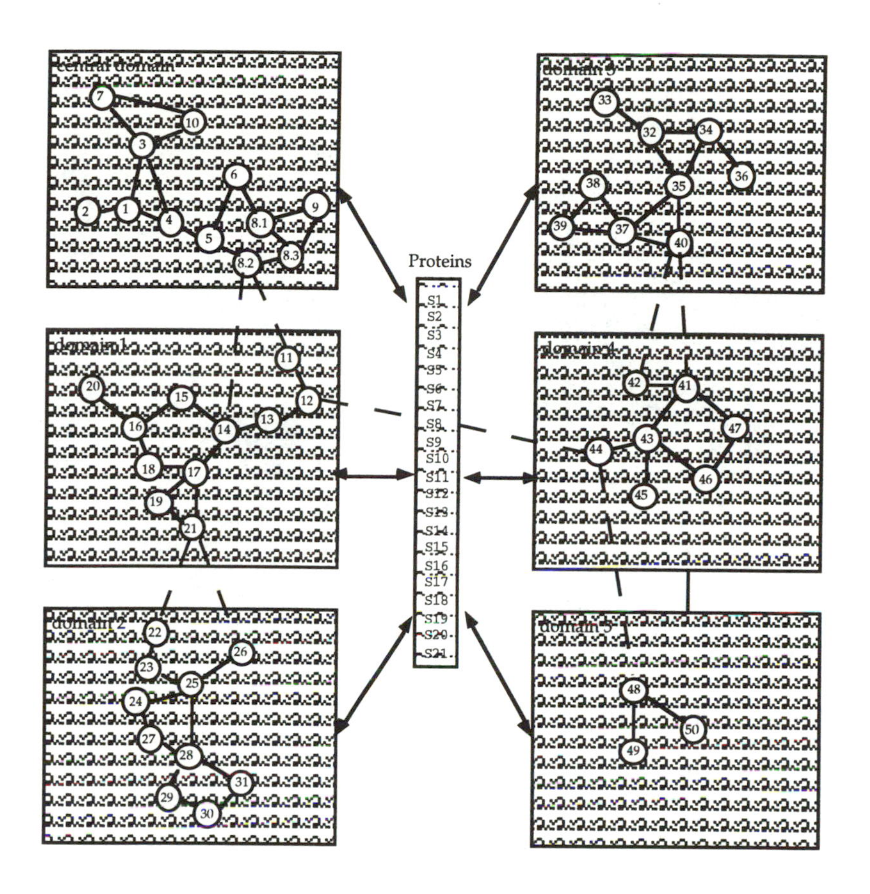

Figure 3: Constraint network for the 16S ribosomal RNA. Circular nodes represent the double-helical regions of the ribosomal secondary structure. We modeled the 52 helices in 16S rRNA. The rectangular node in the center represents the 21 proteins whose positions are fixed in space. Solid lines indicate nodes that have constraints between them. Dotted lines indicate nodes from different domains that have constraints between them. Arrows indicate constraints between domains and proteins. Note that the three helices in domain 5 are not included in our structure because they are severely underconstrained.

consistency checks and can make these computations impractically slow. Domain 5, in fact, was so underconstrained that we chose not to model the three helices which belonging to that domain. Many objects in domain 4 were also underconstrained, but we were able to reduce their location lists to a reasonable size (<1000 locations) for further computations using sampling, as described in the next section.

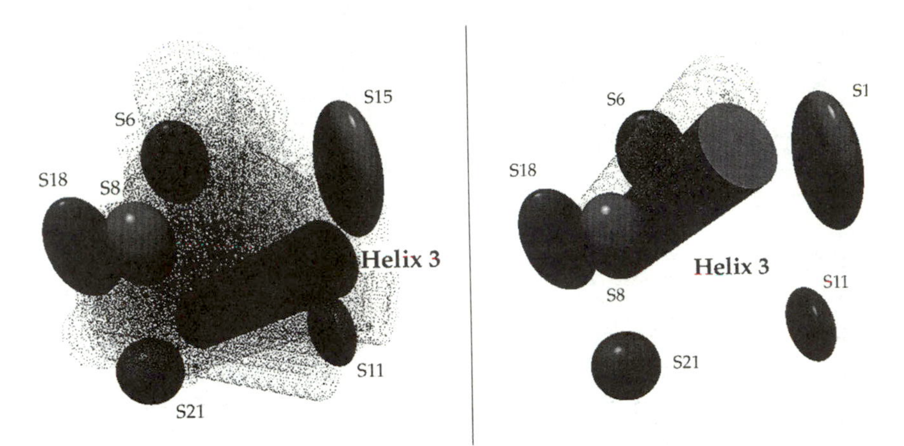

Figure 4: (LEFT) Cloud of valid locations for helix 3 before higher-order yoking operation. Each helix location is marked with some dots scattered on its surface in order to give a visual impression of the volume of space that contains valid locations for the helix. (RIGHT) Cloud of valid locations after higher-order yoking. Notice that the volume of possible locations is considerably reduced by the n-yoking operation.

Coherent instance generation

Having maximally pruned location lists within the individual domains, we used distance constraints between helices in different domains to further reduce the size of locations lists. This represents the first time that information from outside the domains was used to reduce location lists. After such information is introduced, another round of high-order consistency checking is required within the domain, to ensure that the effects of the new information are propagated to all objects in the network. Eventually, the high-order consistency checking mechanisms reach equilibrium, and the location lists for each object remain a constant size.

Using the equilibrium location lists for each object, a list of coherent instances was generated for each domain. Each coherent instance specifies one location and orientation for each object in the domain, such that all the locations and orientations are mutually compatible and consistent with the input distance constraints.

In order to generate aggregate coherent instances, with a valid location and orientation for every helix in the entire 16S rRNA, the coherent instances from the individual domains were combined to find sets of coherent instances (one for each domain) which were mutually compatible. Each helix location within a domain coherent instance is, by definition, compatible with the other helices in that domain. If all the constraints between two domain coherent instances are tested and satisfied, then each helix location in that pair of domain coherent instances must also be compatible. When a set of five domain coherent instances (one for each of the five domains) is found to satisfy all the constraints between domains, then the locations for all the helices must be compatible with the input data, and therefore represent a single valid conformation for the 16S rRNA ensemble. Thus, the task of finding valid overall conformations becomes one of selecting coherent instances for each domain that are mutually compatible.

Methods of reducing computational complexity

As mentioned previously, a major problem encountered during higher-order consistency checks is enormous computational complexity. This computational complexity is directly related to location list size for each object and the number of objects involved in the higher-order consistency check:

$$\text{computational complexity} = (\text{location list size})^{\text{number of objects}}$$

Sparse structural data often result in underconstrained problems. Constraints are not strong enough, or numerous enough, to prune down location lists for some objects. This results in large location lists for these objects, making constraint satisfaction computationally expensive, especially higher-order yoking (which involves more objects). We have already described two methods for reducing computational complexity: performing lower-order checks before high-order checks, and creating relatively independent subproblems (domains) which can be solved separately and recombined.

Another more recent strategy for reducing computational complexity aims to reduce the location list size using intelligent sampling . Since our goal is to have a representative set of conformations that are compatible with the input data, it is not unreasonable to eliminate conformations that are fundamentally the same—with only minor differences in position and orientation. Indeed, there may be a relatively small number of distinct classes of conformations. The idea of grouping conformations into sets that share similar positions/orientations is the basic idea behind intelligent sampling.

Random sampling of conformations can be dangerous because it often results in skewed, uneven representation of the original list of locations—information is lost, and whole families of valid structures may be absent in the final solution. We have developed and tested an intelligent sampling algorithm based on hierarchical clustering, the details of which will be published elsewhere. Figure 5 shows a comparison of the location lists for helix 44 that result from intelligent sampling and random sampling. It demonstrates that intelligent

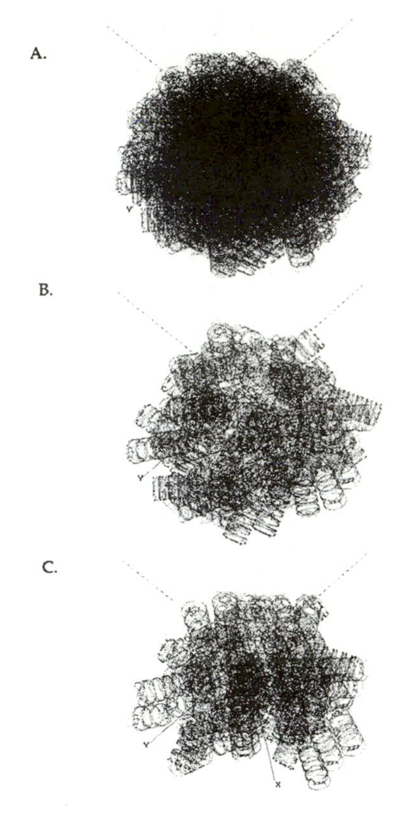

Figure 5: (TOP) A cloud of 571 locations (helix 44) before any sampling. (MIDDLE) The same cloud after intelligent sampling down to 100 locations. (RIGHT) The same cloud after random sampling down to 100 locations. On comparing the two sampling clouds, one can see that intelligent sampling has resulted in a cloud that more closely represents the original cloud. The cloud of locations resulting from random sampling is lopsided and asymmetric, neither of which are characteristic of the original cloud.

sampling provides a set of locations that are more representative of the original list. Intelligent sampling was used during the construction of these models whenever the initial location list lengths were too long (usually greater than 10,000 locations) for tractable processing with standard consistency checks. The sampled lists were subsequently pruned using these consistency checks.

Results

The initial sizes of the location lists for each object, as well as their final sizes are shown in Table 1. Because of the sampling intervals used, the positional uncertainty for helices with only one location is about +/- 7Å. The average excursion for other helices is about +/- 15 to 30 Å, with severely underconstrained helices having uncertainties of more than 4 Å.

Figure 6 shows the final cloud of valid locations for two of the helices. The small clouds for helices H32 and H37 are typical of helices from the central domain and domains 1-3, which proved relatively constrained. Most had fewer than 500 final locations, and 13 of the 31 objects in these domains ended with only 1 valid location (indicating very tight constraints). Helices H6, H20, and H30 predictably had the largest location lists within these four domains, for all

Table 1

Obj	Starting	Final	Obj	Starting	Final
H1	558	1	H24	10719	1
H2	3931	1	H25	1894	1
H3	174	2	H26	5459	475
H4	7460	22	H27	9178	1
H5	1034	93	H28	891	1
H6	27805	6017	H29	1686	38
H7	19409	212	H30	8632	2889
H8.1	6295	543	H31	30281	5
H8.2	1555	22	H32	31356	16
H8.3	3973	752	H33	3010	5
H9	2097	50	H34	40710	5
H10	1171	82	H35	873	6
H11	3227	357	H36	1449	1122
H12	758	80	H37	326	56
H13	16433	25	H38	1968	268
H14	2752	1	H39	3166	267
H15	366	1	H40	29806	428
H16	9936	1	H41	15461	478 (100)*
H17	47	1	H42	33964	18
H18	4105	1	H43	15263	9888 (100)*
H19	19935	13	H44	3876	571 (100)*
H20	93377	26918	H45	3568	3133 (100)*
H21	28320	1	H46	97	5
H22	1930	6	H47	21408	832 (100)*
H23	15647	1			

Table 1: Initial and final number of valid locations per object (helix). The final location lists are consistently several orders of magnitude smaller than the starting location lists. This table shows how achieving higher-order consistency between objects can dramatically reduce the number of valid locations for each object. Note that helices H48-H50 are not included in the above list because they were severely underconstrained, leading to intractably large location lists. The asterisk next to some of the objects from H41-H47 indicates objects whose location lists were intelligently sampled. The number in parentheses indicates the number of locations after intelligent sampling.

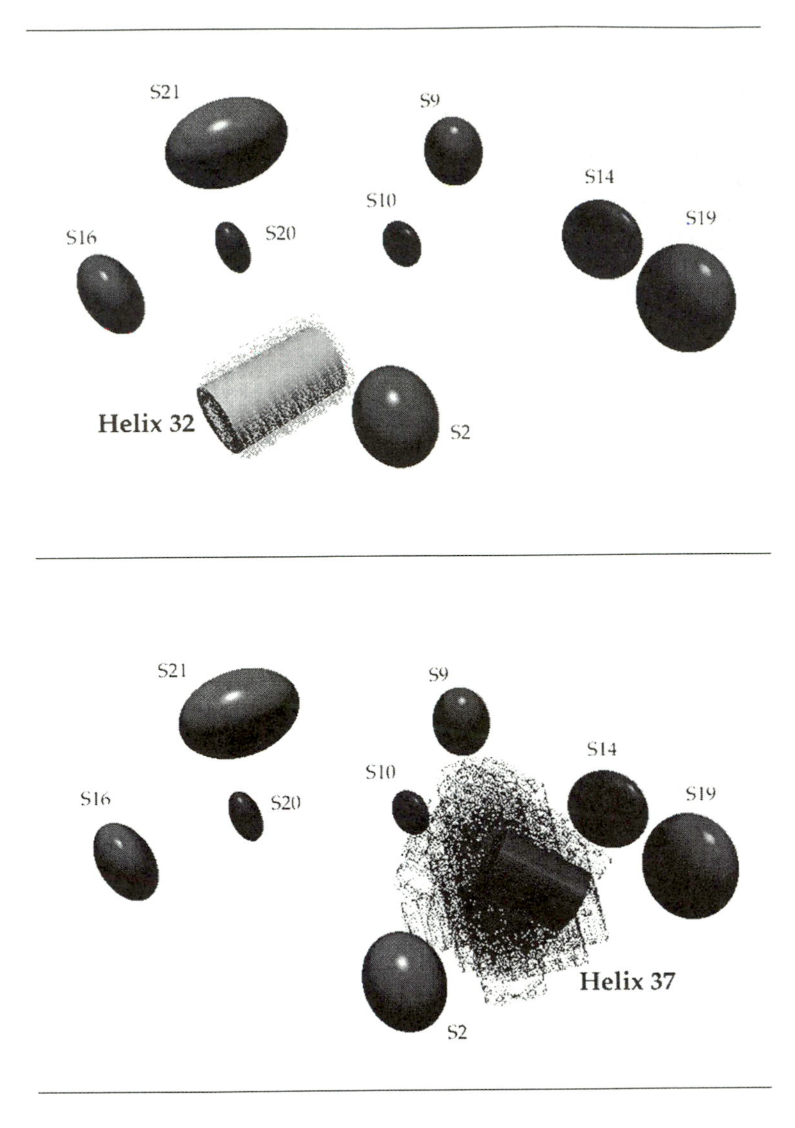

Figure 6: The final cloud of locations for helix 32 (top) and helix 37 (bottom). The locations for both helices are relatively localized indicating the helices are relatively well constrained.

have relatively few constraints to other objects and are more free to move. The largest location lists were consistently in domain 4. Many objects in this domain produced large location lists and required extensive intelligent sampling.

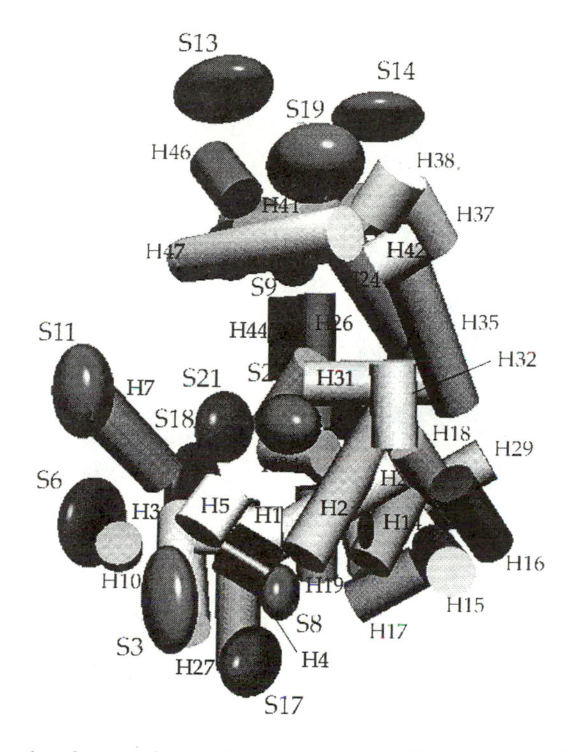

Figure 7: Example of one coherent instance generated by our constraint satisfaction method. Each helix is drawn in one of the locations within its location list, and all locations are compatible with all the distance constraints provided as input data. Proteins are labeled S1 through S21, and helices are labeled H1 through H52.

We generated coherent instances for each domain (except domain 4), and sampled these down to approximately 100 representative conformations for each domain. We then generated roughly 1000 representative combinations of the four domains, some of which are shown in Figure 7 and Figure 8.

Discussion

Examination of the computed structures yielded some surprising results. Because of the large number of valid structures that are produced, we looked for topological similarities among them. The most striking topological feature was a cleft in the 16S rRNA that was ubiquitous among the computed structures. This be seen in the coherent instances in Figure 7 and Figure 8. This cleft corresponds closely to a cleft in the topology of the 16S rRNA that has been determined by electron microscopic (EM) studies (Verschoor *et al.*, 1984). The structures produced by Malhotra *et al.* also have this feature, but their methods

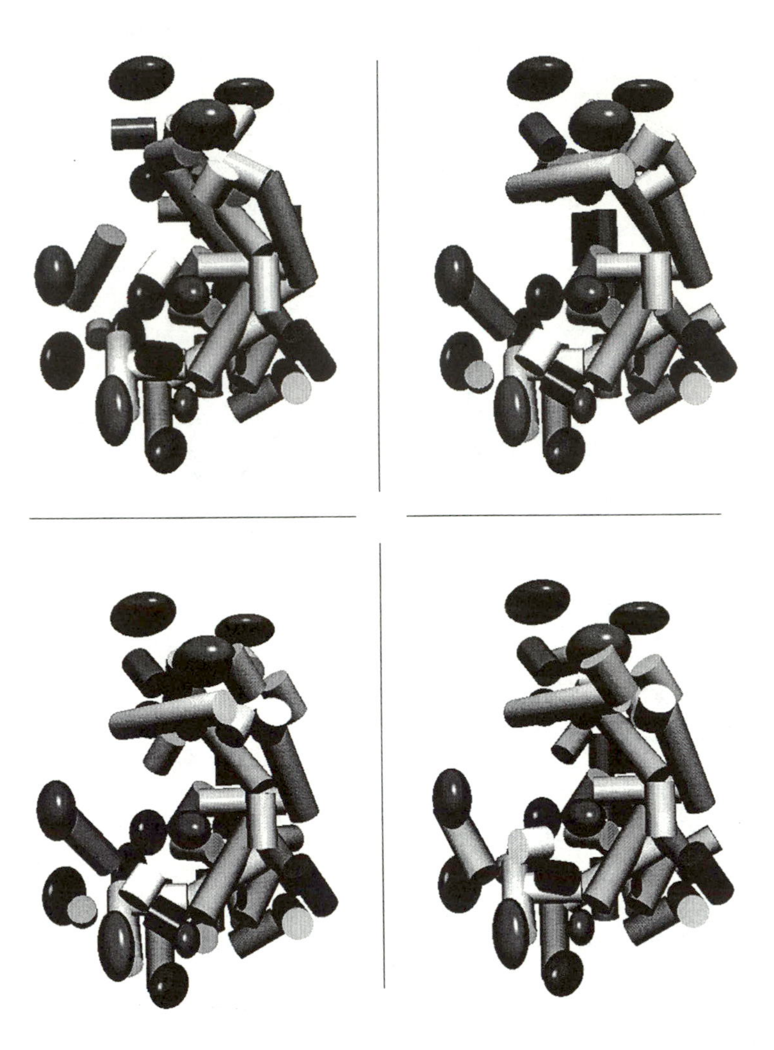

Figure 8: Four coherent instances generated by the constraint satisfaction algorithm. Notice that the cleft on the right is preserved in all four even though there is variation in individual helix positions.

used the topology generated by the EM studies as an explicit constraint on the structure (Malhotra and Harvey, 1994). Our study used no prior knowledge of topology, and still consistently produces this cleft without the EM data. In addition to the cleft, other aspects of the 16S rRNA topology were consistently preserved among candidate structures. For example, there is an additional, smaller cleft on the opposite side of the complex seen on EM that is also preserved in our structures. In addition, the location of residues involved in tRNA binding (1378, 1408, 1492) and the location of residues involved in mRNA binding (1532, 532) are relatively constant, and agree with models previously reported, but based on different data (Brimacombe, 1991; Malhotra and Harvey, 1994).

Because the exact structure of 16S rRNA is not known, we have no direct means for validating our results. Our modeled structures are, of course, consistent with all constraints in our data set. It is also clear that the topology of our model has similarities to the topology determined by EM studies (Verschoor *et al.*, 1984). Our structures can be further validated by comparing them with models derived from other data, and by their ability to predict new experimental results.

Our models may be sensitive to the assumption that all base-paired regions are standard A-form helices. Some of these helices may have significant bends or bulges that make a cylindrical assumption inaccurate. Breaking helices into two connected cylinders, instead of a single cylinder, may provide enough flexibility to solve this problem. For example, Helix 2 was difficult to position, and it may be because our representation of it is too rigid. Other helices that might benefit from fragmentation into smaller parts are H34, H35, H27, and H28. Most of these helices are very long, and have small volumes of valid locations—suggesting that the constraints are difficult to satisfy with current modeling.

Although sampling to reduce the sizes of location lists was used sparingly in most of our calculations, many objects in domain 4 had to be intelligently sampled because they were underconstrained and had very large location lists. The results of experiments such as summarized in Figure 5 suggest that our sampling strategy does not substantially change the range of conformations contained within the location lists, but instead reduces the density of these lists.

Our work has shown that PROTEAN and constraint satisfaction techniques are effective methods for producing a set of structures consistent with sparse structural data that is available for 16S ribosomal RNA. There are several advantages that this method has over optimization techniques. The systematic, exhaustive sampling characteristic of PROTEAN's constraint satisfaction algorithm results in a complete characterization of possible positions and orientations for an object. This allows one to find the full range of valid structures for a molecule that represents, in a sense, an upper bound on the actual solution. Optimization routines, on the other hand, only the find the structures that do best "on average" without indicating how reliable the structures are. For example, structure calculations based on optimization often have a significant number of constraints which are not satisfied (Malhotra and Harvey, 1994). PROTEAN's discrete sampling strategy also allows checks that are sensitive to directional errors in protein position, which may be difficult to implement within an optimization routine. Finally, PROTEAN's step by step process allows sensitivity analysis to different data interpretations. One can, for example, compare the results obtained with liberal distance bounds with those gained from relatively tight bounds used to check distance constraints.

Conclusions

In this paper, we have shown that constraint satisfaction techniques can be used to produce a set of valid structures for 16S rRNA that are compatible with biochemical footprinting data. Each of these structures has an overall shape that is consistent with EM images. In particular two clefts are placed in the appropriate location. These results suggest that the placement of RNA helices within the structure is correct (to within 10 to 30 Å), and allows us to begin localizing functional groups within the 30S subunit.

In the next phase of this work, we will investigate the sensitivity of our models to the use of single cylinders for helices by using multiple cylinders that allow more bends and bulges. We will also test the sensitivity of the method to the specific interpretation of the protection experiment distance bounds (by varying the values of parameters SD and F in Equation 1). In addition, we will undertake a comprehensive comparison of our models with those derived from other data sets, and with other structure-calculation techniques, to look for (and resolve) significant differences between these models.

Acknowledgements

R.B.A. is a Culpeper Medical Scholar, and is supported by NIH grant LM05652. R.C. is supported, in part, by LM07033. The PROTEAN program was developed from 1984-1989 at Stanford as part of a collaboration between Bruce Buchanan (Computer Science) and Oleg Jardetzky (Stanford Magnetic Resonance Laboratory). The images in this paper were created by the program PROTEAND, written by Chris Hughes, and publicly available at ftp://camis.stanford.edu/pub/altman/tar.proteand.

References

Altman, R., and Jardetzky, O. (1989) (N.J. Oppenheimer and T.L. James, Eds.) Academic Press, New York. p. 177.

Altman, R.B., Weiser, B., and Noller, F. (1994) *submitted.*

Altman, R.B., Weiser, B., and Noller, H.F. (1994). in Second International Conference on Intelligent Systems for Molecular Biology, Stanford University, AAAI Press.

Arrowsmith, C.H., Pachter, R., Altman, R. B., Iyer, S. B., and Jardetzky, O. (1990). *Biochemistry* **29**, 6332.

Arrowsmith, C., Pachter, R., Altman, R., and Jardetzky, O. (1991). *European Journal of Biochemistry* **202**, 53.

Brimacombe, R. (1991). *Biochimie* **3**, 927.

Brinkley, J.F., Altman, R. B., Duncan, B. S., Buchanan, B.G., and Jardetzky, O. (1988). *Journal of Chemical Information and Computer Science* **28(4)**, 194.

Capel, M.S., Engelman, D. M., Freeborn, B. R., Kjeldgaard, M., Langer, J. A., Ramakrishnan, V., Schindler, D. G., Schneider, D. K., Schoenborn, B. P., and Sillers, I. Y. (1987). *Science* **238**, 1403.

Capel, M.S., Kjeldgaard, M., Engelman, D. M., and Moore, P. B. (1988). *Journal of Mol. Biol.* **200**, 65.

Dahlberg, A.E. (1989). *Cell* **57**, 525.

Eisenstein, M., Sharon, R., Berkovitch-Yellin, Z., Gewitz, H. S., Weinstein, S., Pebay-Peyroula, E., Roth, M., and Yonath, A. (1991). *Biochimie* **73**, 879.

Gutell, R.R., and Woese, C.R. (1990). *Proc. Natl. Acad. Sci. U.S.A.* **87**, 663.

Hardesty, B., and Kramer, G. (1985) Structure, Function and Genetics of Ribosomes, Springer-Verlag, New York.

Hubbard, J.M., and Hearst, J.E. (1991). *J. Mol. Biol.* **221(3)**, 889.

Mackworth, A.K. (1977). *Artificial Intelligence* **8**, 99.

Malhotra, A., and Harvey, S. (1994). *J. Mol. Biol.* **240**, 308.

Moore, P.B., Capel, M., and Kjelgaard, M. (1985). Structure, Function and Genetics of Ribosomes. (D.M. Engelman, ed.) Springer-Verlag, New York.

Noller, H.F. (1991). *Annu. Rev. Biochem.* **60**, 191.

Noller, H.F. et al., *unpublished.*

Stern, S., Weiser, B., and Noller, H.F. (1988). *J. Mol. Biol.* **204**, 447.

Stern, S., Powers, T., Changchien, L. M., and Noller, H. F. (1989) *Science* **244**, 783.

Verschoor, A., Frank, J., Radermacher, M., Wagenknecht, T., and Boublik, M. (1984). *J. Mol. Biol.* **178**, 677.

Woese, C.R., Gutell, R., Gupta, R., and Noller, H. F. (1983). *Microbiol. Rev.* **47**, 621.

19

Design and Characterization of New Sequence Specific DNA Ligands

D. Wemmer

Department of Chemistry
University of California
Berkeley, CA 94720 USA

During the early 1980s there were two developments which lead to our studies of sequence specific DNA ligands. The first was the development of sequential assignment methods based on 2D NMR spectra which allowed complete assignment of resonances for proteins (Wüthrich, 1986). The assignments in turn allowed determination of many structural restraints through interpretation of NOESY crosspeaks and coupling constants from COSY type spectra. The second advance was the improvement of the chemistry for direct synthesis of DNA oligomers. With multimilligram samples of DNA oligomers available sequential assignment methods for DNA, paralleling those for proteins, were also worked out. Again with assignments came the possibility of determining DNA structures in solution. However for double stranded, Watson-Crick paired DNAs the structure can be reasonably approximated by the standard B-form model derived from fiber diffraction. The accurate determination of local conformational features has been somewhat difficult using NMR since tertiary contacts (as are so valuable in determining protein structures) do not occur. However with careful quantitative analysis some of the local details of structure can be determined.

These NMR methods also offered the possibility of trying to understand the structural basis for binding of ligands to DNA oligomers. In order to make well-defined complexes we wanted to start with a compound that showed some sequence specificity in binding, and selected distamycin (shown below), a polypyrrole antibiotic which was known to have preference for binding to A-T rich DNA sequences. A close relative, netropsin, had been studied by Dinshaw Patel who showed that the binding is in the minor groove by identifying an NOE between a proton of the ligand and an adenosine H2 in the center of the minor groove (Patel, 1982).

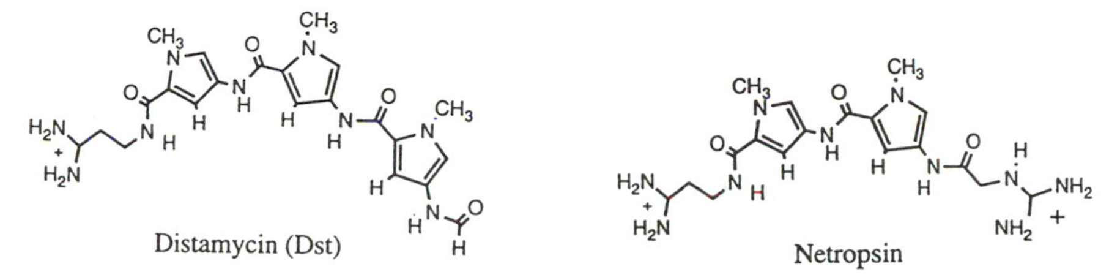

We began by making a complex with the self-complementary DNA oligomer: 5'-CGCGAATTCGCG-3', which had been studied extensively by X-ray crystallography, and also by NMR. Distamycin did form a well-defined complex with this DNA, which was is slow exchange with free DNA during titrations (Klevit *et al.*, 1986). NOESY derived distance restraints were used together with molecular modeling to develop a detailed structure for the complex formed at the 5'-AATT-3' site (Pelton and Wemmer, 1988) (in the following, binding sites will be indicated by the sequence on one of the strands, the complement is always present, and there are always two or three flanking G-C pairs at the ends of the oligomers). We found that there was a very snug fit of the ligand into the A-T region of the minor groove, providing good van der Waals contacts over the full surface of the ligand. In addition there were hydrogen bonds between the amides linking the pyrrole rings and acceptors on the base pairs of the DNA. The charged tail of distamycin was also positioned deep in the groove, where the electrostatic potential is predicted to be maximal, contributing an electrostatic contribution to the stability of the complex. The structure of this unliganded DNA had been determined crystallographically, and the region of minor groove spanning the A-T segment had been found to be very narrow, suggesting that this helped define the A,T preference. At about the same time as our initial studies, a complex of netropsin with the same DNA oligomer was crystallized and the structure was solved using x-ray diffraction by Dickerson and coworkers (Kopka *et al.*, 1985). It showed all of the same basic features contributing to stabilization of the complex. Both distamycin and netropsin covered the 4 A-T pairs defining the binding site.

As the next stage of our work we decided to examine the binding of distamycin to an oligomer containing the sequence: 5'-AAATT-3' to explore how binding occurred on an asymmetric sequence, and whether it would preferentially occupy a single subsite of four base pairs. In titrations we found resonances for two different forms of complex at low distamycin:DNA ratios.

NOESY experiments showed that these corresponded to distamycin bound in the two possible orientations on this site, with a preference of about 3:2 for the formyl end of distamycin being bound to the 5'-AAA-3' sequence. Exchange crosspeaks were seen in the NOESY spectra arising from dissociation and reassociation in the opposite orientation, indicating an off rate of a few per second. When the NOE contacts were studied it was found that there were contacts from each pyrrole of distamycin to two adenosines, rather than one as seen in the complex with 5'-AATT-3'. This was interpreted as arising from sliding of the ligand between the two four base pair subsites in each orientation, with the sliding being rapid (in fast exchange on the NMR timescale) between the subsites (Pelton and Wemmer, 1990). This was supported by later studies of another sequence in which line broadening from this type of sliding was observed (Pelton and Wemmer, 1990). What was much more surprising was the appearance of a new set of resonances from another form of complex, which grew in intensity as a 1:1 stoichiometry of distamycin:oligomer was approached. We found that these resonances continued to grow as more distamycin was added, finally being the only form of complex present when the distamycin:oligomer ratio reached 2:1.

At yet higher ratios resonances from free distamycin were seen in addition to those from the complex. NOESY spectra were again used to assign resonances. Two complete, inequivalent sets of distamycin resonances were observed. One had contacts to sugar protons on one strand of the DNA, while the other made

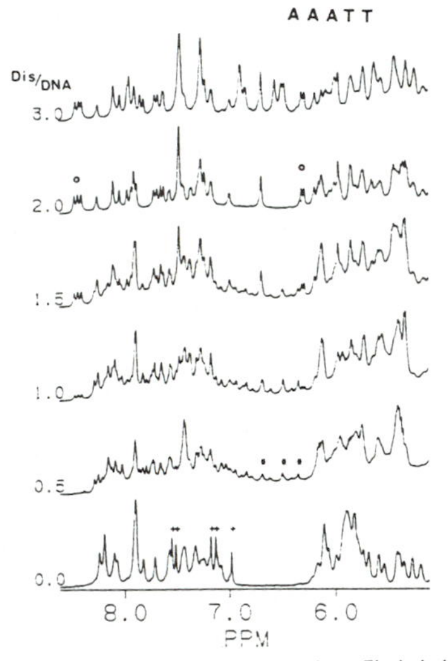

Figure 1: A titration of a DNA with the binding site 5'-AAATT-3' with distamycin is shown, with the stoichiometry labeled at the right. Free DNA H2 resonances are indicated by +, some resonances from the 1:1 complex by *, and some of resonances from the 2:1 complex by o.

contacts only to the opposite strand. Both had contacts to the adenosine H2s in the middle of the groove. Modeling of the complex showed that the two distamycin molecules resided side-by-side in the minor groove (Pelton and Wemmer, 1989), which had been opened by roughly 4 Å relative to narrow groove seen in the AATT structure. The two distamycins were antiparallel, each positioning the formyl end of the ligand against the 5' end of the contacted strand. There is still good contact of each ligand with the wall of the minor groove, but in this case also between the two ligand molecules. There are again hydrogen bonds from the amides of the ligands to base acceptors, and the positively charged tails are deep in the groove.

It is clear that all five A-T pairs are contacted, and hence the binding site in this mode is five base pairs in length rather than the four in the 1:1 binding mode. The power of NMR is apparent in considering these studies, under many conditions there were five different forms of complex present in equilibrium with one another, however by studying complexes under different conditions they could all be identified and characterized.

Both the NMR titrations and subsequent titration calorimetry studies (Rentzeperis *et al.*, 1995) showed that the first molecule binds with affinity slightly higher than the first (about a 10 fold higher dissociation constant for the second molecule), but the second still has quite high affinity (K_d in the μM range). This was surprising because it indicated that the large change in position of the DNA backbone which defines the minor groove can be accomplished at low energetic cost. Subsequently we explored the sequence dependence of the

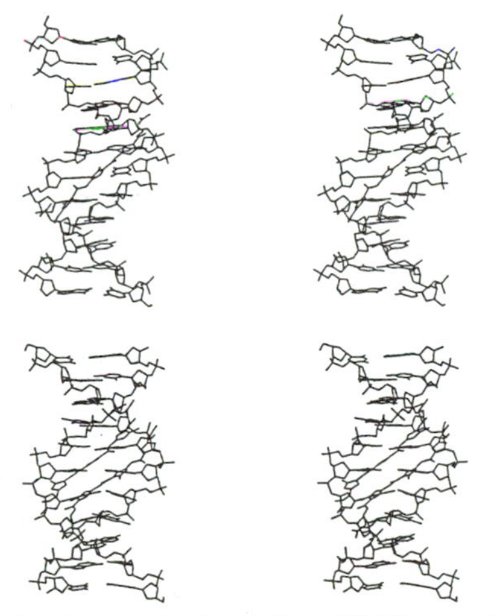

Figure 2: Examples of structures calculated from NMR restrained modeling are shown for 1:1 complex (top) and 2:1 complex (bottom). The change in groove width is quite apparent.

relative affinity of binding of the 1:1 to 2:1 modes, with some interesting results. To try to subtly change the DNA, while leaving the functional groups in the minor groove unchanged, we made A-T -> I-C substitutions. Substitution of a single I-C into the 5'-AAATT-3' site made little difference, but when two sites were substituted (either on the ends or in the middle) an increase in the binding of the second molecule relative to the first was seen. When all five As were substituted with I to give a binding site 5'-IIICC-3', then only 2:1 binding was seen at all stoichiometries (Fagan and Wemmer, 1992). This requires that the dissociation constant for the second molecule be ca. 100 fold lower than that for the first molecule. In NOESY spectra of the free oligomer, cross strand contacts are seen from the inosine H2s to sugar H1's, which are a marker for a fairly narrow minor groove. We also examined the effects of order of A-T pairs in the binding site. For titrations with an oligomer containing the binding site 5'-AAAAA-3' we observed only 1:1 complex until greater than one mole of distamycin had been added per mole of oligomer. Thereafter new resonances were identified which were from the 2:1 complex. This behavior indicates that the dissociation constant for the second ligand binding is ≥100 fold higher than that for the first. However when the binding site was 5'-ATATA-3' only 2:1 complex was observed at all stoichiometries, reflecting a second ligand dissociation constant ≥ 100 fold lower than the first! Thus swapping the orientation of two A-T pairs, converting AAAAA to ATATA, leads to a change of ≥ 10,000 fold in the ratio of 1:1 to 2:1 dissociation constants (K_1 and K_2). Although we do not yet have complete quantitative data yet it appears that the product K_1K_2 is almost the same for these, indicating that the total binding free energy in getting to the 2:1 complex is very similar, but the partitioning between the first and second molecule binding is very different for different sequences. Crystallographic studies have shown that both poly-A type sequences and poly-A-T type sequences have narrow minor grooves, although the former seem slightly narrower. It is also known that poly-A type sequences (runs of A-T pairs with few T to A steps) have an unusual structure which can lead to DNA bending, and gives unusually slow hydrogen exchange for the imino protons. This behavior correlates with the binding behavior we observe: poly-A type sequences are most stable with a very narrow groove, and this makes a particularly good 1:1 binding site, and there is an energetic cost of widening the groove to accommodate the second ligand; alternating poly-A-T sequences have somewhat wider grooves and hence can be 1:1 sites, but open easily to accommodate a second ligand and the full contact with both grooves in this mode leads to full binding affinity. This conclusion is reinforced by the fact that 5'-AAAA-3' and 5'-AATT-3' are good 1:1 binding sites, while 5'-TATA-3' and 5'-ATAT-3' are not.

After solving the crystal structure of netropsin bound to 5'-AATT-3', Dickerson and coworkers suggested a covalent modification which they thought would lead to G-C specific ligands (Kopka *et al.*, 1985). This idea was independently suggested by Lown *et al.* (1986), and later pursued by Dervan as well. The logic was that A-T specificity arose from steric and functional group complementarity of the ligand and the groove of DNA. At sites with a G-C base

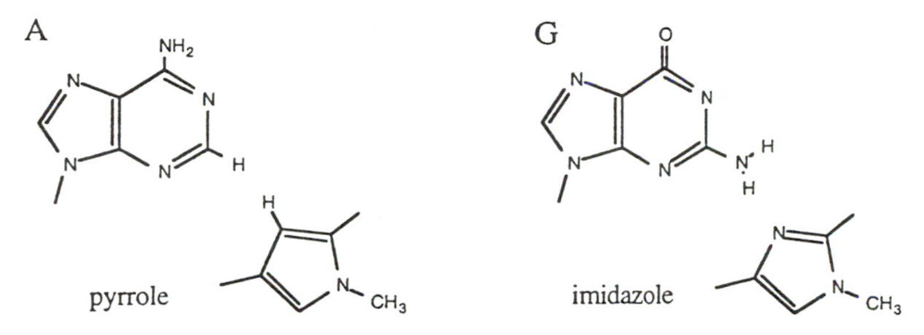

pair, they argued, the groove is not deep enough for good binding because of the amino group of guanosine protruding into the groove. However if the pyrrole group of the ligand were changed to an imidazole the fit would be restored, and the ligand and DNA could hydrogen bond as well. It was thought that these features would lead to G-C specific binding. However as such molecules were made, primarily derivatives of netropsin, it was found that they were more G-C tolerant but were not G-C specific. As our thinking about groove widths and binding modes evolved through looking at various crystal structures of DNA oligomers, we came to the conclusion that the very narrow groove required for good 1:1 binding does not occur in mixed A-T/G-C sequences (Heinemann and Alings, 1989). We also found that netropsin cannot bind in the 2:1 mode, we believe because both ends are charged which would lead to unfavorable electrostatic interactions in the 2:1 mode. Hence we reasoned that a redesigned ligand, incorporating the idea of a hydrogen bond to a G-C pair, but accomplishing it in a 2:1 binding motif might work. Patterned after the 2:1 distamycin complex we had characterized on the binding site 5'-AAATT-3', we thought that imidazole-distamycin (ImD), containing imidazole only in the

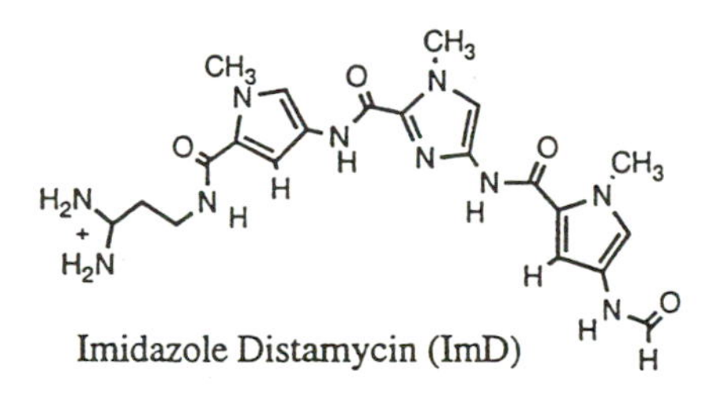

Imidazole Distamycin (ImD)

central ring, should bind to the target sequence 5'-AAGTT-3', with both imidazole rings over the G-C base pair in the center of the site (Dwyer *et al.*, 1992). Y. Bathini in Lown's group made this compound for us and we carried out titrations with both the 5'-AAGTT-3' and 5'-AAATT-3' and found that indeed the binding was good for the G-C containing sequence, forming exactly the complex proposed, but gave a weak complex (fast exchange, extensive line broadening at all stoichiometries during a titration) with the all A-T site. The model for the complex was all but indistinguishable from the 2:1 distamycin complex on 5'-AAATT-3'.

At about the same time Dervan's group had made an analogous compound, but with the imidazole on the end ring rather than the center, which they termed imidazole-netropsin (ImN). With footprinting and affinity cleavage studies they found that it bound at sequences such as 5'-TGACT-3', all of the high affinity sites with two G-C base pairs rather than one as they had expected (Wade *et al.*, 1992). After seeing the description of our 2:1 complexes they concluded that their compound might also binding in that mode, explaining how two G-C pairs could be selectively targeted. They sent samples of their compounds for structural studies, which verified that a complex of virtually the same structure as the 2:1 distamycin and 2:1 2-ImD complexes did form (Mrksich *et al.*, 1992). Interestingly in the 2-ImN case each G-C was being recognized by one imidazole ring and one pyrrole ring. In the spectra of the complex we noted that the G-amino protons, which hydrogen bond to the ligand, gave rise to two separate, fairly sharp resonances rather than the usually very broad lines. We interpreted this as arising from an increased activation energy for rotation about the amino C-N bonds due to the presence of the hydrogen bond to the ligands. Dervan's group also synthesized this compound with ^{15}N labels in the imidazole ring, and very substantial changes in the ^{15}N chemical shifts were observed for both ligands upon binding, again attributable to the presence of the hydrogen bonds.

In looking at our complex with 2-ImD we had noted that the ligand against the G-containing strand was in a good position to hydrogen bond, while that against the C-containing strand was not. We thought that a complex with one 2-ImD and one distamycin might in fact have even higher affinity than the 2:1 2-ImD complex. We decided to test whether such a 'hetero' complex would form by doing competition experiments, titrating distamycin into a sample of the 2-ImD complex to see if one of the 2-ImD ligands would be displaced. Indeed that is what we found, over 90% of the DNA formed a heterocomplex with one 2-ImD and one distamycin, the 2-ImD contacting the G-containing strand in analogy to the 2-ImN complex (Geierstanger *et al.*, 1993). In a control experiment, we also carried out titrations of the 5'-AAGTT-3' DNA with distamycin, not expecting to find a good complex. However, to our surprise a well defined 2:1 complex is formed, again structurally analogous to the complex on 5'-AAATT-3'! In fact, in our competition titrations, we found a small amount of the 2:1 distamycin complex in equilibrium with the heterocomplex, but none of the 2:1 2-ImD. The formation of the 2:1 distamycin complex indicates that the groove depth is not the critical factor for distamycin binding, that probably a match of the width of the ligands matching the width of the groove is much more important.

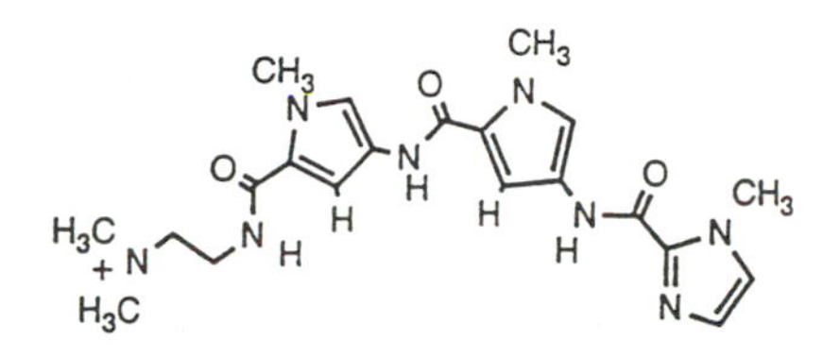

Imidazole Netropsin (ImN)

As we thought about why the heterocomplex was the optimum, the solvation of the imidazole group came to mind. To form the 2:1 2-ImD complex, the ligand against the C-containing strand had to be desolvated, but did not form a hydrogen bond in the complex to compensate, as the molecule against the G-containing strand did. While the importance of solvation could not be tested experimentally in any simple way, this seemed an ideal system to try free energy perturbation calculations since the only differences in the complexes were conversions of pyrrole C-Hs to imidazole N-Hs. We contacted Peter Kollman and described the system, but did not tell him our experimental results, instead challenging him to try to 'predict' the rank order of the complexes which formed (we had experimental data on three of the four possible two ligand complexes, shaded rings indicate imidazole). S. Singh and Ajay in his lab carried out these

A A G T T	A A G T T	A A G T T
T T C A A	T T C A A	T T C A A

calculations, and in fact did very well in predicting not only the rank order but also the relative affinities quite closely (Singh *et al.*, 1994). In the calculations the effects of solvation can be partitioned (at least approximately) and it does seem that the desolvation of the 2-ImD on the C-strand makes the 2:1 2-ImD less favorable than the heterocomplex buy several kcal/mol. On the G-stand the hydrogen bond is sufficiently strong to overcome the desolvation. Subsequently we examined the binding of 2-ImN and distamycin on a 5'-AGAAT-3' site, and with the single G-C the heterocomplex was the optimum complex as we anticipated (Geierstanger *et al.*, 1994).

Since the binding of these ligands at G-C containing sequences always occurred in the 2:1 mode, it seemed possible that the affinity and specificity might be improved through formation of covalently linked 'dimers'. Both our group and Dervan's had considered linking the molecules either through the 'head' of one molecule and the 'tail' of the other (giving a hairpin dimer), or through replacing the N-CH3 of the central pyrrole or imidazole of each ligand with an N-(CH2)3-6-N alkyl linker. Dervan's group carried out syntheses of both types of molecules, and it was found that the specificity was increased (since both ends of the molecule had to bind at the same site), and that there was a small enhancement in affinity (Dwyer *et al.*, 1993). Both homodimeric and heterodimeric molecules were made with imidazole, pyridine and pyrrole rings, and the binding sites were characterized by affinity cleavage, footprinting and the structures by NMR.

At this point we felt that there were 'design rules' which could be applied to generate ligands to recognize specific sequences. These can be summarized: 1) to recognize mixed A-T/G-C sequences use binding in the 2:1 mode; 2) in the binding site each pair of pyrrole/imidazole rings will contact one base pair, and the tails will each contact one base pair; 3) to target a G-C base pair use an imidazole ring for the G-strand and a pyrrole for the C-strand; 4) to target an A-T base pair use two pyrrole rings; 5) the base pairs interacting with the tails must be A-Ts. With these principles we thought that it should be possible to design a ligand which had converted the original specificity of distamycin for all A-T base pairs, to one almost completely reversed in which there are only G-C pair in the center of the binding site contacting the rings. In order to achieve this with a single ligand, we extended the length by one base pair in the design, and alternated imidazole and pyrrole rings to give the ligand designated ImPImP shown below. The target site, following the rules above should then be 5'-(A,T) G C G C (A,T)-3' where the parentheses indicate that either base should be acceptable. Dervan's group synthesized this ligand, and carried out footprinting and affinity cleavage on plasmid restriction fragments, and found the expected binding sites. NMR structural studies verified the nature of the complex, and again provided evidence for the hydrogen bonds to the G aminos (Geierstanger *et al.*, 1994).

The lengthening of the ligand has increased the binding site to be six base pairs. Extrapolating from distamycin analogs it is likely that a further extension to five rings will be possible, all pyrrole compounds with six or more rings start to lose affinity apparently due to a mismatch between the curvature of the DNA and the ligand. We also believe that it should be possible to link these longer ligands in the same ways that the three ring ligands were to enhance specificity and affinity.

The observation of the 2:1 binding motif on the 5'-AAATT-3' site provided critical insight into the coupling of DNA sequence, groove structure and affinity for ligand binding. Exploiting this motif we have been able to design new ligands which have both higher affinity and specificity than distamycin. These

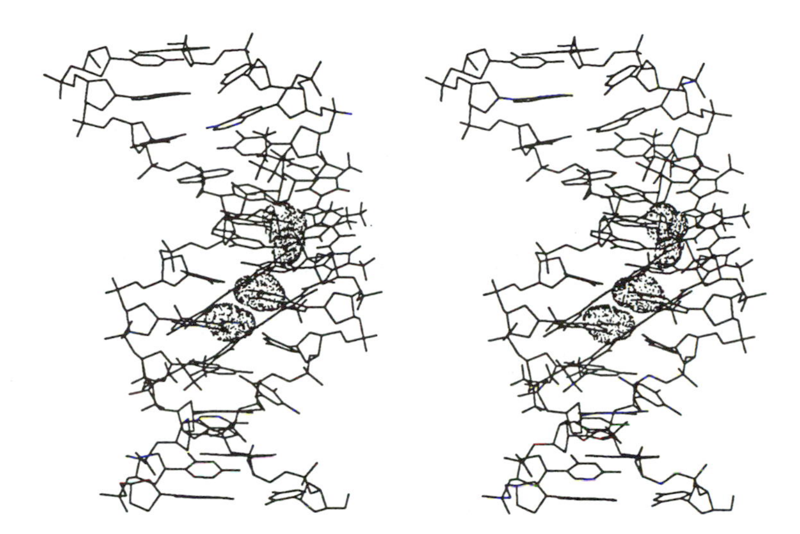

Figure 3: A structure of the 2:1 ImPImP complex is shown derived from NMR restrained molecular modeling. The G-amino protons being recognized in the complex are shown with a stippled surface. (reproduced from Geierstanger *et al.*, 1994)

are capable of recognizing G, C or (A,T) at each position of DNA sites five or six base pairs in length. Extrapolations from the present work suggest that it should be possible to extend the length of the recognition site further by using further end-to-end links.

Acknowledgements

I would like to thank all of the people who contributed to this work. In my lab doing structural studies this includes: Jeff Pelton, Tammy Dwyer, Bernhard

Geierstanger, Patty Fagan Jones, and Jens Peter Jacobsen. The work has benefited enormously through collaborations with the synthetic groups of Peter Dervan, including Warren Wade and Milan Mrksich, and Bill Lown, especially Yadagiri Bathini. The computational work with Peter Kollman also provided insight, and I thank Suresh Singh and Ajay in his group. In addition I would like to thank all the other members of these groups who have contributed both directly and indirectly to the success of these and related studies.

References

Dwyer, T.J., Geierstanger, B.H. Bathini, Y., Lown, J.W., and Wemmer, D.E. (1992). *J. Am. Chem. Soc.* **114**, 5911.

Dwyer, T.J., Geierstanger, B.H., Mrksich, M., Dervan, P.B., and Wemmer, D.E. (1993). *J. Am. Chem. Soc.* **115**, 9900.

Fagan, P., and Wemmer, D.E. (1992). *J. Am. Chem. Soc.* **114**, 1080.

Geierstanger, B.H., Dwyer, T.J., Bathini, Y., Lown, J.W., and Wemmer, D.E. (1993). *J. Am. Chem. Soc.* **115**, 4474.

Geierstanger, B.H., Mrksich, M., Dervan, P.B., and Wemmer, D.E. (1994). *Science* **266**, 646.

Geierstanger, B.H., Jacobsen, J.P., Mrksich, M., Dervan, P.B., and Wemmer, D.E. (1994). *Biochemistry* **33**, 3055.

Heinemann, U., and Alings, C. (1989). *J. Mol. Biol.* **210**, 369.

Klevit, R.E., Wemmer, D.E., and Reid, B.R. (1986). *Biochemistry* **25**, 3296.

Kopka, M.L., Yoon, C., Goodsell, D., Pjura, P., and Dickerson, R.E. (1985). *Proc. Natl. Acad. Sci. U.S.A.* **82**, 1376.

Lown, J.W., Krowicki, K., Bhat, U.G., Skorobogaty, A., Ward, B., and Dabrowiak, J.C. (1986). *Biochemistry* **25**, 7406.

Mrksich, M., Wade, W.S., Dwyer, T.J., Geierstanger, B.H., Wemmer, D.E., and Dervan, P.B. (1992). *Proc. Natl. Acad. Sci. U.S.A.* **89**, 7586.

Patel, D.J. (1982). *Proc. Natl. Acad. Sci. U.S.A.* **79**, 6424.

Pelton, J.G., and Wemmer, D.E. (1988). *Proc. Natl. Acad. Sci. U.S.A.* **86**, 5723.

Pelton, J.G., and Wemmer, D.E. (1988). *Biochemistry* **27**, 8088.

Pelton, J.G., and Wemmer, D.E. (1990). *J. Biomol. Str. Dyn.* **8**, 81.

Pelton, J.G., and Wemmer, D.E. (1990). *J. Am. Chem. Soc.* **112**, 1393.

Rentzeperis, D., Marky, L.A., Dwyer, T.J., Geierstanger, B.H., Pelton, J.G., and Wemmer, D.E. (1995). *Biochemistry* **34**, 2937.

Singh, S.B., Ajay, Wemmer, D.E., and Kollman, P.A. (1994). *Proc. Natl. Acad. Sci. U.S.A.* **91**, 7673.

Wade, W.S., Mrksich, M., and Dervan, P.B. (1992). *J. Am. Chem. Soc.* **114**, 8783.

Wüthrich, K. (1986). NMR of Proteins and Nucleic Acids, Wiley, NY.

20

Determination by ^1H NMR of a Slow Conformational Transition and Hydration Change in the Consensus TATAAT Pribnow Box

C. Milhé, A. Lane, and J.-F. Lefèvre

UPR 9003 du CNRS, ESBS, Pôle API
Boulevard Sébastien Brant
67400 Illkirch-Graffenstaden, France

Laboratory of Molecular Structure
National Institute for Medical Research
The Ridgeway, Mill Hill
London NW7 1AA U.K.

The conformational dynamics and hydration of a DNA 14-mer containing the consensus Pribnow box sequence TATAAT have been measured using rotating frame T_1 measurements and NOESY and ROESY in water. The H2 proton resonances of adenines show fast intermediate exchange behavior which can be attributed to a conformational transition that affects the distances between H2 protons of neighboring adenine residues, both sequential and cross-strand. The relaxation rate constant of the transition was measured at $4000s^{-1}$ at 25°C. Bound water close to the H2 proton of adenines was observed with residence times of >1ns. At low temperature (5°C), the Pribnow box is in a closed state in which hydration water in the minor groove is tightly bound. At higher temperatures, the conformation opens up as judged by the increase in separation between sequential H2 protons of adenines and water exchanges freely from the minor groove. The conformational transition and the altered hydration pattern may be related to promoter function.

The control of gene expression in procaryotes depends on the specific recognition by RNA polymerase of a six base-pair sequence (consensus: TTGACA) located at -35 from the transcription site, and a second one, named the Pribnow box (consensus: TATAAT) at about 10 base-pairs upstream the initiation site (Rosenberg and Court, 1979). It has been shown (Hawley and McClure, 1983) that strong promoters exhibit a high degree of homology with the consensus sequences, separated by an optimum consensus spacer length of 17 base pairs.

The strength of a promoter depends on, among other thing, the rate of the initiation of transcription. This rate depends on the product between the thermodynamic and kinetic constants K_B and k_2 (McClure, 1980). The initial binding of RNA polymerase to the promoter results in the formation of a transcriptionally inactive 'closed' complex, characterized by the association constant K_B. Isomerization to the active 'open' complex then occurs, and is characterized by the first order rate constant k_2. Hence, the frequency of transcription initiation depends both on the strength of the polymerase-promoter interaction, and the ease with which this complex can isomerize to the productive state. Both of these events are likely to depend on the physical properties of the promoter.

The lac promoter governs the transcription of the β-galactosidase operon. The wild type promoter (TATGTT) is weak and can be activated by cAMP receptor protein (CRP) (Malan *et al.*, 1984). Two mutants have been mapped in the Pribnow box region: lac PS (TATATT), which is stronger than the wild type, and lac UV5, which contains the perfect consensus sequence (TATAAT) and which is constitutively very efficient (McClure, 1980).

Conformational transitions have been previously observed in TAA sequence types present in the -10 and -35 trp operator regions (Lefèvre *et al.*, 1985; Lane *et al.*, 1987) and in the -10 lac promoter region (Schmitz *et al.*, 1992; Kennedy *et al.*, 1993). The central adenine H2 proton NMR resonance first broadens as the temperature is raised from 5 to 25 °C, then narrows again as the temperature is raised further to 35 °C. This behavior is typical of a conformational transition giving rise to fast intermediate exchange on the chemical shift time scale. The exchange rate has been quantified by measuring the spin-lattice relaxation rate constant in the rotating frame ($T_{1\rho}$) (Lane *et al.*, 1993). The present study reports similar observations on a 14 base pair oligonucleotide inspired from the lac UV5 sequence (Chart 1) in which the perfect consensus Pribnow box is embedded. Interestingly, the temperature dependence of promoter activity also shows a sharp transition between 25 and 30 °C, depending on the promoter sequence (Buc and McClure, 1985). Below the transition temperature, the promoter-RNA-polymerase complex is unable to initiate transcription of the DNA. Isomerization into an open complex is induced by raising the temperature above the transition point.

1 2 3 4 5 6 7 8 9 10 11 12 13 14
G T C G T A T A A T G T G T
C A G C A T A T T A C A C A
28 27 26 25 24 23 22 21 20 19 18 17 16 15

Chart 1: Sequence and numeration of the 14 base-pair lacUV5 duplex

A spine of hydration, with a first shell of very well ordered water molecules, was first observed in crystal structures of A-T rich sequences by Drew and Dickerson (Drew and Dickerson, 1981). The hydration shell has also been observed by NMR around A-T rich sequences (Liepinsh *et al.*, 1992), and the residence time range was deduced from the sign of the NOE between water and DNA protons. In the major groove, the water molecules reside for less than 500 ps, while in the minor groove, they stay for more than 1 ns. We report here the hydration of the lacUV5 promoter oligonucleotide, and compare the hydration with conformational features measured at different temperatures. It is suggested that the observed conformational transition at low temperature is related to a change in the hydration of the duplex.

Materials and methods

Sample preparation

The oligonucleotides d(GTCGTATAATGTGT) and d(ACACATT ATACGCA) were synthesized chemically (Applied Biosystems, Model 380B) and purified by reverse-phase HPLC on a C18 column (Zymark). The stoichiometric quantities in each strand were determined by absorbance titration. The NMR spectra of the duplex were recorded in 0.5 ml 100 mM phosphate pH 7 solutions of either 99.98% D_2O or 90%H_2O/10%D_2O. The final concentration of the duplex was around 1 mM.

Melting temperature determination

The absorbance of a solution of duplex was monitored at 260nm as a function of temperature (OD_{260} = 1 in 100mM phosphate pH7). Because of the cooperative melting process and the difference between single and double stranded DNA hypochromicity, the absorbance has a sigmoidal dependence on temperature, and the melting temperature Tm can be determined from the maximum of the first derivative curve, dA260 / dT versus T.

NMR experiments

One- and two-dimensional proton NMR spectra were recorded on a Bruker AMX 500 spectrometer or on a Varian Unity 600 and are referenced relative to internal DSS (3-(trimethylsilyl)-1-propanesulfonic acid).

2D spectra were acquired by recording 512 FIDs of 2048 points, in the phase-sensitive mode with time proportional phase incrementation (TPPI) (Wüthrich and Marion, 1983). The relaxation delay was set to 3 s. For experiments recorded in D_2O, the residual signal of HDO was suppressed by low power continuous selective irradiation during the preparation period. NOEs between H5-H6 of cytosine residues were measured at short mixing times using a NOESY experiment with a π-pulse inserted in the mixing time period to avoid peak distortion due to zero-quantum coherence (Rance *et al.*, 1985). The incrementation of the position of the π-pulse shifts the ω_1-frequency of the zero-quantum peaks apart from the exchange peak (Macura *et al.*, 1982). Spectra in H_2O were recorded with a WATERGATE excitation pulse sequence using gradients (Piotto *et al.*, 1992). This sequence allows the use of mixing times as short as 30 ms in NOESY experiments. HOHAHA spectra (Davis and Bax, 1985) were recorded with a radio frequency field strength set to 7800 Hz and a mixing time of 40 ms in order to allow observation of multiple bond correlations. 2D-spectra were recorded with 32 scans for each FID.

2D processing was achieved on a X32 computer using UXNMR software (Bruker) or on an IBM RS6000 workstation using the program FELIX 2.01 (Hare Research Inc.). The final size of the matrices was 2K$\times$2K real points. Prior to Fourier transform, the signal was multiplied in both dimensions by a 80° shifted square sine-bell function. In dimension 1, the signal was zero-filled to produce 2K real points.

Rotating-frame experiments were done in the pulsed Fourier transform mode (Freeman and Hill, 1971) at 600MHz, where the transmitter was used for both low and high power pulses. The spin-lock field strength was varied from 0.8 to 9.4 kHz. The duration of the spin-lock period was increased from 1 to 300 ms in 10 to 12 unequally spaced steps. The data were analyzed by non-linear regression to:

$$M(t) = M_0 \exp(-R_{1\rho}\, t) \qquad (1)$$

where M(t) is the magnetization at time t, M_0 the magnetization at time zero and $R_{1\rho}$ is the spin-lattice relaxation rate constant in the rotating frame. For the weakest B_1 fields, the pulsed spin-lock method was used and the identity $R_2 = R_{1\rho}$ assumed (Farmer *et al.*, 1988).

Fitting of $R_{1\rho}$ data with a two site exchange process

The evolution of the rotating frame relaxation rate constants $R_{1\rho}(\omega_1)$ as a function of the spin lock strength ω_1 were fitted using a two site model

(Deverell *et al.*, 1970):

$$A \underset{k_{-1}}{\overset{k_1}{\longleftrightarrow}} B \tag{2}$$

by non-linear regression to (Lane *et al.*, 1993):

$$R_{1\rho} = R_1^{\infty} + 4\pi^2 \, \Delta\nu_{app}^{2} \, (k / (k^2 + \omega_1^{2})) \tag{3}$$

where p_A and p_B are the fractional populations of conformations A and B, respectively, $\Delta\nu_{app} = 2\Delta\nu(p_A p_B)^{1/2}$, $\Delta\nu$ being the difference in Hz between the resonance frequencies ν_A and ν_B, in site A and B respectively and $k = (k_1 + k_{-1})$ is the exchange rate constant between the two sites.

Relaxation matrix fitting

NOEs at a given mixing time t_M are related to the relaxation matrix G by (Keepers and James, 1984):

$$NOE(t_M) = \exp(-t_M \, \Gamma) \, NOE(0) \tag{4}$$

The relaxation matrix was restricted to the subset spin system comprising the three H2 protons of A8, A9 and A22:

$$\Gamma = \begin{pmatrix} \rho_{A22} & \sigma_{A8\text{-}A22} & 0 \\ \sigma_{A8\text{-}A22} & \rho_{A8} & \sigma_{A8\text{-}A9} \\ 0 & \sigma_{A8\text{-}A9} & \rho_{A9} \end{pmatrix} \tag{5}$$

where ρ and σ represent the longitudinal and the cross-relaxation rate constants, respectively.

The relaxation matrix Γ was determined by a non-linear least-squares method from a set of NOE build up values measured at various mixing times between 10 and 400 ms. A target function :

$$funoe = \Sigma \, (NOE_{cal} - NOE_{exp})^2 \tag{6}$$

is minimized by a simplex method, encoded in the MATLAB software package (The MathWorks, Inc.). NOE_{cal} is the NOE calculated at each step of the minimization process according to equation (4) and NOE_{exp} is the experimental NOE. NOE_{exp} was obtained from the ratio of the volume of the cross-peak to that of the H2 proton diagonal peak at zero mixing time (Kieffer *et al.*, 1993). This initial magnetization was calculated by summing the 2D peak volumes at the frequency of A8H2 which is isolated in the spectrum. Measurement of the intensity of A8H2 resonance line in the first FID of various mixing time NOESY spectra shows that magnetization leakage is negligible up to 400 ms. This is also true for the other H2 protons. In order to avoid rescaling between the various experiments (Kieffer *et al.*, 1993), the initial magnetization volume was calculated from the sum of the cross and diagonal peak volumes at A8H2 frequency in each 2D NOESY map. A complete local 3x3 NOE matrix is obtained by dividing each volume by this initial magnetization. A9H2 and A22H2 diagonal NOE matrix elements have been set to the initial magnetization minus the off-diagonal NOE sum. While the longitudinal relaxation rate for H2 proton of adenines is slow, leading to partial saturation of the signals, it has been shown by extensive simulation and experiment that the errors occurring from this are small (Lane and Fulcher, 1995).

Results and discussion

Melting temperature

The melting temperature of the duplex, determined optically, was 50±1°C. It will be higher in the NMR sample because of the much higher concentration, so that NMR experiments can be safely run on the duplex form up to at least 40 °C.

Assignment of exchangeable and non-exchangeable proton resonances

The non-exchangeable base and sugar protons resonances have been assigned sequentially (Gronenborn and Clore, 1984) from analysis of the connectivities in NOESY and HOHAHA spectra in D_2O buffer. Figure 1 shows part of the NOESY map recorded at 25°C, where cross peaks between the H1´ sugar to H6/8 base protons resonances are found. Sequential connectivities are shown for the non-coding strand of lac UV5. Assignments at 25°C are given in table 1.

The large intra residue NOE observed between H2´ and H8,6 base protons, as well as the strong H1´-H2´ couplings apparent in the HOHAHA spectrum, are characteristic of a B-like conformation for the duplex (Gronenborn and Clore, 1984).

1D spectra recorded in 90% H_2O-10% D_2O, at temperatures ranging from 5 to 35 °C are shown in Figure 2. At 5 °C, imino protons of all five guanines resonate between 12 and 13 ppm. Between 13 and 14 ppm, resonances of only eight out of nine thymine imino protons are observed. Even at 5 °C, the imino proton of thymine 14 is missing in the spectrum. The resonances of T2 (13.95 ppm) and G1 (12.87 ppm) broaden very quickly when the temperature increases.

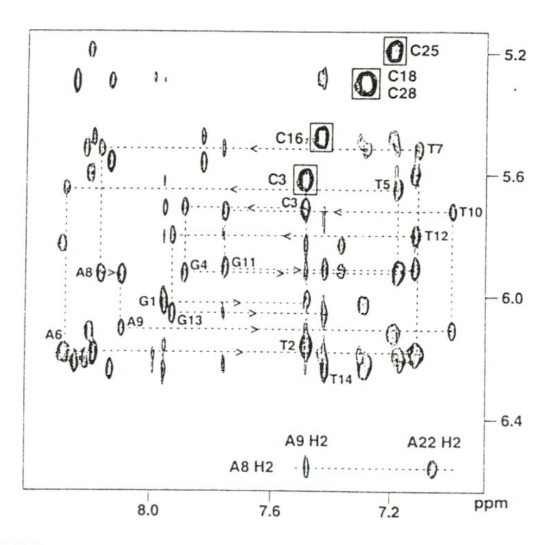

Figure 1: NOESY spectrum of d(GTCGTATAATGTGT)·d(ACACATTATACGCA) at 25℃. Assignments of the base H8-H6 to H1′ protons are shown. The spectrum was recorded at 25℃, 500 MHz with a mixing time of 400 ms.

	H1'	H2'	H2"	H3'	H5	H6	H8	H2	CH3
G1	6.01	2.79	2.69	N.D.	-	-	7.95	-	-
T2	6.15	2.55	2.21	4.91	-	7.48	-	-	1.32
C3	5.71	2.44	2.12	4.86	5.61	7.48	-	-	-
G4	5.93	2.75	2.60	4.95	-	-	7.89	-	-
T5	5.65	2.43	2.07	4.85	-	7.17	-	-	1.47
A6	6.18	2.88	2.62	5.00	-	-	8.28	7.18	-
T7	5.51	2.37	2.00	4.85	-	7.10	-	-	1.39
A8	5.94	2.89	2.67	5.02	-	-	8.16	6.56	-
A9	6.10	2.88	2.52	4.99	-	-	8.08	7.46	-
T10	5.72	2.41	2.03	4.84	-	6.99	-	-	1.27
G11	5.90	2.73	2.53	4.93	-	-	7.75	-	-
T12	5.79	2.32	1.94	4.85	-	7.12	-	-	1.32
G13	6.06	2.68	2.68	4.97	-	-	7.92	-	-
T14	6.23	2.25	2.25	4.54	-	7.42	-	-	1.61
A15	6.18	2.76	2.61	4.83	-	-	8.18	7.98	-
C16	5.27	2.31	2.07	N.D.	5.48	7.42	-	-	-
A17	6.22	2.89	2.71	5.02	-	-	8.23	7.75	-
C18	5.52	2.37	2.00	4.99	5.31	7.28	-	-	-
A19	6.20	2.93	2.64	5.00	-	-	8.21	7.42	-
T20	5.92	2.53	2.01	4.84	-	7.17	-	-	1.31
T21	5.84	2.54	2.21	4.90	-	7.35	-	-	1.58
A22	6.20	2.91	2.64	5.00	-	-	8.29	7.04	-
T23	5.60	2.38	2.00	4.85	-	7.12	-	-	1.35
A24	6.12	2.81	2.62	5.00	-	-	8.21	7.31	-
C25	5.48	2.24	1.85	4.79	5.20	7.19	-	-	-
G26	5.56	2.75	2.65	4.98	-	-	7.82	-	-
A27	6.24	2.88	2.62	4.98	-	-	8.13	7.96	-
C28	6.03	2.10	2.05	4.45	5.30	7.28	-	-	-

Table 1: ^1H NMR assignments of non exchangeable protons of d(GTCGTATAATGTGT)·d(ACACATTATACGCA). Protons were assigned using NOESY and HOHAHA as described in the text. The chemical shifts are given in parts per million (ppm) referenced to DSS resonance at 0 ppm. The data were recorded at 25℃. "N.D." means that the proton has not been assigned.

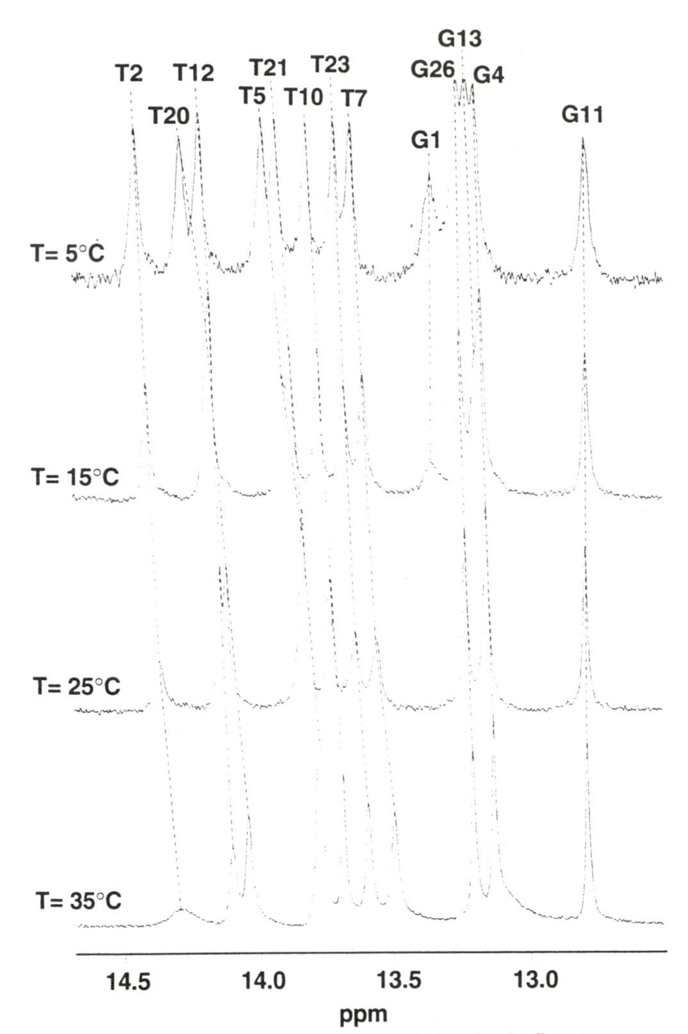

Figure 2: Imino proton region of 1D spectra recorded in H_2O. Spectra were recorded at different temperatures ranging from 5℃ to 35℃.

The assignment of the imino proton (Table 2) is based on a NOESY spectrum recorded at 5 °C (Figure 3). All observed thymine imino protons resonances exhibit a cross-peak with the H2 proton resonance of the paired adenines. All imino protons of guanines, except G1 which is very broad, have NOE with the amino protons of the paired cytosines. Moreover, sequential connectivities between imino protons are observed (peaks labelled from a to j in the box C of Figure 3).

	NH	NH2
G1	12.86	N.O.
T2	13.95	-
C3	-	8.51 / 6.96
G4	12.70	N.O.
T5	13.50	-
A6	-	7.55 / 6.34
T7	13.16	-
A8	-	7.23 / 6.06
A9	-	7.15 / 6.04
T10	13.33	-
G11	12.29	N.O.
T12	13.73	-
G13	12.73	N.O.
T14	N.O.	-
C28	-	8.17 / 6.78
A27	-	N.O.
G26	12.77	N.O.
C25	-	8.10 / 6.52
A24	-	7.60 / 6.41
T23	13.22	-
A22	-	7.47 / 6.33
T21	13.44	-
T20	13.79	-
A19	-	7.46 / 6.12
C18	-	8.11 / 6.61
A17	-	N.O.
C16	-	8.25 / 6.72
A15	-	N.O.

Table 2: ^{1}H NMR assignments of exchangeable protons of d(GTCGTATAATGTGT)·d(ACACATTATACGCA). Protons were assigned using NOESY as described in the text. The chemical shift are given in parts per million (ppm) referenced to DSS resonance at 0 ppm. The data were recorded at 5℃. "N.O." means that no resonance was observed.

All amino protons of the cytosines could be assigned through their dipolar interaction with H5 protons at 5 °C. Strong exchange peaks due to rotation around the C-N bond were observed between the hydrogen bonded and the free amino protons (Figure 3).

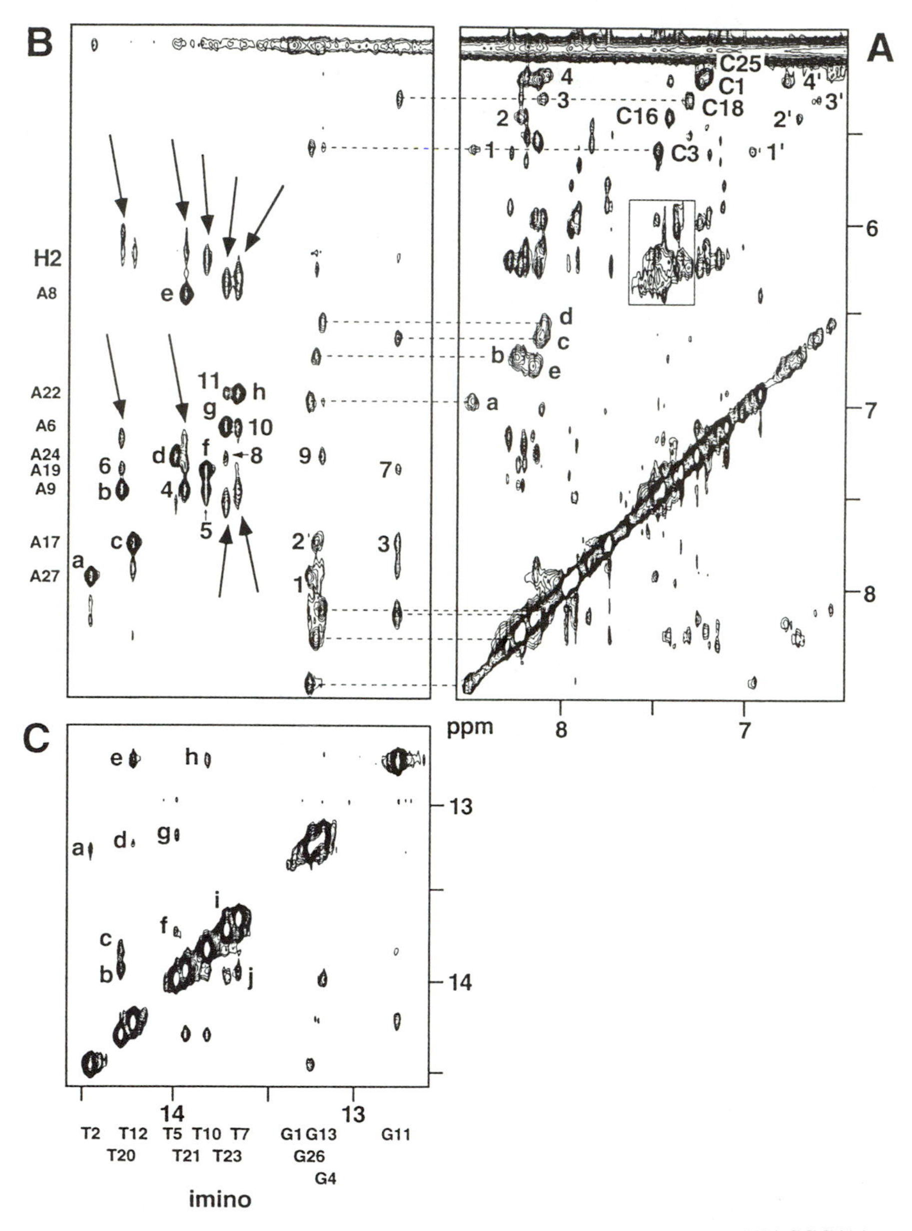

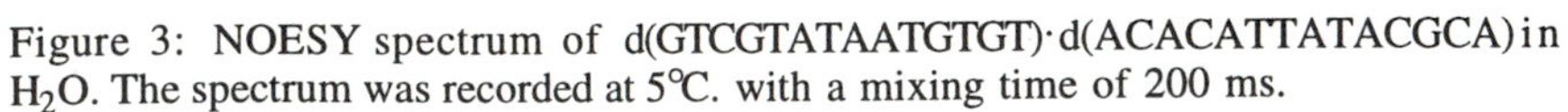

Figure 3: NOESY spectrum of d(GTCGTATAATGTGT)·d(ACACATTATACGCA) in H_2O. The spectrum was recorded at 5℃. with a mixing time of 200 ms.

A: -Peaks labelled with residue name are cross peaks between H5 and H6 protons of cytosines.

-Peaks labelled with letters correspond to exchange peaks between the bound and free amino protons of cytosines.
-Peaks labelled with numbers are cross peaks between respectively bound and free (primed number) amino protons of the cytosines.

NH2 exchange peaks	a	b	c	d	e
NH2-H5 peaks	1-1'	2-2'	3-3'	4-4'	N.O.
Residue name	C3	C16	C18	C25	C28

-The boxed area indicates the very broad exchange peaks between the bound and free amino protons of adenines.

B: -Peaks labelled with letters are cross peaks between the H2 proton of adenines and the imino proton of thymines of the same base pair.

a:	T2-A27	e:	T21-A8
b:	T20-A9	f:	T10-A19
c:	T12-A17	g:	T23-A6
d:	T5-A24	h:	T7-A22

-Peaks labelled with numbers correspond to sequential NOE between H2 protons of adenines and imino protons of thymines or guanines.

1:	A27-G26	5:	A9-T10	9:	A24-G4
2:	A17-G11	6:	A19-T20	10:	A6-T7
3:	A17-G13	7:	A19-G11	11:	A22-T23
4:	A9-T21	8:	A24-T23		

C: -Labelled peaks are sequential NOE between imino protons of guanines and thymines.

a:	T2-G26	e:	T12-G11	i:	T7-T23
b:	T20-T21	f:	T5-T23	j:	T7-T21
c:	T20-T10	g:	T5-G4		
d:	T12-G13	h:	T10-G11		

At 5 °C, some of the imino protons of thymines showed NOEs to unassigned resonances (peaks shown by arrows in the box B of Figure 3). T23(N3H) proton has a dipolar interaction with protons which resonate at 7.55 and 6.34 ppm. T7(N3H), T20(N3H) and T21(N3H) gave smaller NOEs to protons at 7.47 and 6.33 ppm, 7.15 and 6.04 ppm and 7.45 and 6.06 ppm respectively. As all H2 protons of adenines have been assigned, the only remaining unassigned protons sufficiently close to the imino proton of thymines are the amino protons of adenines. These amino protons usually are not detected, because they are broadened due to intermediate rotation rates around their C-NH2 bond (Patel *et al.*, 1987) or fast exchange with solvent, but in the present case, exchange was slow enough so that they could be observed. Interestingly, the exchange of adenine amino protons is slowed down for all AT base pairs located in the Pribnow box sequence. No NOEs between thymine imino proton and adenine amino protons were observed for the other A-T bases in the sequence: T2-A27, T5-A24, T10-A19 and T12-A17.

Chart 2 summarizes the pairings observed by NMR from 5 °C to 35 °C. A thymine imino to adenine H2 proton NOE is indicative of a hydrogen bonded A-T pair. Similarly, an NOE between the guanine imino proton and H5 of cytosine, relayed by its amino protons, indicates that the G is hydrogen bonded to the C.

At 25 °C, the two base pairs at both terminal ends (G1-C28, T2-A27, G13-C16 and T14-A15) are not continuously paired because of fraying, as shown by the disappearance of their resonances. However, integral measurements of thymines 5, 7, 10, 20, 21 and 23 imino proton resonances in the 1D spectra of Figure 2, show that virtually no exchange occurs for these protons up to 25 °C. Strong pairing of the six Pribnow box AT pairs is confirmed by the imino to H2 NOE of comparable intensities.

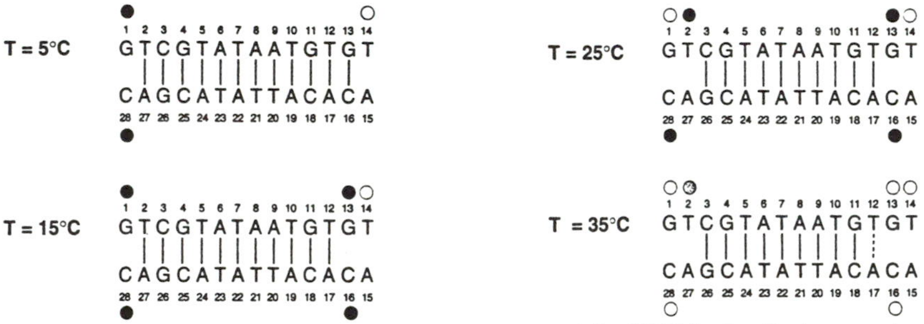

T = 5°C

1 2 3 4 5 6 7 8 9 10 11 12 13 14
GTCGTATAATGTGT
CAGCATATTACACA
28 27 26 25 24 23 22 21 20 19 18 17 16 15

T = 25°C

1 2 3 4 5 6 7 8 9 10 11 12 13 14
GTCGTATAATGTGT
CAGCATATTACACA
28 27 26 25 24 23 22 21 20 19 18 17 16 15

T = 15°C

1 2 3 4 5 6 7 8 9 10 11 12 13 14
GTCGTATAATGTGT
CAGCATATTACACA
28 27 26 25 24 23 22 21 20 19 18 17 16 15

T = 35°C

1 2 3 4 5 6 7 8 9 10 11 12 13 14
GTCGTATAATGTGT
CAGCATATTACACA
28 27 26 25 24 23 22 21 20 19 18 17 16 15

Chart 2: Schematic representation of base pairing of the UV5 duplex for temperatures ranging from 5°C to 35°C.

| strong NOE are observed between protons of the facing bases

| weak NOE are observed between protons of the facing bases

● no NOE is observed between protons of the facing bases, but the exchangeable proton of the base is clearly observed

● no NOE is observed between protons of the facing bases, the exchangeable proton of the base is very broad and difficult to observe

○ no exchangeable proton is observed at all

Conformational transition

Temperature dependence of line-width and chemical shift behavior of proton resonances

Assignment data have been obtained for temperatures from 5 °C to 40 °C. On increasing temperature, the H2 proton resonances of A6, A8, A22 and A24 exhibit a peculiar behavior which is illustrated by the well-isolated resonance of A8H2. The line-width of this resonance first increases from 2.5 Hz at 5 °C to a maximum of about 10 Hz at 25 °C and then decreases (Figure 4). A22H2 is not fully resolved at temperatures over 25 °C, but between 5 and 25 °C its line-width increases by at least 6Hz (Figure 4). H2 protons of adenines 6 and 24 are not resolved at all temperatures in the range 5 to 45 °C, precluding line-width measurements. However, at low temperatures (data not show), the line-width of both protons is much smaller at 5 °C than at 25 °C and it becomes narrower again at temperatures higher than 35 °C. Such behavior, which reflects a conformational exchange in the fast intermediate range, was observed only for these four H2 protons.

Because of strong overlap in 1D spectra, the line-widths of other protons could not be measured. However, visual inspection in the aromatic region of the spectra (data not shown) does not show peculiar broadening around 25-30 °C.

Proton chemical shift variations as a function of temperature are linear between 5 and 45 °C (Table 3). As usually observed, the temperature coefficients are negative for all base protons, except for H2 protons, which exhibit a positive temperature dependence of their chemical shift. The temperature coefficients are very similar for all base protons. However, large variations are observed among H2 protons. A8H2 and A22H2 show a very strong temperature dependence, while A6H2 and A24H2 have temperature coefficients similar to the others.

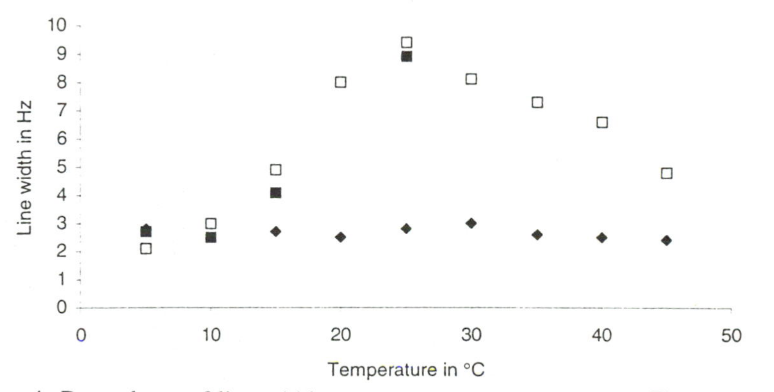

Figure 4: Dependence of line-widths on temperature for the A8H2 (□), A22H2 (■), and DDS (◆) protons.

	$10^3 \frac{d\delta}{dT}$ (ppm / °C)	
	H6/H8	H2
G1	0.6	-
T2	-1.4	-
C3	N.D.	-
G4	-1.3	-
T5	N.D.	-
A6	-2.7	3.2
T7	-2.3	-
A8	-1.0	7.9
A9	-1.7	1.6
T10	-1.1	-
G11	-0.9	-
T12	N.D.	-
G13	-1.2	-
T14	0.8	-
A15	N.D.	2.1
C16	N.D.	-
A17	N.D.	1.4
C18	N.D.	-
A19	N.D.	3.3
T20	N.D.	-
T21	-1.4	-
A22	-2.0	6.8
T23	-1.3	-
A24	-1.7	2.0
C25	N.D.	-
G26	-1.7	-
A27	-1.3	1.1
C28	N.D.	-

Table 3: Temperature coefficients of base proton resonances in d(GTCGTATAATGTGT)·d(ACACATTATACGCA). $d\delta/dT$ was determined by linear regression to the shift versus temperature profiles measured in 5 °C steps from 5 °C to 40 °C.

Rotating frame T1 experiments

Exchange rates in the fast intermediate regime can be determined by measurements of spin-lattice relaxation in the rotating frame ($R_1\rho$) as a function of spin-lock strength (Deverell *et al.*, 1970). Both the rate constant for chemical exchange, k, and the apparent chemical shift difference, $\delta\nu_{app}$, can be determined independently (see Materials and Methods). We have measured $R_1\rho$ at 25 °C, the temperature at which the broadest line widths of H2 resonances are observed. The relaxation of four H2 (A8, A9, A22 and A24) and two H6 (T10 and T21) could be measured using 1D spectra.

As shown in Figure 6, relaxation rates remain independent of the spin-lock field strength for A9H2, T10H6, and T21H6, while they strongly vary for A8H2, A22H2, and A24H2. For these protons, a non-linear fit to the data (Table 4), according to equation (3), gave a mean exchange rate constant of 4,210 s^{-1}. The apparent chemical shift difference (Table 4), $\delta\nu_{app}$, which contains the product of the two site populations (equation (3)), is different from

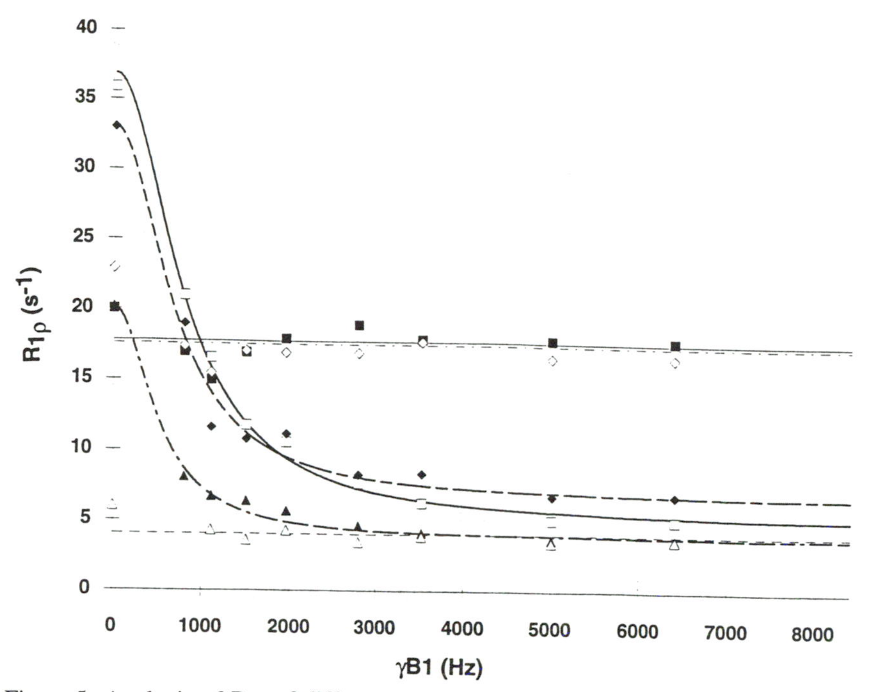

Figure 5: Analysis of $R_{1\rho}$ of different protons. $R_{1\rho}$ is represented in s^{-1} at 25°C as a function of the spin-lock frequency in Hz.

—□— A8 H2
---△--- A9 H2
—■— T10 H6
---◇--- T21 H6
---◆--- A22 H2
---▲--- A24 H2

$R_{1\rho}$ values were determined from magnetization decay curves at each spin-lock strength. Lines are best fit to equation (3).

one proton to the other, and correlates the temperature coefficients, p, reported in Table 3. A linear least squares fit performed on the three points gives:

$$\delta\nu_{app} = 12{,}077.5 \cdot p + 78 \qquad (7)$$

with a correlation coefficient of 0.97. This correlation between p and $\delta\nu_{app}$ is expected, because the temperature coefficient reflects the monotonic change in bulk susceptibility and in conformation seen on other protons, as well as the chemical shift change due to the temperature-dependent conformational exchange.

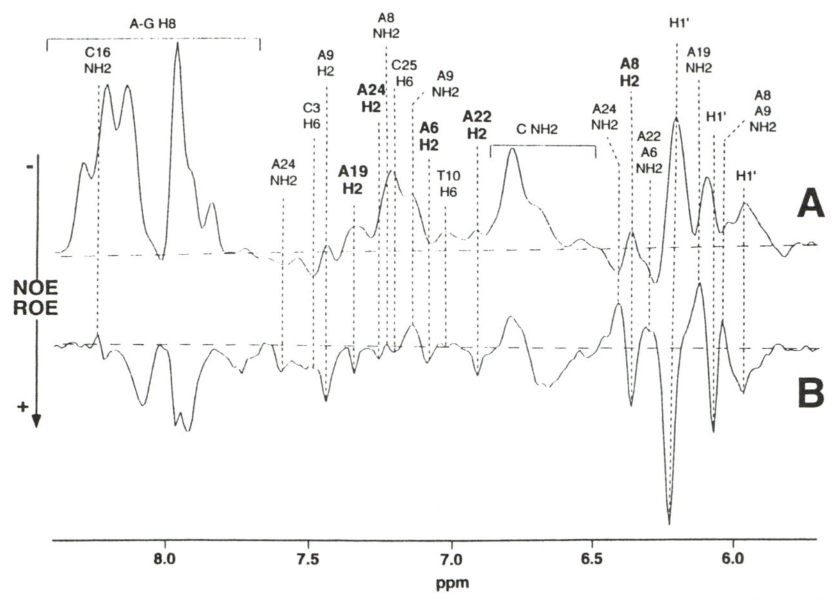

Figure 6: Cross sections through the two-dimensional NOESY (mixing time of 60 ms) and ROESY (mixing time of 30 ms) spectra of lac UV5 taken along ω_2 at the ω_1 chemical shift of the water resonance at 5 °C between 9 and 6 ppm. A. NOESY. B. ROESY

	$R_{1\rho}^{\infty}$ (s^{-1})	$\Delta\nu_{app}$ (Hz)	k (s^{-1})
A8 H2	4.9	180	5000
A9 H2	4.0	-	-
A22 H2	6.6	153	4360
A24 H2	3.8	104	3270
T10 H6	17.8	-	-
T21 H6	17.6	-	-

Table 4: Rate constants and frequency differences from rotating-frame T_1 measurements. $R_{1\rho}^{\infty}$ is the value of the rotating-frame relaxation rate constant at an infinitely large spin-lock field strength. $\Delta\upsilon_{app}$ is the effective frequency difference, and k is the first-order rate constant for the transition. The parameters were obtained by non-linear regression to equation (3). Data were obtained at 25 °C and 600 MHz.

The relaxation rates of A6H2 could not be determined reliably owing to overlap. However, the observed broadening of the line when increasing temperature suggests that this proton experiences the same chemical shift exchange as its neighbors.

Thus, at 25 °C, the four consecutive AT base pairs (from T5-A24 to A8-T21) in the Pribnow box are in conformational exchange with a time constant of 238 μs. As already suggested, the Pribnow box sequence could induce a stable hydration shell. The conformational equilibrium observed at 25 °C correlates with the melting of this hydration shell (*vide infra*).

Change in adenine H2-H2 distances

Cross relaxation rates were measured for the two couples, A8H2-A9H2 and A8H2-A22H2. The three spins lying in the minor groove of the oligonucleotide can be considered as isolated from the others. Dipolar interactions with other protons in the minor groove, especially H1′(i-1), H1′(i+1) and other H2 protons, were neglected because of very small NOE observed compared with those between A8H2 and A9H2, and A8H2 and A22H2. The cross-relaxation between the two extreme protons (A22 and A9) was considered as negligible.

Two sets of 12 2D NOESY experiments, recorded in D_2O, have been carried out at 5 and 40 °C for mixing times ranging from 10 to 400 ms. The relaxation matrices for the A22H2, A8H2 and A9H2 spin subset were calculated as described in Materials and Methods. For data taken at 40 °C, convergence was achieved after 400 steps of minimization and the final value of the target function (funoe, equation (6)) was 10^{-3}. The calculation using data recorded at 5 °C, required 541 steps and ended with a final value funoe equal to 6.10^{-3}. Table 5 gives the σ values obtained for the two H2 protons pairs, A8-A9 and A8-A22, at 5 and 40 °C.

The cross-relaxation rate constant, σ_{ij}, between two protons, i and j, is related to structural and dynamical parameters as follow (Abragam, 1961) :

$$\sigma_{ij} = (a / r_{ij}^6) \cdot (6\, J(\omega_i + \omega_j) - J(\omega_i - \omega_j)) \qquad (8)$$

r_{ij} being the distance between protons i and j, $J(\omega)$ the spectral density function, and $a = 56.92\ 10^{-49}\ m^6 s^{-2}$ for homonuclear proton relaxation. Macromolecules have a long correlation time τ_c, and equation (8) reduces to (Abragam, 1961):

$$\sigma_{ij} = -(a / r_{ij}^6) \cdot J(0) = -(a / r_{ij}^6) \cdot \tau_c \qquad (9)$$

τ_c is related to the viscosity η and the temperature T by the Stokes-Einstein equation:

$$\tau_c = (\eta V) / (k_B T) \qquad (10)$$

where V is the hydrated molecular volume and k_B the Boltzmann constant.

The ratio between the cross relaxation rate constants at two temperatures is:

$$\sigma 40 / \sigma 5 = ((\eta_{40} T_5) / (\eta_5 T_{40})) \cdot ((V_{40} r_5^6) / (V_5 r_{40}^6))$$
$$= 0.38 \cdot (V_{40} r_5^6) / (V_5 r_{40}^6) \qquad (11)$$

If the hydrodynamic volume and the distance do not change (rigid body), the σ_{40}/σ_5 ratio should be equal to 0.38. The ratios calculated for the A8H2-A9H2 and A8H2-A22H2 vectors are much smaller (Table 5). The estimated error in the determination of σ is less than 10%. The observed values for the ratio σ40/σ5 are 0.17±0.03 for A8H2-A9H2 and 0.24±0.05 for A8H2-A22H2. These ratios are about a factor of two lower than the expected ratio of 0.38 and are different from it by 4 to 5 standard deviations. This suggests that either the hydrodynamic volume has decreased from 5 to 40 °C or the distance has increased upon going from low to elevated temperature.

The cross relaxation rate constant for the cytosine H5-H6 vectors of C16 and C18, the H5 and H6 protons of which are not overlapped at 5 and 40 °C, have been measured. The results are given in Table 5. Due to the position of H5 and H6 on the ring of the cytosine, the distance between these two protons does not change with temperature. It is thus possible to determine whether the hydrodynamic volume of the duplex changes with temperature. The σ40/σ5 ratio for the cytosine C16 (0.30 ± 0.06) is smaller than the theoretical value of a rigid body (0.38). This is likely to be due to the fact that at 40 °C, the G13-C16 pair is not fully paired due to fraying (Chart 2). Rapid motion due to fraying decreases the s value at 40 °C and thus the σ40/σ5 ratio. On the other hand, the value found for the C18 cytosine (0.43 ± 0.08) is in agreement with the theoretical value of a rigid body (0.38). Hence, it can be postulated that the hydrodynamic volume of the lacUV5 duplex does not change with the temperature.

	T = 5°C		T = 40°C			
Vector	σ / s^{-1}	r / Å	σ / s^{-1}	r / Å	σ_{40} / σ_5	Δr / Å
A8H2–A9H2	-0.66	3.27	-0.11	3.72	0.17	0.45
A8H2–A22H2	-0.89	3.11	-0.21	3.34	0.24	0.23
C16H5–H6	-3.4	2.46	-1.0	2.46	0.30	-
C18H5-H6	-2.3	2.46	-1.0	2.46	0.43	-

Table 5: Cross-relaxation rate constants for H5-H6 vectors of cytosines and H2-H2 vectors of adenines. Cross-relaxation rates were determined from NOESY experiments at different mixing times and analyzed using the relaxation matrix as described in the text. The distances r were calculated as described in the text.

The correlation time τH5-H6 for the H5-H6 vector can be calculated from equation (9) using a fixed distance of 2.46Å. At 5 °C, it is found to be 9 ns for cytosine C16 and 13 ns for cytosine C18 (τH5-H6 = 13 ns will be used in the following calculations because the correlation time found for C16 may be affected by end effects), and at 40 °C, it is 4 ns. In B-DNA, the H2-H2 vector of sequential adenines is aligned at an angle of about 30° to the helix axis, and therefore the spectral density function will be slightly different from that used to described the cytosine H5-H6 vector. For a 14-mer, the correlation times for the rotation about the long and short axis can be calculated from τH5-H6 as described by Birchall and Lane (1990), assuming an axial rise of 3.38 Å. The effective correlation time for H2-H2 vectors is then approximately 1.3 (τH5-H6). Using the correlation time for overall tumbling as described above, equation (9) can now be used to determine the distance between H2 protons of adenines (Table 5).

The sequential A8H2-A9H2 distance increases from about 3.3 Å at low temperature to about 3.7 Å at high temperature. The value at high temperature is comparable to that for standard B-DNA (ca. 3.8 Å). The cross-strand A8H2 and A22H2 protons remain separated by a short distance, which implies a local structure somewhat different from standard B-DNA. TpA steps seem to be remarkably plastic, because in some instances one finds a narrow minor groove with large propeller twists (Goodsell *et al.*, 1994) and in others the opposite (Kennedy *et al.*, 1993; Liepinsh *et al.*, 1994). It may be that what has been observed in the lacUV5 and in the trp promoters (Lefèvre *et al.*, 1985; Lane *et al.*, 1993) is a manifestation of this plasticity, with high temperatures favoring the flatter propeller twisting and wider minor groove.

Hydration water molecules

Dipolar contacts between protons of water and DNA molecules

The hydration water molecules can be identified through their dipolar interaction with protons of the molecule (Otting *et al.*, 1991). This dipolar interaction has been observed in cross sections taken along ω_2 at the ω_1 chemical shift of the water resonance in two-dimensional NOESY and ROESY recorded in H_2O. Some cross-sections extracted from NOESY and ROESY spectra of lac UV5 at 5°C are shown in Figure 6. Four different cases can be observed :

(i) NOE and ROE are both negative: they correspond to the exchange of imino and amino protons with water.

(ii) NOE are negative and ROE positive: they may correspond to dipolar interaction between H3' protons, which resonate very close to the water proton frequency, and other sugar protons (H1´, H2´, H2´´, H4´, H5´ and H5´´). Such peaks were also observed for some aromatic H6 and H8 protons. Since magnetization transfer from H3' cannot be completely ruled out (in B DNA, these protons lie within 5 Å from H6 and H8 protons), one cannot decide whether these peaks are dipolar interactions between the aromatic protons and water protons lying in the major groove or not.

Negative NOEs and positive ROEs peaks were also observed for H2 of adenines belonging to the Pribnow box sequence. Since these protons are quite isolated from other DNA protons that resonate at the water frequency, we conclude that there are dipolar interactions between these H2 protons and water protons lying in the minor groove. The lifetime of water near H2 protons must be long enough (more than 1 ns (Otting *et al.*, 1991)) to allow for the observation of a negative NOE. It is noteworthy that these NOEs and ROEs are still observable at 15 °C.

(iii) NOE and ROE are both positive: this is indicative of hydration water remaining near these protons for less than 500 ps (Otting *et al.*, 1991). Observation of such NOEs and ROEs for C3H6 and methyl protons is in agreement with previous observations on another DNA sequence by Liepinsh *et al.* (1992).

Since the NOEs and ROEs observed for NH_2 of adenines may involve dipolar interactions with water as well as HOHAHA transfer between the bound and free amino protons and exchange with solvent or arise from rotation about the C-N bond, any interpretation of the peaks observed for amino protons of A6, A22 and A24 (Figure 6), in terms of hydration water, would be too speculative.

At 5 °C, dipolar interactions between water and the H2 protons of A6, A8, A19, A22, and A24 are observed (Figure 6), indicating that stable hydration water molecules are present in the minor groove of the DNA. Interestingly, these stable water molecules are mainly concentrated in the Pribnow box, suggesting that a hydration shell forms around this sequence. The water does not cover the whole oligonucleotide, suggesting that the TATAAT sequence has special hydration properties. The hydration shell is consistent with the fact that at low temperature, the amino protons of adenines in the TATAAT sequence are observed, while they are barely visible in other types of sequence, such as in the rest of the oligonucleotide.

On increasing the temperature, it becomes difficult to identify NOEs between water molecules and non-exchangeable protons. The exchange at 25 °C of all amino protons becomes very fast and huge positive peaks are observed in both the cross sections of the NOESY and ROESY spectra (data not shown), preventing a clear observation of other effects. This means that access of free solvent molecules is increasing for all the amino protons. However, some small, positive ROEs were observed between water protons and A6H2, A19H2 and A22H2, indicating that these water protons still retain a bound state lifetime of the order of the nanosecond.

Conclusion and biological relevance

A two-state conformational equilibrium in the fast intermediate range was found in the lacUV5 duplex as revealed by the temperature dependence of line-widths and by rotating frame experiments. This conformational change has also been observed in other sequences containing the TAA motif (Lefèvre *et al.*, 1985; Schmitz *et al.*, 1992; Kennedy *et al.*, 1993; Lane *et al.*, 1993).

Below the transition temperature, it seems that the Pribnow box region has a narrow minor groove that is stabilized by relatively strongly ordered hydration water molecules. Increasing the temperature induces a conformational transition in which the minor groove widens and from which water can exchange readily.

A stable hydration shell may be a property of AT rich sequences. In fact, bound water molecules were observed mainly at the level of AT base pairs by crystallography (Drew and Dickerson, 1981) as well as by NMR (Liepinsh *et al.*, 1992; Kubinec and Wemmer, 1992). These studies were carried out on DNA fragments containing $d(A_nT_n)$ segments. The main characteristic of these sequences is a narrow minor groove occupied by a spine of hydration. Very recently, studies on DNA fragments containing $d(T_nA_n)$ segments have been carried out by crystallography (Goodsell *et al.*, 1994). In agreement with our results, hydration water molecules were identified in the whole minor groove, especially at the TpA step of the duplex $d(CCATTAATCC)_2$. However, Kennedy *et al.* (1993) have observed, by NMR, an abrupt widening of the minor groove at the TpA step of $d(CGAGGTTTAAACCTCG)_2$ duplex at all temperatures. Only the adenine at the TpA junction was found conformationally mobile and exhibited a low temperature conformational change. Moreover, Liepinsh *et al.* (1994) did not observe any water molecules with long lifetimes in the minor groove of the self-complementary $d(GTGGTTAACCAC)_2$ duplex. They suggested that a wider minor groove, compared to the related duplex containing AATT tract, would not allow for the spine of hydration to be stabilized. They pointed out that the TpA step could disfavor the presence of hydration water because of a large minor groove. Nevertheless, these two results (Kennedy *et al.*, 1993; Liepinsh *et al.*, 1994) are not in contradiction with the fact that the melting of the strong hydration shell around the Pribnow box sequence and the observed conformational change are strongly correlated. Indeed, the hydration state of the $d(CGAGGTTTAAACCTCG)_2$ duplex has not been determined, nor has the dynamic behavior of the d(GTGGTTAACCAC)$_2$ duplex been studied as a function of temperature. These various results suggest that TAAX subsequences are highly plastic; their properties in solution may depend not only on the solution conditions, but also on the flanking sequences.

The stable hydration shell may also explain the exceptional stability of imino protons in sequences containing several consecutive adenines (Leroy *et al.*, 1988). As the base pair must open for the imino proton to exchange (for a review see Guéron and Leroy, 1992), the hydration shell may stabilize the closed form of the pair.

Interestingly, the initiation of transcription by RNA polymerase is also a temperature-dependent process. It has been shown (Buc and McClure, 1985) that at low temperature, the complex between RNA polymerase and lac UV5 promoter does not isomerize and stays in a so called closed complex which cannot initiate transcription. The isomerization toward an open complex, able to initiate transcription, occurs when the temperature increases, with an apparent transition temperature of 25 °C. It is thought that the first closed complex involves mainly the recognition of the -35 consensus sequence (TTGACA) by the polymerase. The consecutive isomerization in an open complex requires a recognition and an unwinding of the Pribnow box (Chan *et al.*, 1990). The results of the present study may explain why the complex can be frozen in the closed form at temperatures below 25 °C. The hydration shell may be stable enough to prevent the Pribnow box from recognition. Nature may have selected AT rich sequences as key sequences in promoters in order to have a temperature dependent switch controlling transcriptional activity.

Acknowledgements

This work was supported by grants from the French 'Ministères des Affaires Etrangères' (Programme ALLIANCE) and the British Council for travel funds. We thank Dr. Bruno Kieffer for suggesting and implementing the experiment for the measurement of the H5-H6 cross-relaxation rates.

References

Abragam, A. (1961). in The Principles of Nuclear Magnetism, Oxford University Press, New York.

Birchall, A. J., and Lane, A. N. (1990). *Eur. Biophys. J.* **19**, 73.

Buc, H., and McClure, W. R. (1985). *Biochemistry* **24**, 2712.

Chan, B., Minchin, S., and Busby, S. (1990). *FEBS Lett.* **267**, 46.

Davis, D. G., and Bax, A. (1985). *J. Am. Chem. Soc.* **107**, 2820.

Deverell, C., Morgan, R. E., and Strange, J. H. (1970). *Mol. Phys.* **18**, 553.

Drew, H. R., and Dickerson, R. E. (1981). *J. Mol. Biol.* **151**, 535.

Farmer, B. T., Macura, S., and Brown, L. R. (1988). *J. Magn. Reson.* **80**, 1.

Freeman, R., and Hill, H. D. W. (1971). *J. Chem. Phys.* **55**, 1985.

Goodsell, D. S., Kaczor-Grzeskowiak, M., and Dickerson, R. E. (1994). *J. Mol. Biol.* **239**, 79.

Gronenborn, A. M., and Clore, G. M. (1984). *Prog. NMR Spectrosc.* **17**, 1.

Guéron, M., and Leroy, J. L. (1992). *Nucleic Acids and Molecular Biology* **6**, 1

Hawley, D. K., and McClure, W. R. (1983). *Nucleic Acids Res.* **11**, 2237.

Keepers, J. W., and James, T. L. (1984). *J. Magn. Reson.* **75**, 404.

Kennedy, M. A., Nuuturo, S. T., Davis, J. T., Drobny, G. P., and Reid, B. R. (1993). *Biochemistry* **32**, 8022.

Kieffer, B., Koehl, P., Plaue, S., and Lefèvre, J. F. (1993). *J. Biomol. NMR*, **3**, 91.

Kubinec, M. G., and Wemmer, D. E. (1992). *J. Am. Chem. Soc.* **114**, 8739.

Lane, A. N., Bauer, C. J., and Frenkiel, T. A. (1993). *Eur. Biophys. J.* **21**, 425.

Lane, A. N., and Fulcher, T. (1995). *J. Magn. Reson.*, *in press*,

Lane, A. N., Lefèvre, J. F., and Jardetzky, O. (1987). *Biochim. Biophys. Acta*, **909**, 58.

Lefèvre, J. F., Lane, A. N., and Jardetzky, O. (1985). *FEBS Lett.* **190**, 37.

Leroy, J. L., Charretier, E., Kochoyan, M., and Guéron, M. (1988). *Biochemistry* **27**, 8894.

Liepinsh, E., Leupin, W., and Otting, G. (1994). *Nucleic Acids Res.* **22**, 2249.

Liepinsh, E., Otting, G., and Wüthrich, K. (1992). *Nucleic Acids Res.* **20**, 6549.

Macura, S., Wüthrich, K., and Ernst, R. R. (1982). *J. Magn. Reson.* **46**, 269.

Malan, T. P., Kolb, A., Buc, H., and McClure, W. R. (1984). *J. Mol. Biol.* **180**.

McClure, W. R. (1980). *Proc. Natl. Acad. U.S.A.* **77**, 5634.

Otting, G., Liepinsh, E., and Wütrich, K. (1991). *Science* **254**, 974.

Patel, D. J., Shapiro, L., and Hare, D. (1987). *Quart. Rev. of Biophys.* **20**, 35.

Piotto, M., Saudek, V., and Sklenar, V. (1992). *J. Biol. NMR* **2**, 661.

Rance, M., Bodenhausen, G., Wagner, G., Wüthrich, K., and Ernst, R. R. (1985). *J. Magn. Reson.* **62**, 497.

Rosenberg, M., and Court, D. (1979). *Ann. Rev. Genet.* **13**, 319.

Schmitz, U., Sethson, I., Egan, W. M., and James, T. L. (1992). *J. Mol. Biol.* **227**, 510.

Wüthrich, K., and Marion, D. (1983). *Biochem. Biophys. Res. Comm.* **113**, 967.

Section 4: *In vivo* Spectroscopy

21

Clinicians Need Localized Proton MRS of the Brain in the Management of HIV-Related Encephalopathies

P.J. Cozzone, J. Vion-Dury, S. Confort-Gouny, F. Nicoli, A.-M. Salvan, and S. Lamoureux

Centre de Résonance Magnétique Biologique et Médicale (CRMBM)
Faculté de Médecine
13005 Marseille, France

The neurological complications of AIDS (neuroAIDS) represent the principal cause of disability and death in HIV-patients (Gray *et al.*, 1993; McArthur, *et al.*, 1993). Several types of lesions affect the brain tissue: direct infection of the nervous tissue by HIV, opportunistic infections (such as toxoplasmosis, cytomegalovirus encephalitis, tuberculosis, progressive multifocal encephalopathy....), and lymphomas. The AIDS-related dementia complex (ADC) affects about 60% of patients in the late stage of AIDS. ADC is characterized by the occurrence of sub-cortical dementia with cognitive, behavioral, and motor decline, psychomotor slowing and apathy. ADC is related to the presence of a diffuse encephalopathy leading to a cortical and sub-cortical atrophy, as well as diffuse white-matter lesions. There is a distinct advantage in diagnosing as early as possible the neurological complications (*e.g.* encephalopathy) of AIDS, since early treatment can improve significantly the quality of life in patients by slowing down or even stopping the neurological and psychological degradation.

Neuroimaging techniques, and mainly magnetic resonance imaging (MRI), constitute so far the best diagnostic tools of neuro-AIDS (Kent *et al.*, 1993; Mundinger *et al.*, 1992). In addition, localized magnetic resonance spectroscopy (MRS) of the brain provides a non-invasive exploration of intracerebral metabolism *in vivo*, and can be performed following a standard MRI examination [for a review see Vion-Dury *et al.* (1994)]. Several key molecules of brain metabolism can be detected, including N-acetyl-aspartic acid which is thought to be a neuronal marker, choline-containing molecules (involved in phospholipid metabolism), glutamate, glutamine, inositol, phosphocreatine and creatine, and lactate.

Recently, significant modifications in the concentration of brain metabolites detected by phosphorus and proton MRS have been described in patients with ADC (Bottomley *et al.*, 1992; Deicken *et al.*, 1991; Menon *et al.*, 1992; Chong *et al.*, 1993; Meyerhoff *et al.*, 1993; Confort-Gouny *et al.*, 1993). In a preliminary study, we have observed that even, if MR images are normal (without atrophy, focal or diffuse lesions) or if the patients are neuro-asymptomatic, the values of metabolic parameters measured by MRS are often modified (Vion-Dury *et al.*, 1994).

We present here, in a synthetic perspective, a pool of results obtained in Marseille by the Neurometabolism Group of CRMBM on HIV-patients with AIDS-related encephalopathy.

Subjects and methods

The details of protocols are described in the reference by Vion-Dury *et al.*, (1994). In summary, this study was performed on nine healthy volunteers who did not belong to any high risk group for HIV infection, and 66 HIV-patients: i) 16 seropositive patients who were asymptomatic with respect to any cerebral dysfunction and had no neurological or psychiatric clinical signs, ii) 47 patients who had AIDS-related dementia complex (ADC), which was sometimes associated with opportunistic infections such as toxoplasmosis (n = 4) or progressive multifocal leukoencephalopathy (n = 3), and iii) 3 children with fœto-maternal contamination and subsequent HIV-related encephalopathy.

Magnetic resonance examinations were conducted on a Siemens Magnetom SP63 (1.5 T) at the Timone Hospital in Marseille. Standard spin-echo (SE; TE = 60 and 120 ms, TR = 3500 ms) or "Turbo" spin-echo imaging (TPSE; TE = 90, TR = 3500 ms) or gradient echo (FLASH-2D, TR = 380 ms, TE = 10 ms, flip angle = 90°) sequences were applied in the transverse plane.
Proton MRS of the brain was performed at 63 MHz using the PRESS sequence (echo time = 135 ms, repetition time = 1.600 s) following a CHESS sequence to suppress the water signal [13]. The spectroscopic volume of interest (VOI) (8 cm^3) was always located in the parieto-occipital region of the brain. In order to document the metabolic consequences of HIV-related encephalopathies, the VOI was always placed outside any focal opportunistic lesion. Typical proton MR spectra of a patient with HIV related encephalopathy and of a control subject are presented in Figure 1.

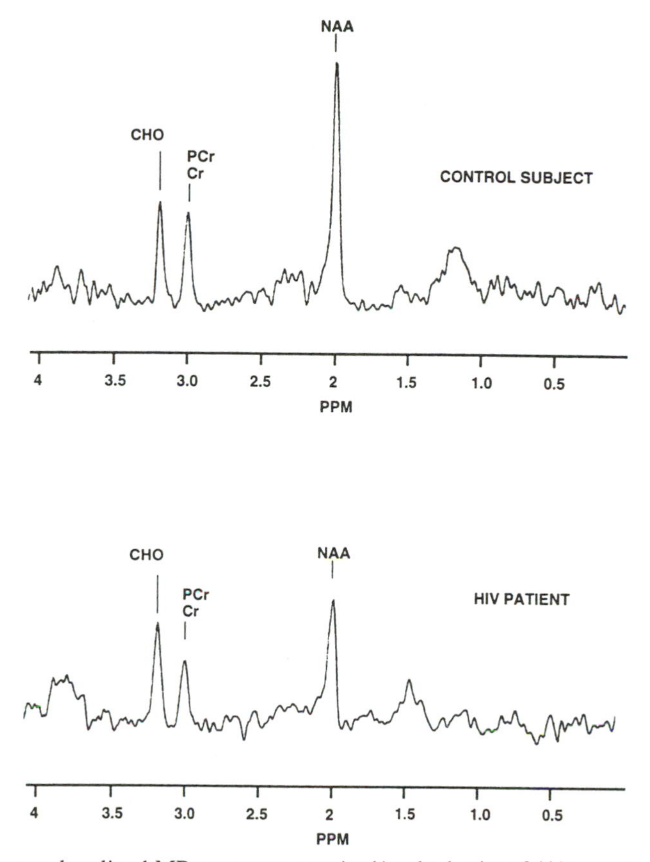

Figure 1: Proton localized MR spectrum acquired in the brain of (A) a control subject and of (B) a patient with ADC. Proton MRS of the brain was performed at 63 MHz using the PRESS sequence (echo time = 135 ms, repetition time = 1.600 s) following a CHESS sequence to suppress the water signal. The spectroscopic volume of interest (VOI) (8 cm^3) was always located in the parieto-occipital region of the brain and placed outside any focal opportunistic lesion. Areas of signals from N-acetyl-aspartate (NAA), choline-containing compounds (Cho) and creatine-phosphocreatine (Cr-PCr) were measured using the Siemens software.

Brain metabolic variations in HIV-related encephalopathy

In a first protocol, the mean values of each metabolic ratio calculated for each statistical population have been compared. Comparison of means was based on a Student's t test or variance analysis followed by a Fisher's test.

This study shows that HIV-related encephalopathy induces significant modifications of brain metabolism as analyzed by proton localized MRS of the

	n	NAA/Cho	NAA/Cr-PCr	Cho/Cr-PCr	CD4 count
Controls	9	2.23 (0.51)	2.23 (0.63)	1.01 (0.07)	nd
limit values		*encephalopathy if < 1.65*	*encephalopathy if < 1.66*	*encephalopathy if > 1.28*	-
Clinical state					
Neuro-asymptomatic patients	16	1.83 (0.32) (§)] *	2.02 (0.30)	1.13 (0.26)	240,4 (229)] &
ADC patients	47	1.50 (0.39) (§)]	1.66 (0.31) (§)	1.16 (0.28)	58.4 (79.6)]
MRI in HIV patients					
Normal	8	1.88 (0.30)] *	2.06 (0.28)] *	1.13 (0.31)	334.5 (189)] *
With non specific WML	9	1.92 (0.31)] *	1.89 (0.28)]	1.18 (0.28)	141.5 (155.3)]
Abnormal (Atrophy or AIDS related WM lesions)	46	1.46 (0.37) (§)]	1.65 (0.32) (§)]	1.00 (0.15)	56.9 (103.3)]

Table 1: Values of the different metabolic ratios calculated from *in vivo* localized proton MR spectra obtained from control and HIV patients. Limit values of each ratio defining the occurrence of a metabolic pattern are shown in italics.

& = Student's t test, significant difference, $p < 0.05$.

* = Variance analysis, significant differences, $p < 0.05$.

§ = Variance analysis, different from controls, $p < 0.05$.

brain. As displayed in Table 1, the mean value of NAA/Cho ratio is significantly reduced in the two sub-groups of neurosymptomatic and neuroasymptomatic patients. In the whole group of HIV patients, mean values of NAA/Cho and NAA/Cr-PCr ratios are significantly reduced, even when MRI is normal. The most sensitive metabolic parameter is the NAA/Cho ratio. The mean value of Cho/Cr-PCr ratio is not significantly altered. The mean CD4 count is significantly reduced when patients present an ADC, and when MRI is abnormal (Table 1).

In spite of a statistical significance, Figure 2A shows a poor correlation between the T4 count and value of brain metabolic ratios, whatever the metabolic ratio used (Figure 2A displays this correlation with NAA/Cho ratio).

Figures 2B and 2C display the variations of NAA/Cho ratio, the most sensitive parameter for the detecting modifications in brain metabolism HIV-related encephalopathies. Interestingly, NAA/Cho value is low when atrophy is large.

Proton MRS is more sensitive than proton MRI in the early detection of brain anomalies in HIV patients

In a second protocol, we have conservatively considered that a spectrum of a patient was abnormal (this means that the patient had a metabolic encephalopathy detected by MRS) when the individual value of at least one of the three metabolic ratios was outside the range of all values found in the

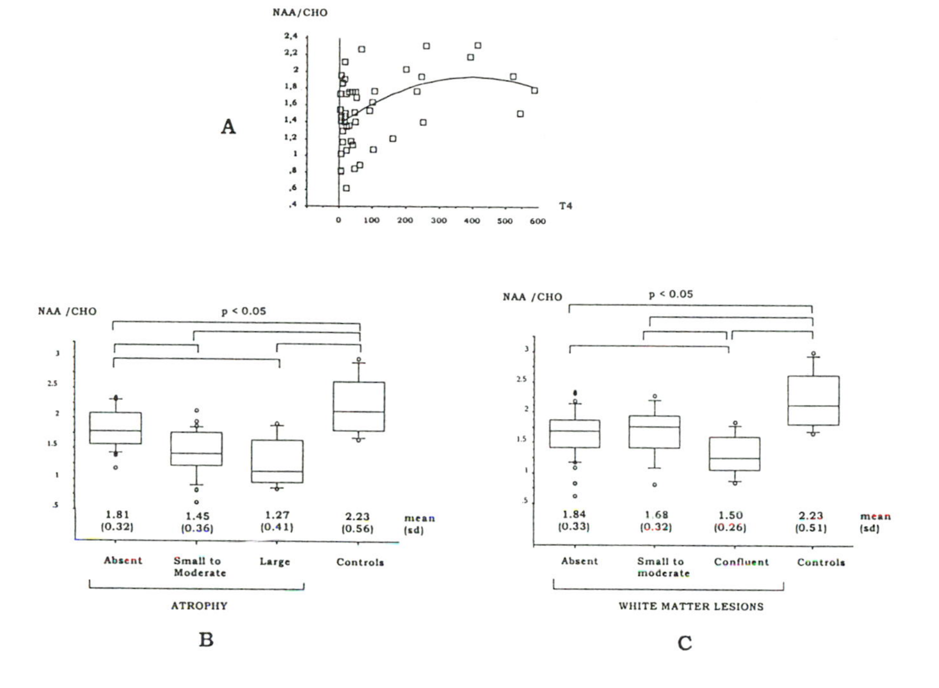

Figure 2: (A) Relationships between T4 count and NAA/Cho ratio (y = 1.378 + 0.0003 x - 3.44 e-6 x2, r = 0,43, p ≤ 0.05). Values of NAA/Cho as a function of (B) the extent of lesion detected by MRI : atrophy and (C) diffuse white mater lesions.

	MRS and MRI Correlated		MRS and MRI Uncorrelated	
	Normal MRI and normal MRS	Abnormal MRI and abnormal MRS (metabolic encephalopathy)	Abnormal MRI and normal MRS	Normal MRI and abnormal MRS (metabolic encephalopathy)
NEURO-ASYMPTOMATIC HIV PATIENTS (n = 16)	7 (43,7%)	3 (18,7%)	2 (12,5%)	4 (25%)
ADC PATIENTS (n = 47)	4 (8,5%)	33 (70,2%)	8 (17%)	2 (4.25%)

Table 2: Comparison of the detection of HIV-related encephalopathies using MRI and MRS. MRS was considered abnormal when one ratio was more than or less than the limit value displayed in Table 1 (see also Figure 4).

control group, *i.e.,* when NAA/Cho or NAA/Cr-PCr ratios were lower than the lowest control value (respectively 1.65 and 1.66) or when the Cho/Cr-PCr ratio was higher than the highest control value (1.28) (Table 1). Under these conditions, the occurrence of an encephalopathy is defined from the cut-off values of metabolic ratios. In this protocol, comparison of frequencies was based on "chi 2 test".

We have found that the correlation between MRS and MRI is good (MRS and MRI are abnormal in the same patient, or MRS and MRI are normal in the same patient) in 47 patients (about 75%) (Table 2). In 4 of the 16 neuro-asymptomatic patients (*i.e.* 25%) a metabolic encephalopathy was found while MRI was still normal. These results confirm and extend those obtained by Chong *et al.* (1993)

In 10 patients, MRI was found abnormal, and no metabolic encephalopathy was detected with MRS. However, among these 10 patients, only one patient had a normal spectrum, and the other 9 patients presented an individual NAA/Cho ratio near the limit value. In fact, one of the two neuro-asymptomatic patients belonging to this group had also a value of the Cho/Cr-Pcr ratio equal to 1.27. Under these conditions, the discrepancy of MRI and MRS results was only apparent and was related to the highly conservative value selected as a cut-off for metabolic ratios in the definition of metabolic encephalopathies. In any event and from a clinical standpoint, MRS examination is still highly valuable to the patients since spectral anomalies (even slight changes in metabolic levels) can alert the physician about the possible existence of a developing encephalopathy.

MRS allows to define metabolic patterns in HIV patients

A third protocol involved a careful examination of each MR spectrum for each HIV patient (whatever the clinical stage) in an attempt to define several metabolic patterns of encephalopathies. We have delineated three metabolic patterns (Figure 3): the first pattern occurs when the value of NAA/Cho only decreases under the limit value (<1.65). This pattern has been called "undifferentiated encephalopathy". The second pattern is characterized by a decrease in the individual value of NAA/Cr-PCr ratio under the limit value (< 1.66). This pattern has been called "NAA encephalopathy". The third pattern is characterized by an increase in the individual value of Cho/Cr-PCr ratio above the limit value (>1.28). This pattern has been called "Cho encephalopathy".

Using this metabolic classification, we have observed the "undifferentiated encephalopathy" pattern (only NAA/Cho is decreased) in 7 patients. The "NAA encephalopathy" pattern was observed in 27 patients. In 19 of them, this pattern was also associated with a decrease in NAA/Cho ratio under the limit value (<1.65). The so-called "Cho encephalopathy" was observed in 14 patients, and

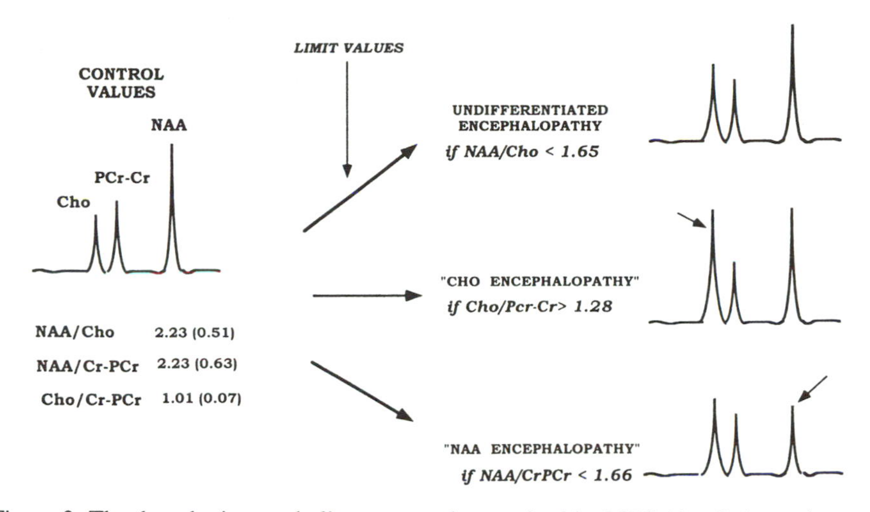

Figure 3: The three brain metabolic patterns characterized by MRS (details in text).

was always found associated to a decrease in NAA/Cho under the limit value, and associated to the "NAA encephalopathy" pattern in 6 cases. Most of neuro-asymptomatic patients (9/16, about 56%) displayed no MR signs of encephalopathy, and the 3 patterns were similar in frequency in the remaining 7 patients (40%). In ADC patients, the most frequent metabolic encephalopathy was the "NAA encephalopathy". We have found that the "NAA encephalopathy" was clearly related to the occurrence of atrophy, since the isolated WM lesions do not seem to induce a preferential metabolic pattern (Vion-Dury *et al.*, 1994). The frequency of "Cho encephalopathy" was constant whatever the type of MRI findings. On the contrary, the "NAA encephalopathy" was more frequent in patients with both atrophy and white matter lesions.

Figure 4 summarizes the different combinations of cerebral metabolic perturbations observed in our patients. The "undifferentiated encephalopathy" is observed mainly in neuro-asymptomatic patients, while the association of "Cho encephalopathy" and "NAA encephalopathy" is observed only in ADC patients.

The NAA/Cho ratio is the most frequently affected metabolic parameter when encephalopathy occurs (34 patients, *i.e.,* 54%). Generally, this ratio is also the first to be affected when a follow-up of brain metabolism by MRS is performed over a period of several months (data not shown). Whatever the nature of the lesions, the "undifferentiated encephalopathy" occurs in about 20% of patients. The NAA/Cr ratio is affected when atrophy is present. But, when atrophy is associated to the WM lesions, the percentage of "undifferentiated

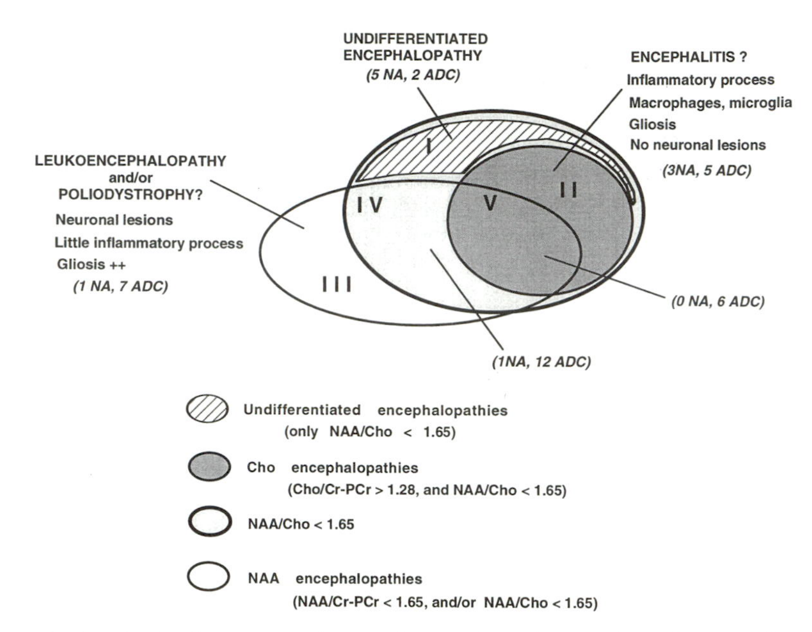

Figure 4: Synthetic diagram displaying the number of patients with the different types of brain metabolic patterns detected by localized proton MRS. NA = neuro-asymptomatic patients, ADC = patients with AIDS-dementia complex.

encephalopathies" falls and the number of "NAA encephalopathies" strongly increases. Since NAA is considered as a neuronal marker (Miller, 1991), it is not surprising that atrophy is well correlated to the reduction of the cerebral amount of NAA. Both mild (cognitive/motor disorders) and severe (dementia complex) disorders have been described in the central nervous system of HIV patients (Janssen, 1991). Cognitive/motor disorders do not seem directly related only to brain neuronal loss, but might be also related to a metabolic dysfunction (Seilhean *et al.*, 1993), because in dementia complex, neuronal injury is clearly documented (Everall *et al.*, 1991, 1993; Lipton, 1992). Consequently, it is reasonable to consider that the decrease of NAA/Cr ratio value reflects the gravity of neuronal suffering, injury, and then death. The Cho resonance includes all choline-containing molecules such as phospholipid metabolites, phosphocholine, and acetylcholine (Michaelis *et al.*, 1991). It has been suggested that the Cho resonance is sensitive to membrane lesions, mainly demyelination (Brenner *et al.*, 1993). In a recent work (Preece *et al.*, 1993), it has been demonstrated that betaine might also contribute to the Cho resonance. Inflammatory processes, without demyelination, can also induce an increase of betaine, phosphocholine, and choline levels, in relation to the presence of interleukin-2 activated lymphocytes. The occurrence of "Cho encephalopathy", might be related either to a demyelinating process, or to an inflammatory process, or to a combination of both. At the final stage of HIV-related encephalopathy, when both diffuse inflammatory processes and atrophy are present, and when the ADC is complete, it is not surprising to observe the maximal alteration of neuronal viability. Figure 4 also illustrates the relationships between the occurrence of ADC and the gravity of the metabolic lesion.

An important issue raised in this study pertains to the two main metabolic patterns detected by MRS in AIDS-related encephalopathies which might describe the histological entities of HIV-induced lesions of the brain (Gray, 1993). The first one is HIV-related encephalitis, with reactive gliosis, macrophages, and lymphocytic infiltrates, in which neurons and axons seem preserved. This encephalitis affects predominantly the WM, and is characterized by an increase of signal in T2-weighted images. This encephalitis would be well described by the "Cho encephalopathy", when this pattern is not associated to a decrease of NAA/Cr-PCr ratio. The second one is the HIV-related progressive leukoencephalopathy, and/or poliodystrophy, which creates diffuse lesions, with little or no inflammatory processes, and leads to neuronal loss. This leukoencephalopathy might be reflected by the decrease of the NAA/Cr-PCr ratio. Between these two neuropathological forms, all the combinations of lesions can be found, explaining the "undifferentiated" pattern, and the simultaneous occurrence of the "Cho encephalopathy", and "NAA encephalopathy" (Figure 4). The increase in the relative number of ADC patients from "undifferentiated encephalopathy" group to the group displaying simultaneous occurrence of the "Cho encephalopathy", and "NAA encephalopathy" provides an argument in favor of the construction of a tentative grading of the gravity of metabolic encephalopathy on the basis of MR spectra.

Treatment by Zidovudine (AZT) can reverse brain metabolic alterations detected by MRS in patients with AIDS dementia complex

Using the same protocol, we have performed a follow-up in three patients with ADC. Magnetic resonance examinations were performed at the time of diagnosis and after anti-retroviral treatment by AZT (Vion-Dury *et al.*, 1995). The spectroscopic volume of interest (VOI = 8 ml) was carefully positioned in the same parieto-occipital white matter area in the two examinations performed on each patient.

All the patients showed, at first examination, clinical signs of HIV-related encephalopathy, with alteration of neuropsychological tests. In all cases, NAA/Cho and NAA/PCr-Cr ratios calculated from localized MR spectrum were

	NAA/Cho normal if >1.65	NAA/PCrCr normal if >1.66	Cho PCr/Cr normal if <1.28	NAA$	Cho$	PCr-Cr$
Patient 1						
-1st *examination*	0.75*	0,90*	1.20	49.75	66.12	54.99
-2nd *examination* (4 months later) **AZT = 900mg/j**	2.27	1.71	0.75	116.96	51.49	68.41
Patient 2						
-1st *examination*	1.58*	1.47*	0.93	63.75	40.37	43.21
-2nd *examination* (4 months later) **AZT = 1g/j**	1.70	1.68	0.98	73.86	43.32	44.65
Patient 3						
-1st *examination*	1.00*	1.14*	1.14	81.27	81.27	71.29
-2nd *eaxmination* (2 months later) **AZT = 950mg/j**	1.47*	1.36*	0.92	114.68	77.76	84.05

$: absolute values expressed in arbitrary units; *: abnormal values

Table 3: Follow-up of AZT therapy. Values of NAA/Cho, NAA/Cr, and Cho/Cr ratios were calculated from the proton MR spectra. The absolute intensities (in arbitrary units) of N-acetyl-aspartate (NAA), phosphocreatine-creatine (PCr-Cr) and choline (Cho) signals were calculated by multiplying the area of resonances by the actual reference pulse amplitude, used as an internal calibration (Michaelis *et al.*, 1993). This procedure allows sequential spectra recorded on the same patient to be compared.

abnormal (lower than control values) (Table 3). After 3 or 4 months of a 900-1000 mg/day AZT treatment neurological and neuropsychological examination were strongly improved. Diffuse WM lesions detected by MRI had regressed in intensity and area only in one case. The most striking result was that both NAA/Cho and NAA/PCr-Cr ratios returned to normal values in two patients and improved in the third. This normalization is mainly reflected to a large increase in NAA signal and a small decrease in Cho signal.

The results presented here shed new light on several issues: 1) the modifications of brain metabolism associated with HIV-related encephalopathy can be reversed by an anti-retroviral treatment. Reversion was observed in the three patients and was based on the increase of the NAA signal, and, to a lesser extent, on the increase of the PCr-Cr signal and the decrease of the Cho signal; 2) there exists a good correlation between the evolution of brain metabolic status and the neurological/neuropsychological findings; correlation between brain metabolic status and MR imaging is poor; 3) since the depressed level of NAA in the brain can be reverted to normal values (NAA is considered as a neuronal marker), one can propose that cognitive/motor disorders are not only related to a reduction of the number of neuronal cells occurring in HIV-related encephalopathies (Seilhean *et al.*, 1993; Everall *et al.*, 1991, 1993), but can be also related to a metabolic dysfunction of existing neurons. The decrease of

NAA/Cho and NAA/PCr-Cr ratios reflects several stages of brain damage, with at first, the occurrence of neuronal suffering or injury and later, when metabolic lesions cannot be reversed, neuronal death.

HIV-Related encephalopathies in children

HIV-related encephalopathies observed in children after materno-foetal contamination are less frequent than in adults and clinically different (Burns, 1992). The most dramatic forms, the so-called "progressive encephalopathies", are characterized by a predominant motor syndrome with pyramidal tract abnormalities, cognitive impairment, developmental retardation or regression, and ultimately spastic paraparesis or hemiparesis.

In this part of the study, we have examined three children (a 3.5 year-old boy, a 2.5 year-old girl and a 4 year-old boy) with progressive HIV-related encephalopathy, using a STEAM sequence (stimulated echo time = 20 ms, repetition time = 1.6 s) (Confort-Gouny *et al.*, 1993). The spectroscopic volume of interest (VOI = 8 ml) was positioned in the white matter part of centrum ovale corresponding to the pyramidal tract. The data are presented as the ratio of area of the signal of each metabolite of interest (X) over the sum of areas of all signals of metabolites of interest (S) (Confort-Gouny *et al.*, 1993).

The three children had a maternal-infant transmission of HIV. Antiretroviral and immunoglobulin therapies were initiated at the time of the diagnosis of contamination. At the time of MR spectroscopy, CD4 counts were respectively 73, 53, and 376 cells/mm^3. The first patient had microcephaly, spastic quadriparesis with pyramidal rigidity and facial diplegia. The second patient had microcephaly and progressive motor regression with spastic diplegia. The third patient had microcephaly, spastic quadriparesis, and facial diplegia. He could neither speak nor eat. In the three patients, MR images of brain showed a moderate atrophy and hyperintensities of periventricular and deep white matter (WM) in T_2-weighted images. No cortico-spinal tract degeneration was observed.

As displayed in Figure 5, MR spectra of these patients are characterized by a very moderate decrease of NAA/S, and a slight increase in PCr-Cr/S and INS/S. Surprisingly, the brain metabolic profile recorded in the pyramidal tract is not dramatically modified, in spite of a pyramidal syndrome and diplegia or quadriparesis observed in these very disabled patients.

The brain metabolic status of children with severe HIV-related progressive encephalopathies is only slightly altered when compared to adult patients at the terminal stage of encephalopathy. The decrease in relative NAA concentration (NAA/S) in all three young patients is less than 25%, whereas the same parameter is often decreased by more than 50% in the spectrum recorded in WM of adults with HIV-related encephalopathy. These results suggest that the

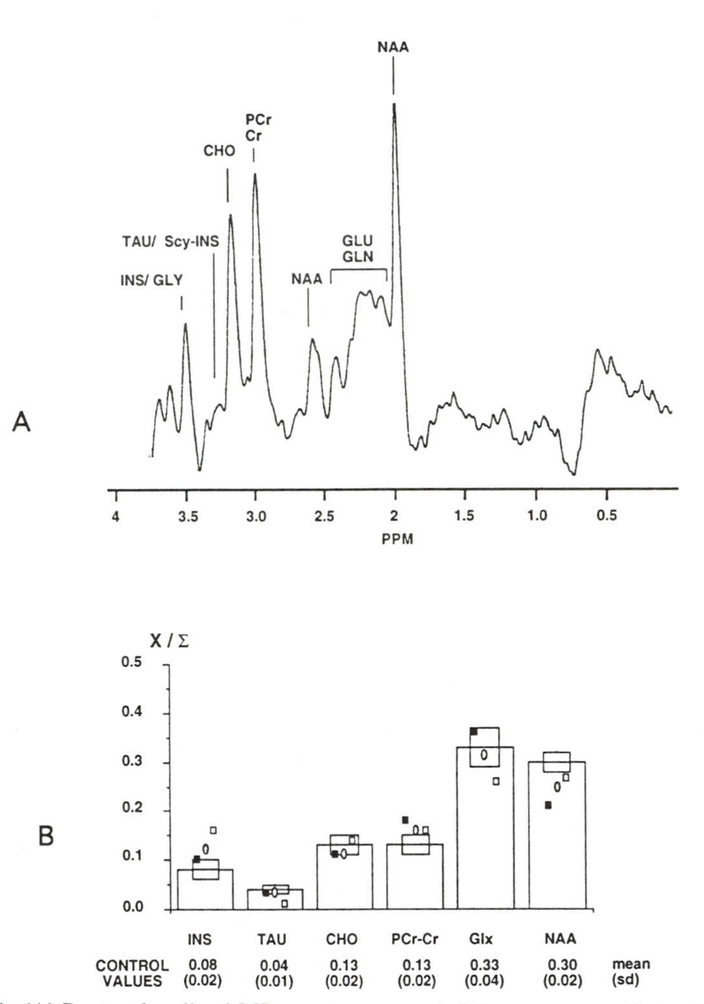

Figure 5: (A) Proton localized MR spectrum recorded from corona radiata of a child (patient # 1) with progressive HIV-related encephalopathy. (B) For each metabolite, the ratio [area of the signal of each metabolite of interest (X) / sum of areas of signals of all metabolites of interest (S)] is given. Patient 1 (dark square), patient 2 (grey circle), and patient 3 (white square). Histograms = control values (mean ± sd) of X/ S for each metabolite (6 children of matching age). Assignments of resonances are in Kent *et al.* (1993).

alteration of pyramidal neurones is weak or moderate in pediatric progressive HIV-related encephalopathy. They also constitute a strong argument, based on *in vivo* results, in favor of an indirect lesion of pyramidal cells secondary to abnormal myelination or Wallerian degeneration related to lesion of brain stem or spinal cord (Gray, 1993).

Why do clinicians need localized brain proton MRS in HIV-related encephalopathies?

In brain lesions due to HIV infection the functional impairment of the patient is constant and often very extensive, constituting the dementia complex. It is very critical to perform an early diagnosis of encephalopathy in HIV patients in order to treat as soon as possible. Early treatment can lead to preservation of the capital of neurons, to reduction in the dementia complex and to improvement in the quality of life of the patient (motricity, interpersonal relations, memory...). The high sensitivity of localized proton MRS to the metabolic brain anomalies induced by HIV permits the diagnosis of the occurrence of encephalopathy before MRI. In addition, localized MRS is very useful when neuropsychological anomalies are observed in HIV patients, and when MRI is still normal. In this case, MRS helps differentiate the beginning of an HIV-related encephalopathy (MR spectra are abnormal) from drug ingestion and/or a depressive syndrome (MRS is normal). MRS also allows a quantitative monitoring of treatment.

MRS offers a new and unique modality for the early diagnosis and follow-up of HIV-related encephalopathies. It is currently used as a routine diagnostic procedure at the Timone University Hospital in Marseille.

Acknowledgements

This work is supported by CNRS (URA 1186), APM (Assistance Publique à Marseille), SIDACTION and the Programme Hospitalier de Recherche Clinique (Ministère de la Santé). We thanks Catherine Dhiver, Jean-Albert Gastaut, Jean-Louis Gastaut and Gérard Michel for their fruitful discussions and patients referral.

References

Bottomley, P.A. (1984). US patent 4 480 228.

Bottomley, P.A, Cousins, J.P., Hardy, C.J., Pendrey, D.L., Wagle, W.A., Hardy, C.J., Eames, F.A., McCafrey, R.J., and Thomson, D.A. (1992). *Radiology* **183**, 695.

Brenner, R.E., Munro, P.M.G., Williams, S.C.R., Bell, J.D., Barker, G.J., Hawkins, C.P., Landon, D.N., and Donald, W.I. (1993). *Magn. Reson. Med* **29**, 737.

Burns, DK. (1992). *J. Child Neurol.* **7**, 332.

Chong, W.K., Sweeney, B., Wilkinson, I.D., Paley, M., Hall-Craggs, M.A., Kendall, B.E. Shepard, J.K., Beecham, M., Miller, R.F., Weller, I.V.D., Newman, S.P., and Harrisson, M.J. (1993). *Radiology* **188**, 119.

Confort-Gouny, S., Vion-Dury, J., Nicoli, F., Dano, P., Donnet, A., Grazziani, N., Gastaut, J.L., Grisoli, F., and P.J., Cozzone. (1993). *J. Neurol. Sci.* **118**, 123.

Deicken, R.E., Hubesch, B., Jensen, P.C., Sappey-Marinier, D., Krell, P., Wisniewski, A., Vanderburg, D., Parks R., Fein, G., and Weiner, M.W. (1991). *Arch. Neurol.* **48**, 203.

Everall, I.P., Luthert, P.J., and Lantos, P.L. (1991). *The Lancet* **337**, 1119.

Everall, I.P., Luthert, P.J., and Lantos, P.L. (1993). *J. Neurol. Neurosurg. Psych.* **56**, 481.

Gray, F. (1993). Atlas of the neuropathology of HIV infection. Oxford University Press, Oxford, pp 290.

Janssen R.S. (and group of merican academy of neurology AIDS task force). (1991). *Neurology* **41**, 778.

Kent, T.A., Hillman, G.R., Levin, H.S., Guinto, F., Kaye, A.R., and Casper, K. (1993). *Adv. Neuroimmunology* **3**, 129.

Lipton, S.A. (1992). *Trends Neurosci.* **15(3)**, 75.

McArthur, J.C., Hoover, D.R., Bacellar, H., Miller, E.N., Cohen, B.A., Becker, J.T., Graham, N.M.H., McArthur, J.H., Selnes, O.A., Jacobson, L.P., Visscher, B.R., Concha, M., and Saah A. (1993). *Neurology* **43**, 2245.

Menon, D.K., Ainsworth, J.G., Cox, I.J., Coker, R.C., Sargentoni, J., Coutts, G.A., Baudouin, C.J., Kocsis, A.E., and Harris, J.R.W. (1992). *J. Comp. Ass. Tom.* **16(4)**, 538.

Meyerhoff, D.J., MacKay, S., Bachman, L., Poole, N., Dillon, W.P.,Weiner, M.W., and Fein, G. (1993). *Neurology* **43**, 509.

Michaelis, T., Merboldt, K.D., Hänicke, W., Gyngell, M.L., Brunh, H., and Frahm, J. (1991). *NMR Biomed.* **4**, 90.

Michaelis, T., Merboldt, K.D., Bruhn, H., Hänicke, W., and Frahm, J. (1993). *Radiology* **187**, 219.

Miller, B.L. (1991). *NMR Biomed.* **4**, 47.

Mundinger, A., Adam, T., Ott, D., Dinkel, E., Beck, A., Peter, H.H., Volk, B., and Schumacher, M. (1992). *Neuroradiology* **35**, 75.

Preece, N.E., Baker, D., Butter, C., Gadian, D.G., and Urenjak, J. (1993). *NMR Biomed.* **6**, 194.

Seilhean, D., Duyckaerts, C., Vazeux, R., Bolgert, F., Brunet, P., Katrlama, C., Gentilini, M., and Hauw, J.J. (1993). *Neurology* **43**, 1492.

Vion-Dury, J., Meyerhoff, D.J., Cozzone, P.J., and Weiner, M.W. (1994). *J. Neurol.* **241**, 354.

Vion-Dury, J., Confort-Gouny, S., Nicoli, F., Dhiver, C., Gastaut, J.A., Gastaut, J.L., and Cozzone, P.J. (1994). *C. R. Acad. Sci.* **317**, 833.

Vion-Dury J., Nicoli F., Confort-Gouny S., Salvan A.M., Dhiver C., and Cozzone P.J. (1995). *The Lancet* **8941**, 60.

22

Nuclear Magnetic Resonance Spectroscopy Studies of Cancer Cell Metabolism

O. Kaplan and J.S. Cohen

Department of Surgery
Tel-Aviv Medical Center
Tel-Aviv, Israel 69978

Cancer Pharmacology Section
Georgetown University Medical Center
Washington D.C. 20057 USA

Nuclear magnetic resonance spectroscopy (NMR) is a powerful technique that provides information on biochemical status and physiological processes both in-vitro and in-vivo. The metabolism of intact cells and tissues can be studied in a continuous manner, and thus, NMR is a unique non-invasive research tool enabling detection of the metabolic changes as they occur (Cohen *et al.*, 1983; Morris, 1988; Daly and Cohen, 1989). The first NMR study of cellular metabolism was done some 20 years ago, when Moon and Richards reported on the diphosphoglyceric acid (DPG) and pH shifts in erythrocytes (Moon, and Richards, 1973). NMR studies of metabolism of tumor cells were initiated by Navon *et al.* who investigated phosphorylated compounds in Ehrlich ascites cells (Navon *et al.*, 1977).

The choice of the element and isotope for a specific study of metabolism depends on its NMR properties, and the required data. The proton has the highest NMR sensitivity, and is the most abundant nucleus in biological molecules. However, this may cause difficulties in the interpretation and assignment of the ^{1}H NMR spectrum. Moreover, since metabolic studies are usually performed in aqueous solutions, the huge signal from the water protons should be suppressed.

Similarly, the wide signals arising from proteins and membrane components should be suppressed. These problems can be addressed now by several innovative NMR methods (Daniels *et al.*, 1976; van Zijl and Cohen, 1992). The most widely used nucleus in NMR studies of metabolism has been ^{31}P (see reviews Cohen (1988); Kaplan *et al.* (1992)). Phosphorous NMR spectroscopy can provide data on energy metabolism and substrate utilization, phospholipid pathways, precise intracellular pH, and membrane permeability and ion and water distribution. The spectrum is easy to interpret, but the number of compounds which are detectable is limited. Carbon NMR is also useful for NMR studies of metabolism since it is found in most biological compounds; however, ^{13}C has a natural abundance of only 1.1%, and ^{13}C enrichment is necessary. Other nuclei which are used less often in NMR studies of cellular metabolism are ^{23}Na (Gupta *et al.*, 1984), ^{19}F (Malet-Martino, *et al.*, 1986), and rarely ^{15}N (Legerton *et al.*, 1983) and ^{39}K (Brophy *et al.*, 1983).

NMR studies of metabolism can be performed with cellular extracts, cell suspensions, and perfused intact cells, which is the most physiological, and, therefore, the recommended method. Since NMR is a relatively insensitive method, it is necessary to have a large number of cells in each experiment (10^7-10^9), and this somewhat limits its use to cell lines that can be grown in culture conditions. Not surprisingly, many of the studies of cell metabolism have focused on cancer, which is an area of major biochemical interest. Moreover, many immortal human, as well as experimental, neoplastic cell lines have been established, so they can be grown in large numbers necessary for NMR experiments.

Methods in cell metabolism studies

Cellular extracts

Extraction of cellular metabolites is often essential for the interpretation of results of NMR studies of intact cells. The excellent resolution of spectra of extracts enables assignments of proximate signals with the aid of pH titration curves, and by reference to the addition of known compounds to the extract solution (Navon *et al.*, 1977; Kaplan *et al.*, 1990). In contrast to living cells, there are no time constraints, and with prolonged data accumulation, compounds present at low concentrations may be observed and quantified (Glonek *et al.*, 1982; Evanochko *et al.*, 1984). The principal disadvantages of extraction studies are that metabolic processes cannot be continuously monitored and that the intercellular milieu is profoundly altered. Data obtained from extracts should be evaluated with caution, since they can represent artifacts due to the extraction procedure, and may be misleading (Kushnir *et al.*, 1989). Perchloric acid extraction, containing water soluble metabolites (Lowry and Passonneau, 1972), has been used for most NMR studies. The final concentration of perchlorate is

Compound		WT	ADR	p value
		nmole per mg protein		
^{31}P NMR				
	PCho	22.4±0.7	18.0±0.6	0.025
	PEth	17.8±0.6	2.8±0.3	0.0002
	GPC	9.8±0.5	2.3±0.3	0.0033
	GPE	4.2±0.2	ND	ND
	βATP	10.6±0.6	16.2±0.6	0.003
	αADP	3.9±0.3	5.5±0.4	0.027
	PCr	0.2±0.1'	4.6±0.7	0.007
	DPDE	38.8±1.3	32.1±1.8	0.059
^{1}H NMR				
	Ch/PCho/GPC	61.4±1.1	47.9±1.1	0.002
	Cr/PCr	12.5±0.8	30.9±1.2	0.0002
	Lactate	12.5±0.8	24.8±1.0	0.002
Calculated				
	Ch	29.2±1.2	27.6±1.9	0.192
	Cr	12.3±0.9	26.3±1.0	0.0008

Table 1: Concentrations of metabolites in perchloric acid extracts of WT and ADR MCF-7 cells. The metabolite contents were determined from the integrals of NMR signals relative to references. Means and SDs from three sets of data. Abbreviations used in the table: PCho, phosphocholine; Ch, choline; PEth, phosphoethanolamine; GPC, glycerophosphocholine, GPE, glycerophosphoethanolamine; PCr, phosphocreatine; Cr, creatine; DPDE, diphosphodiester; ND, not determined.

critical; it was shown that 0.5 M is the optimal concentration to minimize artifacts (Askenasy *et al.*, 1990). In order to prevent chemical changes, the perchlorate should be neutralized as quickly as possible, metal cations should be removed, the solution should be cold, and for storage the extracts should be dried and kept frozen. Lipid extraction can be performed by chloroform/methanol treatment of cells (Folch *et al.*, 1951). This extraction may be useful for studying membrane-bound phospholipids and glycoproteins (Bines *et al.*, 1985; Meneses and Glonek, 1988).

Cell suspensions

NMR studies of metabolism of suspensions of cells preceded those of cell perfusion (see reviews Shulman *et al.* (1979), Boddie *et al.* (1989), Cohen *et al.* (1988)). Although these studies provided useful information, and demonstrated the value of NMR spectroscopy for biomedical research, they have many drawbacks. Packed cells in the NMR tube experience shortage of oxygen and nutrients, accumulation of toxic waste products, and pH changes. Metabolic studies often involve chemical manipulation that require the capability for reversibility of environmental conditions, not attainable in cell suspensions. Moreover, for most cell types free suspensions are not their normal physical state. Sedimentation of suspended cells at the bottom of the NMR tube aggravates their metabolic status, and may lead to erroneous data. Several techniques to overcome these obstacles have been proposed (Ugurbil *et al.*, 1982; Balaban *et al.*, 1981; Jentoft and Town, 1985; Labotka *et al.*, 1985); the simplest and most useful is oxygen bubbling (Navon *et al.*, 1977), but none is significantly efficacious. Also, most cellular suspension studies were conducted at ambient temperature to alleviate metabolic instability; however, it is clearly superior to study metabolism at 37 °C. All these considerations clearly show that cellular perfusion is much preferred for NMR studies of metabolism. Cell suspensions are still warranted in a limited number of experimental circumstances, particularly when using scarce compounds or isotopically labeled substrates in short-term experiments (Navon *et al.*, 1989; Kaplan *et al.*, 1990).

Perfusion of cells

Perfused intact cells represent perhaps the best approach to the non-invasive study of metabolism. Therefore, our research has focused on perfused cells (mostly breast cancer cells), and herein we review our studies of their metabolism. In contrast to the in-vivo situation, perfused cells are homogeneous, and there are no "artifact" data from connective tissues and blood vessels. The cells are metabolically stable for prolonged periods during perfusion under physiological conditions. Thus, the effects on metabolism following manipulation with nutrients (Lyon *et al.*, 1988; Ronen and Degani, 1989; Ronen *et al.*, 1991; Daly *et al.*, 1987), drugs (Kaplan *et al.*, 1990, Daly *et al.*, 1990; Jaroszewski *et al.*, 1990; Ben-Horin *et al.*, 1993), hormones (Neeman and Degani, 1989; Ruiz-Cabello *et al.*, 1993) and growth factors (Kaplan *et al.*, 1990) can be monitored. During the last decade, several methods of restraint and perfusion of cells for NMR studies have been developed (Ugurbil *et al.*, 1981; Foxall *et al.*, 1984; Daly *et al.*, 1988; Shankar-Narayan *et al.*, 1990; Gonzalez-Mendez *et al.*, 1988; Freyer, 1988). There are now enough methods that an appropriate one can be adjusted to almost all cell types and experimental requirements.

Perfusion cannot be performed in cells freely suspended in the NMR tube, since the flow would wash them away, or the cells would block filters if used; therefore, the cells should be restrained. There are some major differences between the various methods used for cell perfusion studies. Table 1 presents a comparison between the various perfusion techniques for NMR studies of cellular metabolism. Some methods are appropriate only for anchorage-dependent cells, while in others also anchorage-independent cells can be studied; in some, cells cannot multiply and cellular growth cannot be monitored; the technical aspects and difficulties in maintaining experimental conditions are markedly variable. The essentials of cellular perfusion are that metabolic events are unhampered; thus, substrates and nutrients should be continuously furnished,

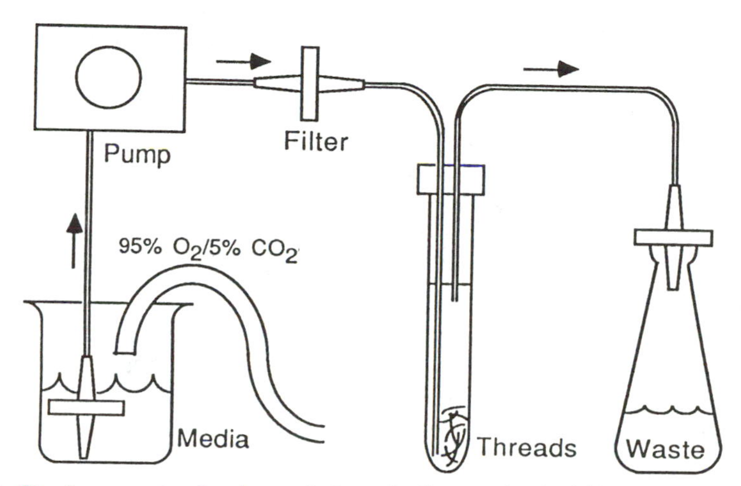

Figure 1: Simple apparatus for the perfusion of cells restrained within a matrix.

and waste products removed, while stable pH levels and temperature of 37 °C are maintained. It is recommended that perfusion is done with the cells growth medium, at an appropriate rate, using a peristaltic pump (Figure 1). Recirculation perfusion is sometimes necessary, for example when scarce labeled isotopes are used, but is adequate only as long as nutrients are not depleted, and waste materials are not accumulated. Another problem that should be addressed is whether to perfuse with phosphorus-containing solution in ^{31}P experiments. While phosphate is essential for many biochemical reactions, its signal (from the medium) interferes with intracellular pH determinations. In all prolonged experiments sterility should be maintained.

The two most widely used perfusion techniques for NMR studies of cellular metabolism are bead attachment and embedding in gels. Attachment of anchorage-dependent cell to dextran microcarrier beads was initially introduced by Ugurbil *et al.* (1981). The cells multiply on the beads, and coat them in monolayers. Besides dextran, beads of other compositions and sizes are now available: *e.g.*, treated polystyrene, polyacrylamide, agarose polyacrolein. Since the cells actually grow on the beads, they can be studied in several stages of development, and effects on cell growth can be monitored. Also, the effects of large compounds (for example proteins) can be evaluated, since the cells are in direct contact with the perfusion solution. The disadvantages of the bead techniques are that anchorage-independent cells cannot be studied, that the "dead" volume taken by the beads reduces the amount of cells and the S/N, that the field homogeneity is far from optimal, and that measures to prevent overgrowth of cells on the beads, beads sedimentation, and cell detachment should be taken. The gel methods differ from the bead methods in two fundamental features: a) in the gels the cells are inside the matrix, and therefore the porosity of the matrix and the ease of nutrients diffusion are of critical importance; b) since attachment to the gel is not essential, both anchorage-dependent and -independent cells can be studied. In the first gel technique, cells were embedded in low-temperature gelling agarose threads (Figure 2A). With this method the matrix occupies a relatively small volume, and large number of cells can be maintained in a favorable metabolic status (Figure 2B). The major disadvantage of this method is the limited growth of cells in the threads. The basement membrane (Matrigel) method overcomes this obstacle (Daly *et al.*, 1988). Cancer cells grow in the Matrigel (Figure 3), and are morphologically and functionally identical to their in-vivo counterparts. The alginate capsules method, which was recently applied to human cells, seems to be very promising (Shankar-Narayan *et al.*, 1990). Cells are mixed with low viscosity sodium alginate, and small capsules are formed by dropping this suspension into a 0.1 M calcium chloride solution (Figure 4). This method is applicable for both small and large anchorage-dependent and -independent cells. The cells multiply in the alginate, and the technique is very simple and quick.

Other perfusion methods which are used less often are hollow fibers and dialysis membranes (Gonzalez-Mendez *et al.*, 1988), and spheroids (Freyer, 1988).

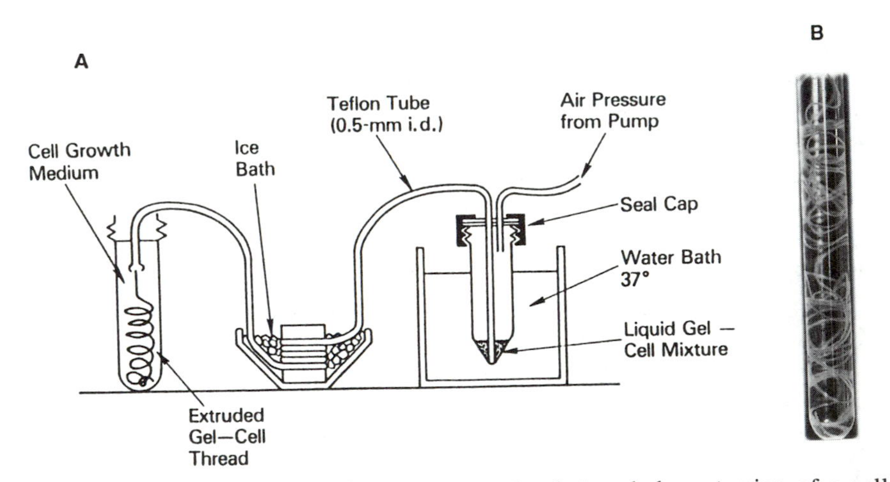

Figure 2: (A) Apparatus for the preparation of gel threads by extrusion of a cell suspension in a liquid gel through a cooled capillary. (B) Agarose gel threads (0.5 mm diameter) containing cells; note that the threads would be transparent without the cells.

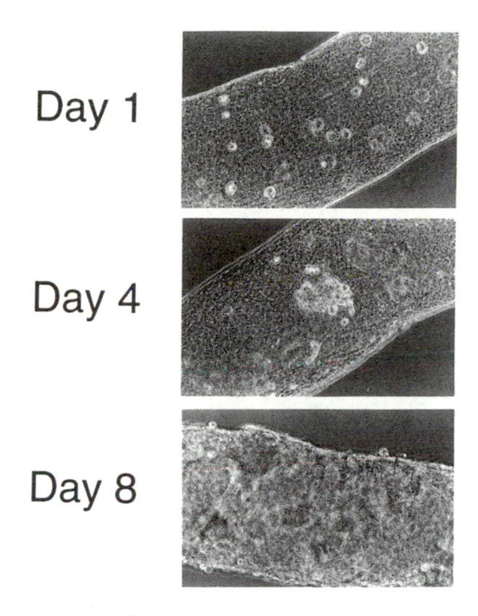

Figure 3: MB231 breast cancer cells growing in Matrigel threads. Note that the cells grow in spaces in the thread, and they do not cause a change in diameter (0.5 mm).

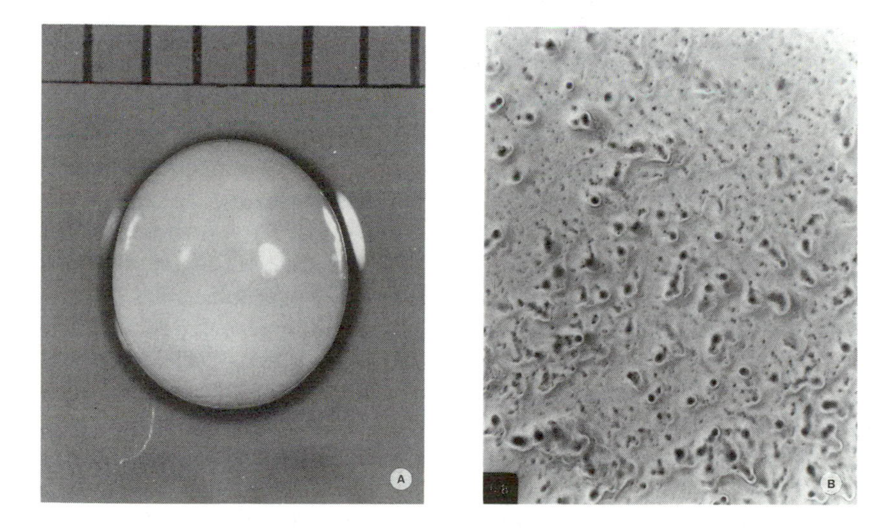

Figure 4: Alginate capsule containing embedded lymphocytes. (A) whole capsule unenlarged (scale is in mm), and (B) cut surface (enlarged x400) showing lymphocytes (small particles) and large channels (black pores).

	Suitable for anchorage-dependent cells	Suitable for anchorage-independent cells	Suitable for cellular growth studies	Metabolic stability of exp. preparations	Suitable for studies of large molecules	Cellular density[2]	S/N (general)	Field homgeneity (general)	Perfusion rates	Technical aspects
A) Micro carrier beads	yes	no	yes	[3]good	yes	1-3 x 10^7 cells/ml low density	fair	fair	1.5-15 ml/min for 2.5-30 ml of sample	simple, cheap
B) Gels										
agarose threads	yes	yes	no	good	[4]yes	0.5-1 x 10^8 cells/ml high density	good	good	0.5-1 ml/min for 2 ml of sample in tube	very simple, cheap
matrigel	yes	yes	yes	good	[5]N/D		good	good		simple
alginate crystals	yes	yes	yes	good	[4]yes		fair	fair		very simple. cheap
C) Hollow fibers and membranes	yes	[5]N/D	yes	good	no	low density	fair-low	fair-low	5-200 ml/min	elaborate complex apparatus, special probes relatively expensive
D) spheroids	yes	no	yes	low (necrosis of inner parts)	[5]N/D (probably no)	medium	fair-low density	fair-low	10-25 ml/min	simple, however means to prevent clumping should be used

Table 2: Comparison of perfusion techniques for NMR studies of cellular metabolism. Footnotes: [1]For references, see appropriate section of text. [2]Cell number in 1 ml of [cells + matrix perfusion solution]. The volume taken up by cells depends on their number, size and shape. Therefore, the calculations are only approximate, and one needs to consider also, the specific cell type. [3]Metabolic stability might deteriorate whenever cells overgrow on their attaching matrix. [4]Established for molecules of up to 45kD. [5]N/D - not determined.

Applications to metabolism of cancer cells

Effects of nutrients/metabolites

NMR studies of cancer cells were initiated by assignments of their ^{31}P spectra, and though the search for metabolic differences between normal and neoplastic cells (Daly *et al.*, 1987, Foxall and Cohen, 1983; Merchant *et al.*, 1988). No qualitative differences in ^{31}P or ^{1}H NMR spectra were noted. Relatively increased levels of PME compounds in some tumor cell lines were found, and this was hypothesized to be associated with intensified cell membrane synthesis and rates of cells replication (Maris *et al.*, 1985; Radda *et al.*, 1987). The decline in PME content after therapy is consistent with this theory (Sijens *et al.*, 1988).

NMR is an excellent method for studying glucose and energy metabolism in cancer cells. In our laboratory, phosphorylated glucose compounds were detected and followed by ^{31}P, and glycolysis products by ^{13}C NMR. Glucose deprivation caused ATP depletion to approximately 40% of the original concentration (Lyon *et al.*, 1988) in perfused MCF-7 breast cancer cells. When the cells were challenged with azide, an inhibitor of oxidative phosphorylation, there were only minor ATP changes, which is in accordance with the very well known theory that glycolysis is the main source of energy in cancer cells (Warbug, 1956). The ATP concentration decreased further when azide was added to glucose-deprived cells, an indicator of some oxidative phosphorylation, presumably using glutamate as the substrate. The cells were perfused with ^{13}C enriched glucose, and ^{13}C NMR, measuring lactate production, provided glycolysis rates (Figure 5). Azide had no effect on the levels of lactate production, but inhibited the conversion of lactate into pyruvate.

Using the glucose analog, 2-DG, we have obtained additional information on metabolism (Navon *et al.*, 1989; Kaplan *et al.*, 1990; Kaplan *et al.*, 1990), and on the effects of drugs and growth hormones (see below). 2-DG competes with glucose on the same uptake mechanism, and simultaneous monitoring of transport and phosphorylation were done by recording ^{13}C NMR of MCF7 and MDA-468 breast cancer cells, using 2-DG labeled with ^{13}C in the 6-position. The advantage of this method is that both the 2-DG and 2-DG-6P signals are clearly resolved, and quantitative measurements of reaction rates can be performed (Figure 6).

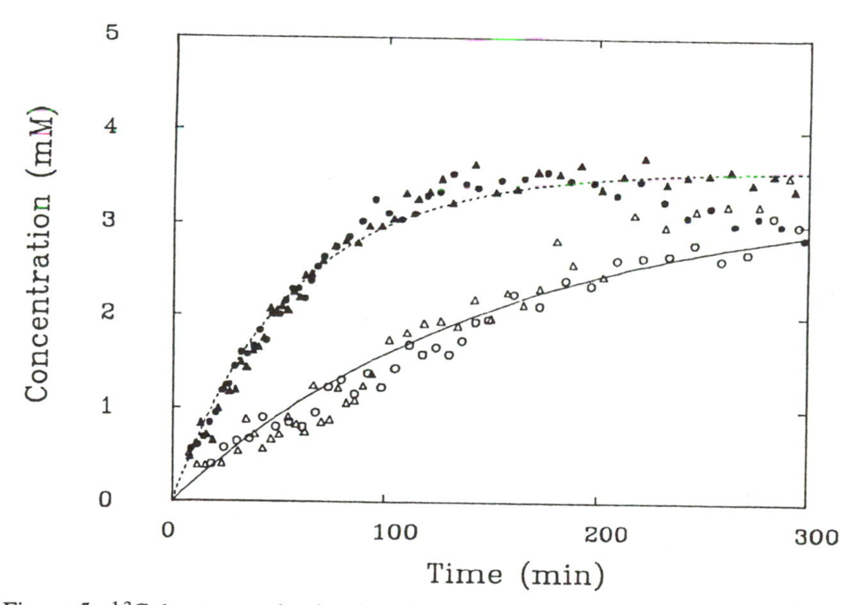

Figure 5: ^{13}C lactate production by glycolysis of perfused wild-type (WT, open symbols) and multi-drug resistant ADR, filled symbols) MCF-7 cells, following addition of $^{13}C_1$-glucose in a closed perfusion system.

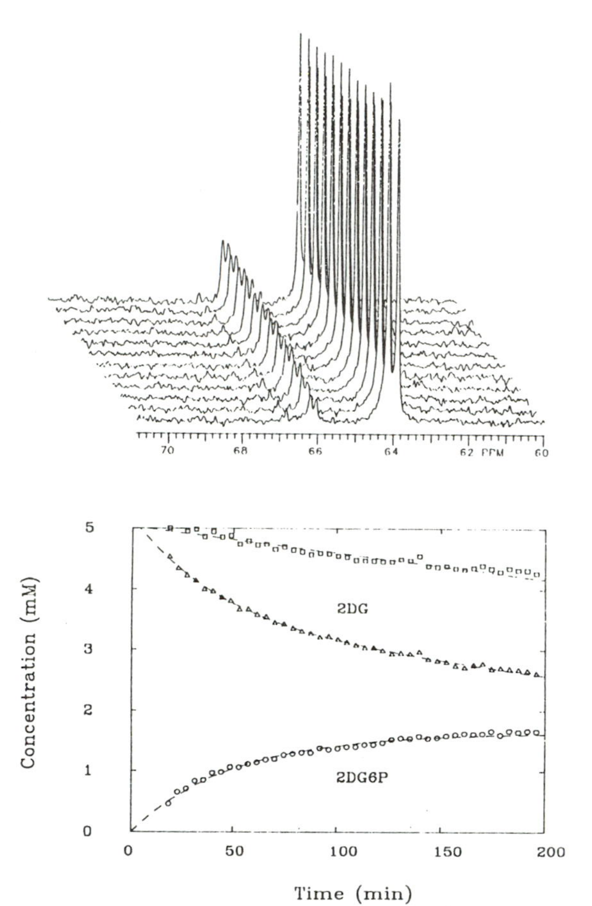

Figure 6: (A) Metabolism of $^{13}C_6$-2-deoxyglucose (2-DG, right peak) to $^{13}C_6$-2-DG (left peak) in a suspension of MCF-7 cells; $^{13}C\{^1H$ decoupled$\}$ NMR spectra obtained at 100 MHz. (B) Plot of time course showing exponential decrease in 2-DG (triangles), exponential increase in 2-DG6P (circles), and overall linear decrease (squares).

Spheroids constituted a "tumor model" for NMR investigations of biochemical and metabolic changes (Lim *et al.*, 1987). The ATP signal furnished information on cell function; it initially increased, and when the spheroids approached 350 mm started to decrease, indicating necrosis of the innermost cells (Ronen and Degani, 1989). Also, the rates of lactate production, as measured by ^{13}C NMR when the cells were perfused with [1-^{13}C]-glucose, decreased as the spheroids became "older" and larger.

Phospholipid metabolism was studied extensively in our laboratory by ^{31}P NMR of perfused MDA-MB-231 human breast cancer cells embedded in agarose threads and Matrigel (Daly *et al.*, 1987; Daly *et al.*, 1990). Most notable are the effects of substrates and inhibitors on specific enzymatic processes, such as ethanolamine kinase, which in the presence of extracellular ethanolamine resulted in the formation of PE, while a specific inhibitor of choline kinase, hemicholinium-3 resulted in the reduction of concentration of PC (Figure 7). The exponential decay implied that the PC and PE levels influenced the forward reaction rates of cytidyltransferase enzymes intracellularly. Furthermore, when perfused with choline and ethanolamine, rapid large increases in PC and PE occurred, whereas no appearance of the CDP-choline and CDP-ethanolamine peak was seen. These studies showed that the second step of the three-step process of PL biosynthesis, namely choline-phosphate cytidyltransferase activity, is the rate limiting step. Inhibition of GPC phosphodiesterase by ethanolamine, allowed estimation of the flux through phospholipid degradative pathways.

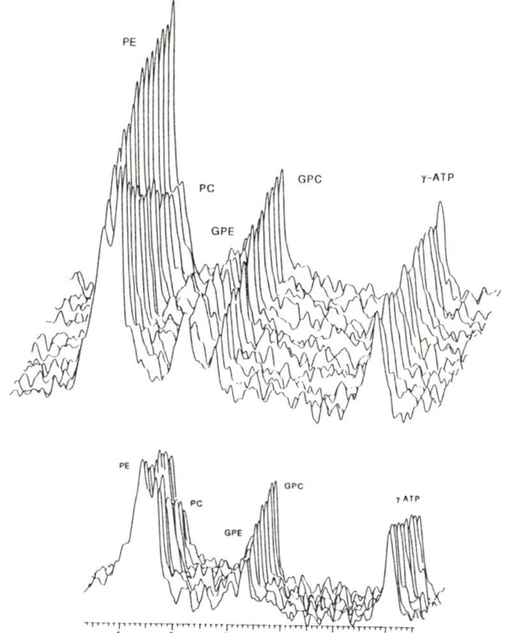

Figure 7: (A) Effect of ethanolamine (2 mM) on the ^{31}P spectra at 162 MHz of perfused breast MDA-231 cancer cells at 37 °C (each spectrum is one hour accumulation). Note the growth of the phosphoethanolamine (PE) peak. (B) Effect of hemicholinium-3, a specific inhibitor of choline kinase. Note the reduction of the phosphocholine (PC) peak, while the PE remains constant (GPC and GPE are the corresponding glyceryl di-esters). Note also that the γ-ATP peak is an internal control in each case.

Phospholipid synthetic pathways were investigated by ^{13}C and ^{31}P NMR studies of T47D breast cancer cells grown as spheroid and perfused with ^{13}C labeled choline and ethanolamine (Ronon *et al.*, 1991). Results of studies of both nuclei showed that PC and PE production rates were on the order of 1.0 fmol/cell per hour, and that the kinetics of choline incorporation did not alter in the presence of ethanolamine, indicating that they have non-competing pathways.

The concentrations of PL metabolites were found to correlate with growth rates and cell cycle in NMR studies of extracts of breast tumors (Smith *et al.*, 1991). PC was highest, and GPC lowest, in the S-phase. We have also shown that PME levels decreased as the cells approached confluency, and it seems that PME is more an indicator of growth rather than of nutritional status, and represents enhanced biosynthesis of PL in rapidly proliferating cells (Daly *et al.*, 1987; Daly *et al.*, 1990). Another NMR indicator of cellular overgrowth was the pH; near confluency it became acidic.

Novel NMR techniques may contribute significantly to the delineation of cellular metabolic processes. Thus, complete separation of intracellular and extracellular information was obtained in cancer cells using diffusion-weighted spectroscopy (van Zijl *et al.*, 1991). These techniques also furnish information on membrane permeability and uptake of metabolites, and the ratio of extra- and intracellular water volumes.

Effects of drugs

The effects of drugs on cancer can be investigated by NMR spectral changes, both *in vitro* and *in vivo* (see below), most notably by ATP and PME depletion, and pH changes. In cancer cell studies the principal use of NMR spectroscopy is in the development and evaluation of new therapeutic modalities, and for studying the mechanisms of action of the drugs.

As indicated above, many cancer cells are characterized by high NMR signals of the PL precursors, PC and PE. In the human lymphoma cell line MOLT-4 the levels of these precursors are lowered by 40% after 6 h of perfusion with 1-beta-arabinofuranosylcytidine, and the cells lysed after 8-10 h (Daly *et al.*, 1990). Human breast cancer cells MDA-MB-231 are insensitive to this drug, and following perfusion with it, there were no ^{31}P NMR spectral changes. Thus, NMR can be used as an indicator for the efficacy of an anti-neoplastic agent.

Multidrug-resistance phenomenon is an important factor in treatment failure in cancer patients. Initially, we described substantial ^{31}P NMR spectral differences between sensitive and resistant MCF-7 human breast cancer cells (Cohen *et al.*, 1986). The resistant cells had higher levels of PCr, and lower levels of PL metabolites, mainly PDE, compared to their sensitive counterparts (Figure 8). These findings were corroborated by other studies (Evelhoch *et al.*,

1987). However, since the number of compounds detected by ^{31}P NMR is limited, the nature of these differences could not be evaluated fully. Proton NMR could not provide these data by itself, due to the fact that ^{1}H signals of some phosphorous-containing compounds and their precursors resonate at almost identical chemical shifts. The resolution to this problem was to perform combined ^{31}P and ^{1}H NMR analyses of extracts (Kaplan *et al.*, 1990), which provided quantitative data on metabolite concentrations (Figure 9). In the resistant cells, the high energy compounds, ATP and PCr and also their precursors, ADP and creatine, respectively, were elevated compared to the sensitive cells, while the latter showed higher PME and PDE concentrations than the resistant cells, but choline levels were similar (Table 2). These studies, which exemplify the usefulness of extracts for specific well-defined circumstances, delineated the differences in control of metabolic pathways between drug-sensitive and -resistant cancer cells.

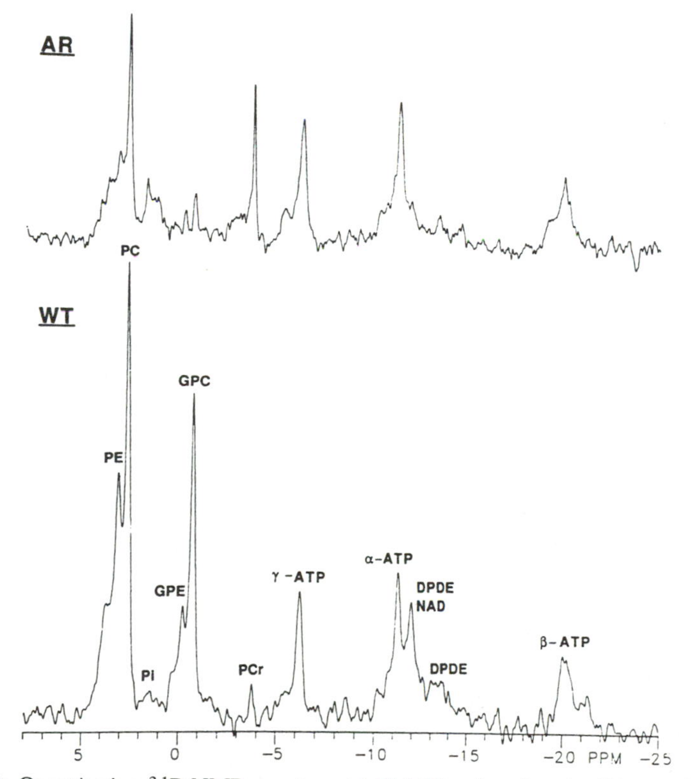

Figure 8: Quantitative ^{31}P NMR spectra at 162 MHz of perfused wild type (WT) and adriamycin resistant (AR) MCF-7 cells embedded in agarose gel threads (200 scans were accumulated with a 40 sec recycle time). Peak assignments are: 1, PME; 2, Pi; 3, PDE; 4, PCr; 5, α-ATP; 6, γ-ATP, NAD and UDPG; 7, UDPG; 8, β-ATP.

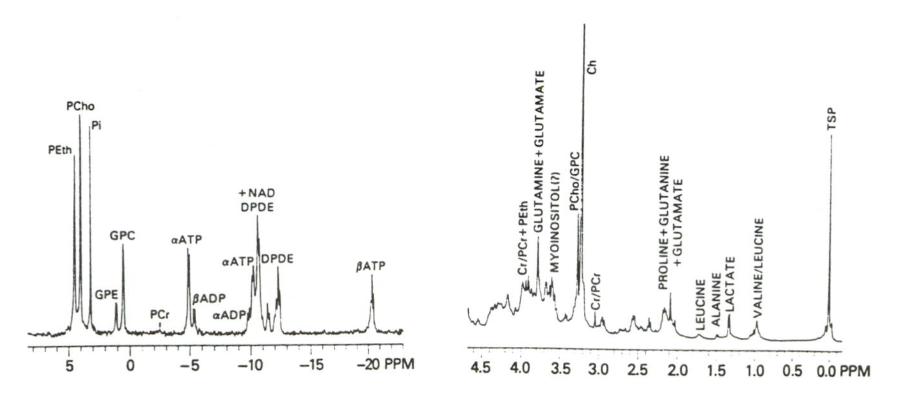

Figure 9: Examples of extract spectra. (A) ^{31}P NMR spectrum at 162 MHz of a perchloric acid extract of MCF-7 breast cancer cells, and (B) ^{1}H NMR spectrum of the same extract. The ^{1}H NMR spectrum was recorded in 1 ml of D_2O and for the ^{31}P spectrum an additional 1 ml was added.

^{13}C NMR studies of perfused MCF-7 cells enabled quantitative comparison in glycolysis rates between sensitive and drug-resistance cells, and it was much faster in the resistant cells, probably due to their increased energy requirements. On the basis of these findings, glycolysis inhibitors which might overcome the MDR phenomenon, were examined. Upon entering the cell 2-DG is phosphorylated through an ATP-consuming reaction, and undergoes no further metabolism (Wick *et al.*, 1957). Moreover, the phosphorylated product, 2-DG-6P, inhibits hexose phosphate isomerase and further interferes with utilization of glucose in the glycolytic pathway (Horton *et al.*, 1973), leading to cell starvation. The therapeutic applications of 2-DG were tested in various malignant disorders (Bessel *et al.*, 1973; Jain *et al*, 1979; Purohit and Pohlit, 1982; Kern and Norton, 1987), but it was found impractical as a single agent. However, 2-DG may have a role in combination with other therapeutic means, and especially whenever there are inherent abnormal energy requirements. Therefore ^{31}P and ^{13}C NMR studies of perfused drug-sensitive and -resistant MCF7 cells, embedded in agarose threads, were performed (Kaplan *et al.*, 1990). ^{31}P NMR kinetic studies demonstrated that the resistant cells accumulated 2-DG-6P faster, and to a greater extent, than the sensitive cells, concomitant with faster and more profound depletion of ATP (Figure 10). Phosphorylation rates were measured by ^{13}C NMR, using 2-DG enriched with ^{13}C at the 6-position, and were 11.2×10^{-4} and 6.5×10^{-4} mmol/min/mg protein for resistant and sensitive cells, respectively (see Figure 6). Toxicity studies revealed that 2-DG is 15-fold more toxic to the resistant compared to the sensitive cells, and that 2-DG has an additive, but not synergistic, effect to adriamycin toxicity.

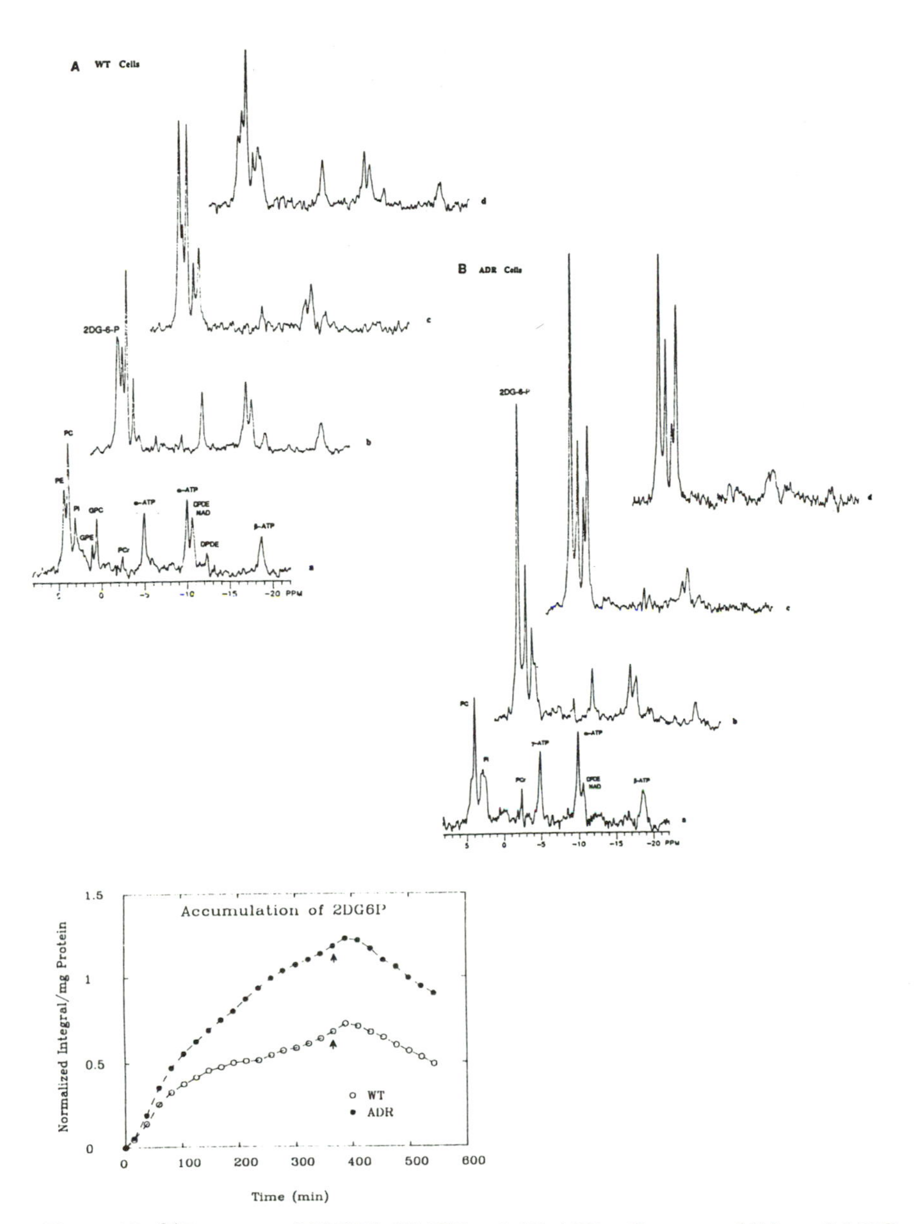

Figure 10: ^{31}P spectra of MCF-7 (A) WT and (B) ADR cells upon addition of 2-DG; (spectrum a) control, (spectrum b) 2-DG perfusion for 7 hr; (spectrum c) 2-DG perfusion for 21 and 18, respectively; (spectrum d) wash-out after 4 and 3.5, respectively. (C) Comparison of accumulation of 2-DG6P in WT and ADR MCF-7 cells following addition of 2-DG (5 mM) at time zero; the arrow represents the wash out with normal medium.

In order to investigate whether MDR is associated with specific NMR features, ^{31}P spectra of several resistant cancer cell lines (ovarian, cervix, melanoma, breast, dog kidney, hamster lung) were recorded (Kaplan *et al.*, 1991). The sensitive cells were either selected to MDR with various drugs, or transfected with the MDR1 gene, and had variable resistances. In all cell lines, including those transfected with the MDR1 gene and low level of drug resistance, there were NMR spectral changes upon acquiring the MDR trait; however, these changes were not consistent for all cell lines. The toxicity to 2-DG was higher in all MDR, as compared to the drug-sensitive cells, and in several sublines of MCF7 cells, the toxicity was highest for the most resistant lines. These results support the hypothesis that MDR is associated with increased energy requirements, and that therapeutic means that attack energy production mechanisms may play a role in preventing and/or negating the development of the MDR phenotype.

Gossypol, a polyphenolic bisnaphthalene aldehyde from the cotton seed, has attracted much attention as a potential male anti-fertility drug, and also as an anti-neoplastic agent (Band *et al.*, 1989). Rhodamine-123, a dye used in photographic industry, is supposed to have anti-mitochondrial properties, and was also suggested as an anti-cancer drug (Modica-Napolitano and Aprille, 1987). Previous non-NMR studies postulated that the actions of both gossypol and rhodamine-123 might be through energy production inhibition (Benz *et al.*, 1987). However, our ^{31}P NMR studies showed that the mechanisms of their effects were different (Jaroszewski *et al.*, 1990). Gossypol induced ATP depletion, markedly increased levels of pyridine nucleotides, and decreased GPC signals, while rhodamine-123 caused only ATP depletion. There were also differences regarding glucose metabolism; both drugs induced elevation of glucose uptake, but an increase in lactate production exceeding that of glucose consumption, and indicating inhibition of oxidative phosphorylation, was observed only in the case of rhodamine-123. Moreover, the resistant cells exhibit cross-resistant to rhodamine-123, but remained sensitive to low concentrations of gossypol. The resistance to rhodamine-123 and sensitivity to gossypol were observed in both selected and transfected MDR cells, pointing to the potential use of gossypol against MDR tumors.

The effects of lonidamine, a relative new anti-neoplastic agent, were studied by combined ^{13}C and ^{31}P NMR spectroscopy (Ben-Horin *et al.*, 1993). When introduced, it was postulated that its mechanism of activity is both respiration and glycolysis inhibition (Paggi *et al.*, 1987). NMR spectra of perfused MCF7 cell in alginate capsules demonstrated a pronounced decrease of intracellular pH, concomitant with increased intracellular and decreased extracellular lactate content. Indeed, the toxicity of lonidamine was greater in acidic environment. The results lead to the conclusion that lonidamine inhibits the efflux of lactate from the cells. NMR thus provided unique data on the mechanism of action of lonidamine, which was unattainable by conventional biochemical methods.

Hormonal and growth factors effects

Hormonal manipulation may have an important role in the treatment of malignant diseases, especially in those of the reproductive organs. Breast cancer may progress from a phenotype, which is responsive to endocrine manipulations and chemotherapy, to a more malignant form, which is resistant to all therapeutical approaches (Clarke *et al.*, 1990). Estrogen receptor status, and responsiveness to hormonal treatment is important in determining prognosis and considering therapy. Therefore, studies were performed trying to correlate NMR spectral changes with hormonal receptors status and hormonal effects (Degani *et al.*, 1988; Ruiz-Cabello *et al.*, 1991). In a study of extracts of human malignant breast tumors low GPC, and high PCr, were associated with high content of estrogen receptors (Barzilai *et al.*, 1991).

The effects of estrogens and anti-estrogens on metabolism of T47D human breast cancer cells were studies by Neeman and Degani (1989). Cells were attached to microcarrier beads and perfused with either 17-b-estradiol or tamoxifen and ^{31}P and ^{13}C spectra were recorded. ^{31}P NMR showed that PC and nucleoside diphosphate were higher in cells exposed to tamoxifen, compared to estrogen treated cells. Glucose utilization and glycolysis products were monitored by ^{13}C NMR using [1-^{13}C]-glucose. Estrogen administration was followed by elevated rates of lactate and glutamate production. This enhanced glucose metabolism was inhibited by actinomycin D or cycloheximide, suggesting that estrogen stimulation requires synthesis of mRNA and/or proteins (Neeman and Degani, 1989).

We have recently investigated the effects of 17-β-estradiol and tamoxifen on the metabolic/bioenergetic spectra of a series of agarose embedded and perfused human breast cancer cell lines, with variable estrogen dependency (Ruiz-Cabello *et al.*, 1991, 1993). A comparison of baseline ^{31}P NMR spectra associated higher PDE and UDPG, and lower PC/GPC and PC/PE ratios, with the acquisition of estrogen independent growth in estrogen receptors expressing cells. No metabolic changes were clearly associated with the metastatic phenotype. Whilst estrogen treatment induced no consistent spectral changes in all cell lines, the estrogen-independent and -responsive MCF7/MIII cells responded to tamoxifen treatment by significantly increasing all spectral resonances 30%-40% above baseline values (Figure 11). This may reflect a tamoxifen induced change to a more differentiated or apoptotoc phenotype, or an attempt by the cells to reverse the inhibitory effects of the drug.

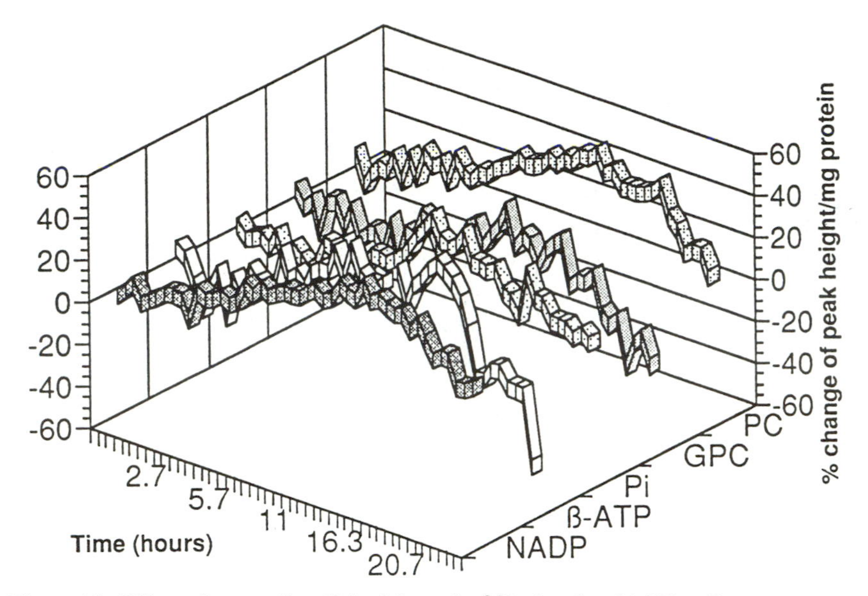

Figure 11: Effect of tamoxifen (0.5 μM) on the ^{31}P signals of MIII cells.

^{31}P NMR studies, including monitoring the phosphorylation rates of 2-DG, provided a plausible explanation for the "paradoxical" effects of epidermal growth factor on MDA-468 human breast cancer cells (Kaplan *et al.*, 1990). These cells are characterized by large number of EGF receptors, but are inhibited by high levels of EGF. EGF accelerated glucose consumption in these cells, but when we perfused them with EGF and high glucose concentrations during the NMR measurements, ATP levels remained stable, and they survived. Thus, the inhibitory EGF effect may be confined to cultured cells in-vitro, and to glucose depletion situations in-vivo.

Clinical relevance

The high resolution of NMR spectra of cancer cells and the facility by which experimental conditions can be controlled has made it possible to assign signals from metabolites and to determine their biochemical profiles. As indicated above, there were no qualitative differences between normal and malignant cells, and NMR can not provide diagnostic biochemical "fingerprints" of cancer. However, the typical quantitative changes associated with the malignant process can be of value for clinical purposes.

The differences between NMR of cells, and those of tissues or organs in clinical NMR, should be emphasized. These differences may be due to the heterogeneity of the whole tumor tissues, which include connective tissues and blood vessels; blood flow and topographical parameters which interfere with signal resolution and assignments; and the constituents of the growth medium, trypsinization of cells, and restraining methods for cell perfusion. Therefore, the relevance of experiments involving cells should be proven in each case, and sometimes the results are found wanting.

In order to test the significance of the results obtained from cell studies we performed similar experiments in animals. Thus, MCF7 human breast cancer cells were injected into nude mice and grew into subcutaneous tumors. ^{31}P, and also ^{13}C, NMR spectroscopy studies of these tumors, performed with surface coils in specially built NMR probes (Lyon *et al.*, 1988), indeed showed the same characteristics and metabolic events (e.g. glycolysis rates) detected by NMR of the perfused cells. Tumor-bearing mice were treated with 2-DG, and NMR results obtained in vivo with surface coils, were similar to the those obtained in cellular studies (Navon *et al.*, 1989).

NMR spectroscopy of human tumors in whole-body magnets is now available; it should be mentioned, however, that their lower magnetic field strength means lower sensitivity and resolution than in NMR of cells.

Perhaps the most promising clinical application of NMR spectroscopy in oncology is in evaluating new therapeutic modalities, and quantization of the effects of treatment. To date, the techniques most often used have been *in-vivo* ^{31}P NMR spectroscopy of human tumors before and following their therapy (Ng *et al.*, 1989; Glaholm *et al.*, 1989; Degani *et al.*, 1986). The decline of PME signals was found to be an indicator for cancer response to chemotherapy, as well as endocrine treatment. Since the PDEs are involved in PL catabolism, it is plausible that the ratio PME/PDE may reflect membrane structure and rates of synthesis versus catabolism, and indeed this ratio was suggested as an indicator for treatment success (Ng *et al.*, 1989).

NMR spectroscopy can yield data on prognostic parameters of tumors, which influence treatment policy. The correlation between NMR features and estrogen receptors status of breast cancer may be used in considering endocrine manipulation, and it was even suggested that by combined MRI and NMR spectroscopy invasive procedures may be avoided (Barzilai *et al.*, 1991). It seems to us that this conclusion is rather premature.

The differentiation process of malignant cells can also be followed by NMR spectroscopy (Smith *et al.*, 1991). PC signals correlated strongly with cell proliferation and differentiation as measured by number of cells in S-phase, rate of DNA synthesis, and pathological grading. PE did not correlate with these parameters, whereas GPC and GPE showed a strong and weak correlations, respectively. This finding may be useful in developing new effective therapeutic means. Care should be taken when analyzing data concerning PL levels, and the ratios of their metabolites by NMR, since PL metabolism is very complex, the data may be species-dependent, and opposite results were reported.

Conclusions

NMR spectroscopy is an established research tool for studies of metabolism of cancer cells. The main advantages of this method are that it is non-invasive and that metabolic activities can be monitored continuously. Energy metabolism, PL synthetic and degradation pathways, and intracellular pH changes are accurately delineated. Valuable data on the effects of substrates and nutrients, drugs, hormones and growth factors on cancer cells have been obtained. It seems that the major potential of NMR spectroscopy of cancer is in the development and evaluation of new therapeutic modalities. While MRI has become an integral and important component of the armamentarium of diagnostic techniques, NMR spectroscopy of human tumors is still in the early stages of development and applications. Further NMR spectroscopy studies of cancer cells can be expected to pave the way for wider and more tangible uses of NMR spectroscopy in clinical practice.

Acknowledgments

JSC would like to take this opportunity to acknowledge the dedication of those who worked in his laboratory in NCI (1983-90) and GUMC (1990-93) and who did the work described here: Robbe Lyon, Pat Faustino, Peter Daly, Jerzy Jaroszewski, Ofer Kaplan, Peter van Zijl, Kirsten Berghmans, and Jesus Ruiz Cabello.

References

Askenasy, N., Kushnir, T., Kaplan, O., and Navon, G. (1990). *NMR Biomed.* **3**, 220.

Balaban, R.S., Gadian, D.G., Radda, G.K., and Wong, G.G. (1981). *Anal. Biochem.* **116**, 450.

Band, V., Hoffer, A.P., Band, H., Rhinehardt, A.P., Knapp, R.C., Matlin, S.A., and Anderson, D.J. (1989). *Gynecol. Oncol.* **32**, 273.

Barzilai, A., Horowitz, A., Geier, A., and Degani, H. (1991). *Cancer Res.* **67**, 2919.

Ben-Horin, H., Kaplan, O., and Navon, G. (1993). *Cancer Res. Submitted.*

Benz, C., Hollander, C., Keniry, M., James, T.L., and Mitchell, M. (1987). *J. Clin. Invest.* **79**, 517.

Bessel, E.M., Courtenay, V.D., Foster, A.B., Jones, M., and Westwood, J.H. (1973). *Eur. J. Cancer* **9**, 463.

Bines, S.D., Tomasovic, S.P., Frazer, J.W., and Boddie, A.W. (1985) *J. Surg. Res.* **38**, 546.

Boddie, A.W., Frazer, J.W., Tomasevic, S.P., and Dennis, L. (1989). *J. Surg. Res.* **46**, 90.

Brophy, P.J., Hayer, K.M., and Riddell, F.G. (1983). *Biochem. J.* **210**, 961.

Clarke, R., Dickson, R.B., and Brunner, N. (1990). *Ann. Oncol.* **1**, 401.

Cohen, J.S., Knop, R.H., Navon, G., and Foxall, D. (1983). *Life Chem. Rep.* **1**, 281.

Cohen, J.S., Lyon, R.C., Faustino, P.J., Batist, G., Shoemaker, M.,

Rubalcaba, E., and Cowan, K.H. (1986). *Cancer Res.* **46**, 4087.

Cohen, J.S. (1988). *Mayo Clin. Proc.* **63**, 1199.

Cohen, J.S., Lyon, R.C., and Daly, P.F. (1988). *Methods. Enzymol.* **48**, 435.

Daly, P.F., Lyon, R.C., Faustino, P.J., and Cohen, J.S. (1987). *J. Biol Chem.* **262**, 14875.

Daly, P.F., Lyon, R.C., Straka, E.J., and Cohen, J.S. (1988). *FASEB J.* **2**, 2596.

Daly P.F., and Cohen, J.S. (1989). *Cancer Res.* **49**, 770.

Daly, P.F., Zugmaier, G., Sandler, D., Carpen, M., Myers, C.E., and Cohen, J.S. (1990). *Cancer Res.* **50**, 552.

Daniels, A., Williams., R.J.P., and Wright, P.E. (1976). *Nature* **261**, 321.

Degani, H., Horowitz, A., and Itzchak, Y. (1986). *Radiology* **161**, 53.

Degani, H., Victor, T.A., Neeman, M., Itzchak, Y., Horowitz, A., and Kaye, A.M. (1988). *Progr. Cancer Res. Ther.* **35**, 378.

Evanochko, W.T., Sakai, T.T., Ng, T.C., Krishna, N.R., Kim, H.D., Zeidler, R.B., Ghanta, V.K., Brockman, R.W., Schiffer, M., Braunschweiger, P.G., and Glickson, J.D. (1984). *Biochim. Biophys. Acta* **805**, 104.

Evelhoch, J.L., Keller, N.A., and Corbett, T.H. (1987). *Cancer Res.* **47**, 3396.

Folch, J., Ascoli, L., Less, M., Meath, J.A., and Lebaron, N. (1951). *J. Biol. Chem.* **191**, 833.

Foxall, D.L., and Cohen, J.S. (1983). *J. Magn. Reson.* **52**, 346.

Foxall, D.L., Cohen, J.S., and Mitchell, J.B. (1984). *Exp. Cell Res.* **154**, 521.

Freyer, J.P. (1988). *Cancer Res.* **48**, 2432.

Glaholm, J., Leach, M.O., Collins, D.J., Mansi, J., Sharp, J.C., Madden, A., Smith, I.E., and McCready, J.R. (1989). *Lancet.* **1**, 1326.

Glonek, T., Kopp, S.J., Kot, E., Pettigrew, J.W., Harrison, W.H., and Chen, M.M. (1982). *J. Neurochem.* **39**, 1210.

Gonzalez-Mendez, Wemmer, D., Hahn, G., Wade-Jardetsky, N., and Jardetsky, O. (1982). *Biochim. Biophys. Acta* **720**, 274.

Gupta, R.K., Gupta, P., and Moore, R.D. (1984). *Ann. Rev. Biophys. Bioeng.* **13**, 121.

Horton, R.W., Meldrum, B.S., and Bachelard, H.S. (1973). *J. Neurochem.* **21**, 507.

Jain, V.K., Kalia, U.K., Gonipath, P.M., Naqvi, S., and Kucheria, K. (1979). *Ind. J. Exp. Biol.* **17**, 1320.

Jaroszewski, J., Kaplan, O., and Cohen, J.S. (1990). *Cancer Res.* **50**, 6936.

Jentoft, J.E., and Town, C.D. (1985). *J. Cell Biol.* **101**, 778.

Kaplan, O., Navon, G., Lyon, R.C., Faustino, P.J., Straka, E.J., and Cohen, J.S. (1990). *Cancer Res.* **50**, 544.

Kaplan, O., van Zijl, P.C.M., and Cohen, J.S. (1990). *Biochem. Biophys. Res. Commun.* **169**, 383.

Kaplan, O., Jaroszewski, J., Faustino, P.J., Zugmaier, G., Ennis, B.W., Lippman, M.C., and Cohen, J.S. (1990). *J. Biol Chem.* **265**, 13641.

Kaplan, O., Jaroszewski, J., Clarke, R., Fairchild, C.R., Schoenlien, P., Goldenberg, S., Gottesman, M.M., and Cohen, J.S. (1991). *Cancer Res.* **51**, 1638.

Kaplan, O., van Zijl, P.C.M. and Cohen, J.S. (1992). NMR Basic Princpl. Progr. 28, 3 (Diehl, P., Fluck, E., Gunther, H., Kosfeld, R., Seelig, eds.) Springer-Verlag, Berlin.

Kern, K.A., and Norton, J.A. (1987). *Surgery* **102**, 380.

Kushnir, T., Kaplan, O., Askenasy, N., and Navon, G. (1989). *Mag Reson. Med.* **10**, 119.

Labotka, R.J., Warth, J.A., Winecki, V., and Omachi, A. (1985). *Anal. Biochem.* **147**, 75.

Legerton, T.L., Kanamori, K., Weiss, R.L., and Roberts, J.D. (1983). *Biochemistry* **22**, 899.

Lim, P.-J., Blumenstein, M., and Mikkelsen, R.B. (1987). *J. Magn. Reson.* **73**, 399.

Lowry, O.H., and Passonneau, J.V. (1972). A flexible system of enzymatic analysis. Academic Press, New York, p 120.

Lyon, R.C., Cohen, J.S., Faustino, P.J., Megnin, F., and Myers, C.E. (1988). *Cancer Res.* **48**, 870.

Lyon, R.C., Tschudin, R.G., Daly, P.F., and Cohen, J.S. (1988). *Mag. Reson. Med.* **6**, 1

Malet-Martino, M.C., Lame, F., Ialaneix, J.P., Palevedy, C., Hollande, E., and Martino, R. (1986). *Cancer Chemother.* **18**, 5.

Maris, J.M., Evans, A.E., McLaughlin, A.C., D'Angio, G.J., Bolinger, L., Manos, H. and Chance, B. (1985). *New Engl. J. Med.* **312**, 1500.

Meneses, P., and Glonek, T. (1988). *J. Lipid Res.* **29**, 679.

Merchant, T.E., Gierke, L.W., Meneses, P., and Glonek, T. (1988). *Cancer Res.* **48**, 5112.

Modica-Napolitano, J.S., and Aprille (1987). *Cancer Res.* **47**, 4361.

Moon, R.B., and Richards, J.H. (1973). *J. Biol. Chem.* **248**, 7276.

Morris, P.G. (1988). *Ann. Rep. NMR Spectrosc.* **20**, 1-60.

Navon, G., Ogawa, S., Shulman, R., and Yamane, T. (1977). *Proc. Natl. Acad. Sci. U.S.A.* **74**, 87.

Navon, G., Lyon, R.C., Kaplan, O., and Cohen, J.S. (1989). *FEBS. Lett.* **247**, 86.

Neeman, M., and Degani, H. (1989). *Cancer Res.* **49**, 589.

Neeman, M., and Degani, H. (1989). *Proc. Natl. Acad. Sci. U.S.A.* **86**, 5585.

Ng, T.C., Grundfest, S., Vijayakumar, S., Baldwin, N.J., Majors, W., Karalis, I., Meaney, T.E., Shin, K.H., Thomas, F.J., and Tubbs, R. (1989). *Mag. Reson. Med.* **10**, 125.

Paggi, M.G., Zupi, G., Fancuiulli, M., Del Carlo, C., Giorno, S., Laudonio, N., Silvestrini, B., Caputo, A., and Floridi, A. (1987). *Exp. Mol. Pathol.* **47**, 154.

Purohit, S.C., and Pohlit, W. (1982). *Int. J. Radiat. Oncol. Biol. Phys.* **8**, 495.

Radda, G.K., Oberhaensli, R.D., and Taylor, D.J. (1987). *Ann. N.Y. Acad. Sci.* **508**, 300.

Ronen, S., and Degani, H. (1989). *Mag. Reson. Med.* **12**, 274.

Ronen, S., Rushkin, E., and Degani, H. (1991). *Biochim. Biophys. Acta* **1095**, 5.

Ruiz-Cabello, J., Berghmans, K., Clarke, R., Andrews, P., Simpkins, H., Kaplan, O., and Cohen, J.S. (1991). 10th Ann. Meeting SMRM.

Ruiz-Cabello, J., Berghmans, K., Kaplan, O., Lippman, M.E., Clarke, R., and Cohen, J.S. (1993). *Cancer Res. Submitted.*

Shankar-Narayan, K., Moress, E.A., Chatham, J.C., and Barker, P.B. (1990). *NMR Biomed.* **3**, 23.

Shulman, R.G., Brown, T.R., Ugurbil, K., Ogawa, S., Cohen, S.M., and den Hollander, J.A. (1979). *Science* **205**, 160.

Sijens, P.E., Wijrdeman, H.K., Moerland, M.A., Bakker, C.J.G., Vermeulen, J., Wa, H., and Luyten, P.R. (1988). *Radiology* **169**, 615.

Smith, T.A.D., Eccles, S., Ormerod, M.G., Tombs, A.J., Titley, J.C., and Leach, M.O. (1991). *Br. J. Cancer* **64**, 821.

Ugurbil, K., Uernsey, D.L., Brown, T.R., Glynn, P., Tobkes, N., and Edelman, I.S.

(1981). *Proc. Natl. Acad. Sci. U.S.A.* **78**, 4843.

Ugurbil, K., Rottenberg, H., Glynn, P., and Shulman, R.G. (1982). *Biochemistry* **21**, 1068.

van Zijl, P.C.M., Moonen, C.T.W., Faustino, P.J., Pekar, J., Kaplan, O., and Cohen, J.S. (1991). *Proc. Natl. Acad. Sci. U.S.A.* **88**, 3228.

van Zijl, P.C.M. and Moonen, C.W.T. (1992). NMR Basic Principle Progr. 26, 67. (Diehl, P., Fluck, E., Gunther, H., Kosfeld, R., Seelig, eds.) Springer-Verlag, Berlin.

Warburg, O. (1956). *Science* **123**, 309.

Wick, A.N., Drury, D.R., Nakada, H.I., and Wolfe, J.B. (1957). *J. Biol Chem.* **224**, 963.

23

Ex Vivo Multinuclear NMR Spectroscopy of Perfused, Respiring Rat Brain Slices: Model Studies of Hypoxia, Ischemia, and Excitotoxicity

L. Litt, M.T. Espanol, Y. Xu, Y. Cohen, L.-H. Chang, P.R. Weinstein, P.H. Chan, and T.L. James

Departments of Anesthesia, Pharmaceutical Chemistry, Neurosurgery, Radiology, and The Cardiovascular Research Institute
University of California
San Francisco, CA 94143 USA

We believe there are important roles for *in vivo* NMR spectroscopy techniques in studies of protection and treatment in stroke. Perhaps the primary utility of *in vivo* NMR spectroscopy is to establish the relevance of metabolic integrity, intracellular pH, and intracellular energy stores to concurrent changes occurring both at gross physiological levels (*e.g.*, changes in cerebral blood flow, or blood oxygenation), and at microscopic or cellular levels. It has long been known that the brain is exquisitely sensitive to deprivations of oxygen, glucose, and cerebral blood flow. Routine human surgery on a limb takes place every day with tourniquets stopping all blood flow for up to two hours. In contrast, the deprivation of all blood flow to the brain (global ischemia) for approximately 5 minutes can result in severe, permanent brain damage. Research has gone on for more than 30 years to understand why the brain's revival time is so much shorter, and to discover brain biochemical interventions that might dramatically extend the brain's intolerance beyond 5 minutes, and therefore be relevant to protection and treatment of stroke. (Kogure and Hossmann, 1985; 1993)

Stroke, defined as a permanent neurologic deficit arising from the death of brain cells, kills ~150,000 people in the U.S.A. each year, and is the third leading cause of death (Feinleib *et al.*, 1993). It is the next malady to escape, once one has dodged death from cardiovascular disease and cancer. Many, if not most, U.S.A. stroke victims will receive neurological clinical care not substantially different from what was provided 30 years ago. Most stroke patients will be put in intensive care units where blood pressure will be regulated and kept in a "safe" range, with the body given supportive care and the brain given an opportunity to heal itself. The problem of stroke is actually quite complex because there are several different kinds of stroke (ischemic, hemorrhagic, *etc.*), and because numerous systemic physiological factors are of relevance. Nevertheless, exciting advances in brain biochemistry suggest that stroke therapy and prophylaxis are likely to improve dramatically in the near future (Zivin and Choi, 1991). The proper design of experimental NMR studies of stroke is complex, because one must properly control systemic physiological factors if one wants to determine intracellular mechanisms. However, because intracellular energy failure is of rapid onset, and because NMR spectroscopy quickly reveals changes in brain energy, *in vivo* and *ex vivo* NMR spectroscopy are useful for assessing initial pharmacological effects during and after stroke. Recent studies from our laboratory are reviewed below. We have tried to reference important works of others in this field, and apologize in advance to any groups that we might be inadvertently omitting!

In vivo ^{31}P NMR spectroscopy of low intracellular pH

In addition to yielding relative intracellular concentrations of adenosine triphosphate (ATP), phosphocreatine (PCr), inorganic phosphate (Pi), and other high energy phosphates, *in vivo* ^{31}P NMR spectroscopy provides a measure of intracellular pH (Adler, 1990). Figure 1 shows a typical *in vivo*, ^{31}P NMR brain spectrum from a healthy, anesthetized rat. Intracellular pH (pHi) is determined from the chemical shift separation, δ, between P_i and PCr. The physical basis for NMR determinations of pHi comes from the fact that monobasic and dibasic inorganic phosphate are in rapid chemical exchange during the acquisition phase of the NMR experiment. Thus one NMR resonance peak represents two P_i species: monobasic P_i ($H_2PO_4^-$) and dibasic P_i (HPO_4^{2-}). The observed ^{31}P resonance peak represents a time average for phosphorous nuclei within P_i as hydrogen ion binding or release causes transitions between mono- and dibasic P_i. Because the time spent as mono- or dibasic P_i is proportional to their relative concentrations, the observed P_i resonance peak occurs at a chemical shift that is the weighted average of chemical shifts for mono- and dibasic P_i. Thus the pH variation of the P_i chemical shift provides a measure of the ratio $H_2PO_4^-$ to HPO_4^{2-}, which, via the Henderson-Hasselbalch equation, provides a determination of pH. Of course the subject is actually more complex than just described, the Henderson-Hasselbalch equation is never actually written out, and carefully obtained calibration data are instead obtained to guide the use of theoretical conceptions. A careful study regarding determination of brain intracellular pH was conducted by Petroff and colleagues (1985). Among other things, they showed that the P_i-PCr chemical shift difference, given as δ in ppm units, was most affected by changes in intracellular magnesium concentrations ($[Mg^{++}]_i$), but that the effect was negligible *in vivo* so long as $[Mg^{++}]_i$ was less than 2.5 mM. The phenomenological formula decided upon by Petroff and collaborators for brain intracellular pH in the brain is: $pH_i = 6.77 + \log [(\delta\text{-}3.29)/(5.68\text{-}\delta)]$, which has the same functional form as that used by earlier investigators (Seo *et al.*, 1983). Other assumptions to the whole procedure are that all NMR-detectable P_i and PCr are intracellular, and that *in vivo* buffering conditions resemble those in the calibration experiment. More significantly, it should be noted that the formula is both phenomenological and *ad hoc*, with constants coming from a least squares fit. The functional form, reminiscent of the Henderson-Hasselbalch equation, was chosen because it is mathematically convenient. In the actual titration curves, *in vivo* measurements of δ stop changing as pH is lowered below ~5.8 or raised above ~8.0. Mathematically, however, pH_i values become infinite if δ is found to be near 3.29 ppm or 5.68 ppm. Values of δ near 3.29 can occur *in vivo* during severe tissue acidosis in deteriorating ischemic tissue. At such times the quality of *in vivo* ^{31}P NMR spectra is usually degraded, with the PCr peak tending to disappear, and the P_i peak tending to become multicompartmental, or just plain messy. The point here is that *in vivo* pH_i determinations might be severely inaccurate, even meaningless, if values produced by the experimentally determined δ fall outside the calibration range: 5.8 to 8.0. *In vivo* brain pH_i determinations can be tricky!

Low pH_i, particularly during hypoxia or ischemia, has for some time been viewed as bad. Indeed in the heart there is a worsening of muscle contractility as extracellular pH decreases below 7.2. *In vivo* pH_i often falls only as a result of increased CO_2 (hypercapnia), with hydrogen ions coming from the carbonic anhydrase reaction: $CO_2 + H_2O \leftrightarrow H_2CO_3 \leftrightarrow H^+ + HCO_3^-$. Clearly, intracellular physiology is different when cytosolic hydrogen ion concentrations increase during hypercapnia, instead of during uncompensated ATP hydrolysis during periods of oxygen deprivation. The normal adult human turns over ~150 moles of hydrogen ions per day (Alberti and Cuthbert, 1982). ATP hydrolysis (energy utilization), plus inadequate recycling of hydrogen ions into ATP resynthesis (as a result of oxygen deprivation) rapidly produces intracellular acidosis. However, the existence of comparable low pH_i values in dissimilar physiological states, hypercapnia and hypoxia/ischemia, permits an inquiry about the extent to which low pH_i, all by itself, is sufficient to cause cellular injury. In more simple terms, does low pH_i accompany the cerebral injury that results from oxygen deprivation, or does it accelerate it? Does brain injury occur because of acidosis, or simply with it? To what extent is intracellular acidosis an incidental epiphenomenon? In non-NMR studies there has been some evidence that low extracellular pH is good, because it inactivates NMDA-type glutamate receptors (Kaku *et al.*, 1993), but only so long as decreases in pH extend no lower than ~6.7. Other non-NMR studies of hypercapnia and ischemia find low intracellular pH to be bad (Katsura *et al.*, 1994; Kristian *et al.*, 1994).

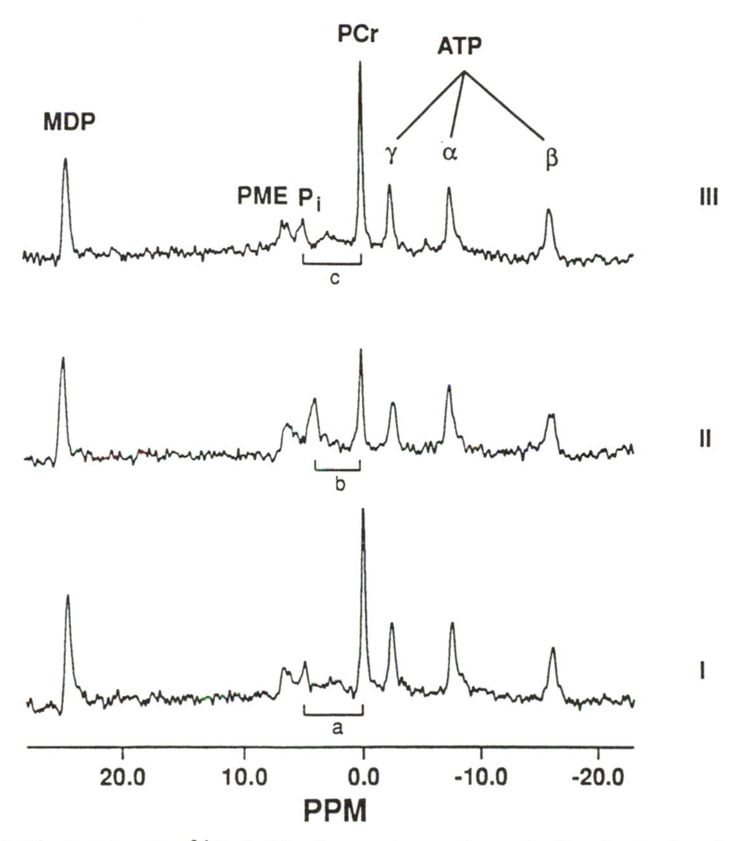

Figure 1: Typical *in vivo* ^{31}P adult brain spectra, each acquired in ~5 minutes, from an anesthetized rat in the study by Xu *et al.* (1990). Reprinted with permission. Data were acquired at 4.7 Tesla in the nonmagnetic hyperbaric chamber described by Litt *et al.* (1993). The P_i-PCr chemical shift difference, indicated by "a" in the bottom tracing (I), was measured normocapnia. In the severe acidosis of supercapnia ($PaCO_2$ ~750 mm Hg), the middle tracing (II) shows that the P_i-PCr chemical shift difference, indicated by "b", has decreased by ~1 ppm. In the top tracing (III) the P_i peak has recovered its original chemical shift value. All animals woke up and appeared normal after this perturbation.

In our 4.7 Tesla *in vivo* NMR studies, CO_2 administration was used to lower intracellular pH under conditions that avoided oxygen deprivation. In this regard it must be mentioned that increasing PCO_2 *in vivo* has numerous physiological side effects (Litt *et al.*, 1985). *In vivo* CO_2 does not restrict its perturbations to one chemical reaction, only in cells viewed by our NMR brain coil. Increased cerebral and systemic blood flow, and increased sympathetic tone are initially produced by increased CO_2. Curiously, at high concentrations, CO_2 is also a general anesthetic agent, much like nitrous oxide, halothane, or even nitrogen (which can produce "narcosis of the deep" in divers). An estimate of CO_2's anesthetic potency is calculable from its lipid solubility, which is very close to that of nitrous oxide. However, CO_2 is three times as potent an anesthetic as expected from such a calculation, with approximately one third of an atmosphere (~250 mm Hg PCO_2) being equivalent to approximately 1% of halothane (ED50, or MAC).

The results of our *in vivo* hypercapnia studies in rats were: 1) at one atmosphere we could administer 50% CO_2 for several hours, decreasing intracellular pH to ~6.6, with no change in ATP during that time. We could then have the rats wake up, having no gross neurobehavioral changes and no histological injury (Cohen *et al.*, 1990); 2) at a barometric pressure of ~2 atmospheres we could administer 50% CO_2 in oxygen for 15 minutes, causing PCO_2 to increase to ~1 atmosphere, and causing intracellular pH decrease to around 6.2. As in the first case, we could then have the rats wake up, having no gross neurobehavioral changes and no histological injury. The low values of brain intracellular pH that were reached, ~6.2, are comparable to those that occur during stroke (Xu *et al.*, 1991). After 15 minutes of cerebral ischemia, and thus after 15 minutes of comparable low brain intracellular pH, there is devastating neurologic injury. Thus low intracellular pH (~6.2) can be obtained in the presence of oxygen adequate to sustain ATP levels. Under such conditions low pH does not appear injurious.

Some details of the hyperbaric studies are worthy of mention. After fifteen minutes of PCO_2 near 1 atmosphere, cardiac output would decrease to levels causing cerebral ischemia. To do the NMR experiments, a special polycarbonate hyperbaric chamber was designed and built that would fit inside our 4.7 Tesla animal magnet, and permit remotely controlled ventilation (Litt *et al.*, 1993).

It would be desirable to perform similar *in vivo* studies during hypoxia and ischemia, because low pH issues during this state occur in a different biochemical context. However, the performance of similar *in vivo* studies during brain ischemia would be considerably more complex and difficult. In order to untangle some of the complexities, as well as to obtain better NMR spectral resolution, we switched, after conducting the above *in vivo* studies with a surface coil, to an *ex vivo* NMR system having respiring ("live") brain slices inside a solenoidal coil.

Ex Vivo NMR spectroscopy of neonatal brain slices

The switch to studying "live" brain slices in an NMR tube brought numerous advantages. In addition to being able to benefit from the better spectral resolution that comes with the homogeneous B_1 field of a solenoidal NMR coil, our arrangement has: (a) the elimination of contaminating NMR signals from nearby extracerebral regions of soft tissue and muscle; (b) the ability to select specific regions of the brain for study; (c) the ability to control extracellular fluid concentrations, while maintaining functional integrity of molecular and cellular structures found *in vivo*; (d) the capacity to accomplish pharmacological interventions with controlled drug concentrations, without interference from the blood-brain barrier or hepatic metabolism; (e) the absence of complications due to systemic physiologic responses that occur *in vivo*, such as hypotension; and (f) the absence of an anesthetic requirement or limit during experimental perturbations. Additionally, because brain slices in NMR experiments are easily removed during NMR protocols, accompanying non NMR studies (*e.g.*, histology, immunohistochemistry, etc.) are possible, permitting investigations of associations between NMR spectroscopy results and cell-specific changes. In such studies slices are obtained more easily than in an *in vivo* biopsy. Because slices are only 350 μ thick, metabolites are faithfully preserved when slices are rapidly frozen or fixed. We validated our use of the brain slice system by conducting hypercapnia studies in slices, and finding the same results as occur *in vivo* (Espanol *et al.*, 1992).

We were assisted in the development of our NMR slice system by publications and personal communications from two groups that did pioneering NMR spectroscopy work with perfused, respiring brain slices: 1) a Cambridge/Nottingham group led by Professor Peter G. Morris (Ben Yoseph *et al.*, 1993; Badar-Goffer *et al.*, 1990a; Badar-Goffer *et al.*, 1990b); and 2) a Université de Paris/Gif-sur-Yvette collaboration led by Drs. Champagnat and Beloeil (Jacquin *et al.*, 1989). In addition to the above investigators, other groups have been carrying out interesting NMR spectroscopy studies in respiring brain slices (Kauppinen *et al.*, 1994; Pirttila *et al.*, 1993; Pirttila *et al.*, 1994; Schanne *et al.*, 1993). Distinct differences between our slice technique and those of others have included: 1) our use of the Chan-Fishman slicing technique, in which cortical slices are cut by hand with a blade, so that slices are harvested rapidly (less than 30 seconds) and with only one side having an injury layer; and 2) our use of a larger NMR tube containing ensembles of 20-80 slices, or ~3 gm, of brain tissue. Given that another group, from Université de Paris/Gif-sur-Yvette, regularly conducts high-field, 2D-NMR studies *in vivo* in rats, using chronically implanted RF coils (Barrère *et al.*, 1990; Loubinoux *et al.*, 1994), 2D and 3D NMR studies in slices would also seem feasible.

When neonatal slices are studied, one can use ^{31}P NMR to simultaneously determine intracellular and extracellular pH (Espanol *et al.*, 1992). The chemical shift of the phosphoethanolamine (PE) resonance is used to determine the former (Corbett *et al.*, 1990; Corbett *et al.*, 1992), while the chemical shift of extracellular P_i (in the buffer) is used to determine the latter. Both in *ex vivo* brain slices and *in vivo* brain, ^{31}P NMR determinations of intracellular pH are phenomenological and not cell-specific. Recently developed light microscopy techniques permit the examination of pH and different intracellular compartments in a single cell (Bright *et al.*, 1987; Chacon *et al.*, 1994). In contrast, ^{31}P NMR measurements of pH (and all other metabolites) come from an average over ~10^8 cells, without distinguishing neurons from endothelial cells, glia, blood, or extracellular compartments.

Methodologies for animal studies and NMR techniques have been described elsewhere in more detail than will be given here (Espanol *et al.*, 1992). All animal protocols have been approved by the UCSF Committee on Animal Research. Our artificial cerebrospinal fluid (ACSF) generally consists of a modified Krebs balanced salt solution containing 124 mM NaCl, 5 mM KCl, 1.2 mM KH_2PO_4 1.2 mM $MgSO_4$, 1.2 mM $CaCl_2$, 26 mM $NaHCO_3$, and 10 mM glucose. It is prepared by the UCSF Cell Culture Facility, and administered after being warmed to 37 °C and equilibrated with an appropriate gas mixture, *e.g.*, 95% O_2 /5% CO_2 for hyperoxia 95% N_2 /5% CO_2 for hypoxia. When Mg^{2+}-free or other special ACSF is prepared, ionic composition is adjusted to obtain an osmolarity of ~300 mOsm.

To date, all NMR slice studies have been performed on a Nalorac Quest 4400 4.7 Tesla NMR instrument, operating at 81 and 200 MHz, respectively, for ^{31}P and ^1H. After being loaded with slices, and with the perfusion system in operation, the 20 mm NMR tube containing eighty tissue slices is positioned inside a custom-made, double-tuned, 4-turn, 23 mm × 15 mm solenoidal coil. The entire probe is then placed in the magnet, where B_o homogeneity is optimized by adjusting room-temperature shim currents until the water proton linewidth is less than 0.06 ppm. Typically, the total acquisition time one set of interleaved $^{31}P/^1H$ spectra is ~5 minutes. Acquisition consists of 2048 complex data points for both ^{31}P and ^1H. For ^{31}P, the time duration of the radiofrequency excitation pulse is typically ~27 ms for a 45° tip angle. The broad hump from phospholipids in the ^{31}P spectra is removed by a convolution difference method. Typical ^{31}P brain slice spectra are shown in Figure 2. NMR signal intensities for ^{31}P metabolites are determined by numerical integration of optimal computer fits to corresponding NMR resonance peaks in the spectra (Nalorac Quest 4400 Curve Fitting Program). ^{31}P metabolite concentrations are measured relative to corresponding signal intensities in the control run. Relative ATP levels are determined from the β-ATP peak at 16.3 ppm. Fully relaxed spectra (20 second interpulse delay) were obtained in special studies to obtain relaxation time corrections for different metabolites. As discussed below, intracellular pH is determined from the chemical shift separation between phosphocreatine (PCr) and inorganic phosphate (P_i), or from the chemical shift separation between PCr and phosphoethanolamine (PE). The two methods, if used to determine pH_i in the same experiment, need to be calibrated against each other, as explained below on and in Figure 3.

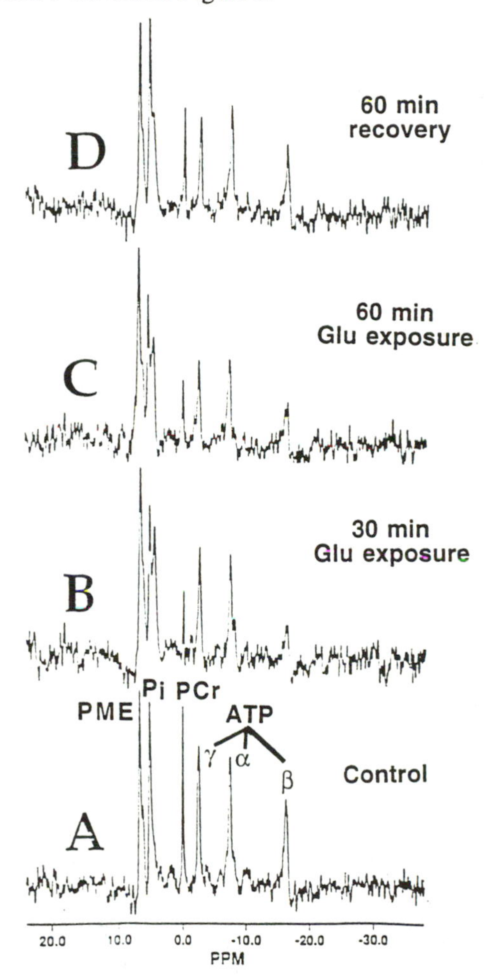

Figure 2: Four ^{31}P spectra from a typical excitotoxicity study in neonatal brain slices. Reprinted with permission from Espanol *et al.* (1994b). Chronologically, the order is from bottom to top, A to D. Each spectrum was acquired in interleaved mode, as described in the text, in ~5 minutes. After 30 minutes of glutamate exposure, the P_i peak can be seen to be split into intracellular and extracellular portions. The perfusate contained P_i. In such circumstances, pH_i could be determined from both the PME peak (phosphoethanolamine) and the intracellular P_i peak.

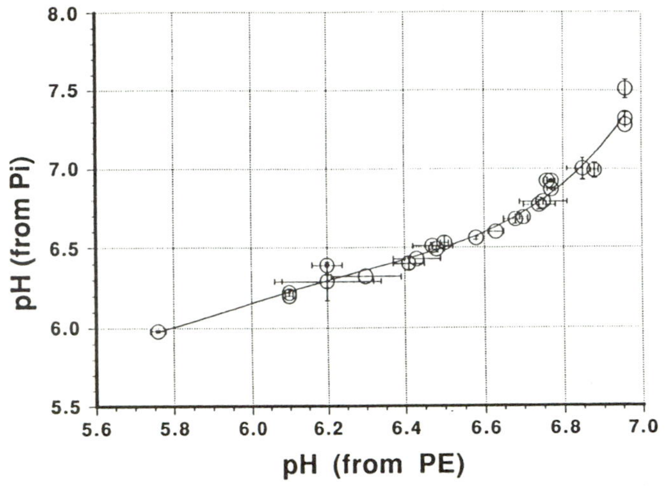

Figure 3: Comparison of pH_i determinations from PE and P_i peaks. Reprinted with permission from Espanol *et al.* (1992). In this study, hypercapnia was used to pull apart the P_i peaks, and pH_i formulae were used from studies by others. At pH_i values above 6.8, the use of PE for determining pH_i was less accurate.

For spin-echo ^1H NMR spectroscopy, experiments are initiated with a ~100 ms low power presaturation pulse centered on the water resonance, and followed with a 1- O AC(1,–) RF pulse having frequency maxima and minima located for selective excitation of metabolites with chemical shifts near lactate (1.32 ppm), and for selective non-excitation of water (4.70 ppm) (Hetherington *et al.*, 1985; Williams *et al.*, 1988). The spin-echo delay (TE) for refocusing pulses is ~136 milliseconds (~J). Having a long spin-echo delay permits substantial discrimination against more rapidly relaxing lipid signals that resonate near the lactate resonance peak at 1.32 ppm. Because lactate is a doublet with splitting (1/J) of ~7.3 Hz, choosing TE to be an integral multiple of J caused lactate to have a maximally positive spectral amplitude. Refocusing pulses were phase-cycled using the EXORCYCLE sequence (Bodenhausen *et al.*, 1977). Other typical ^1H NMR spectroscopy parameters are: 50 ms for a 90° nutation, +1,500 Hz spectral width, 500 millisecond repetition time, 862 millisecond interpulse delay. Typically, one ^1H spectrum requires 192 acquisitions. Additionally, each ^1H spectrum undergoes Lorenzian-to-Gaussian transformation by -12 Hz exponential and 7 Hz Gaussian multiplication. Relative changes in intracellular lactate levels are quantitated relative to the N-acetyl-aspartate (NAA) peak, an internal neuronal reference (Chang *et al.*, 1987). Figure 4 shows ^1H NMR brain slice spectra from one study.

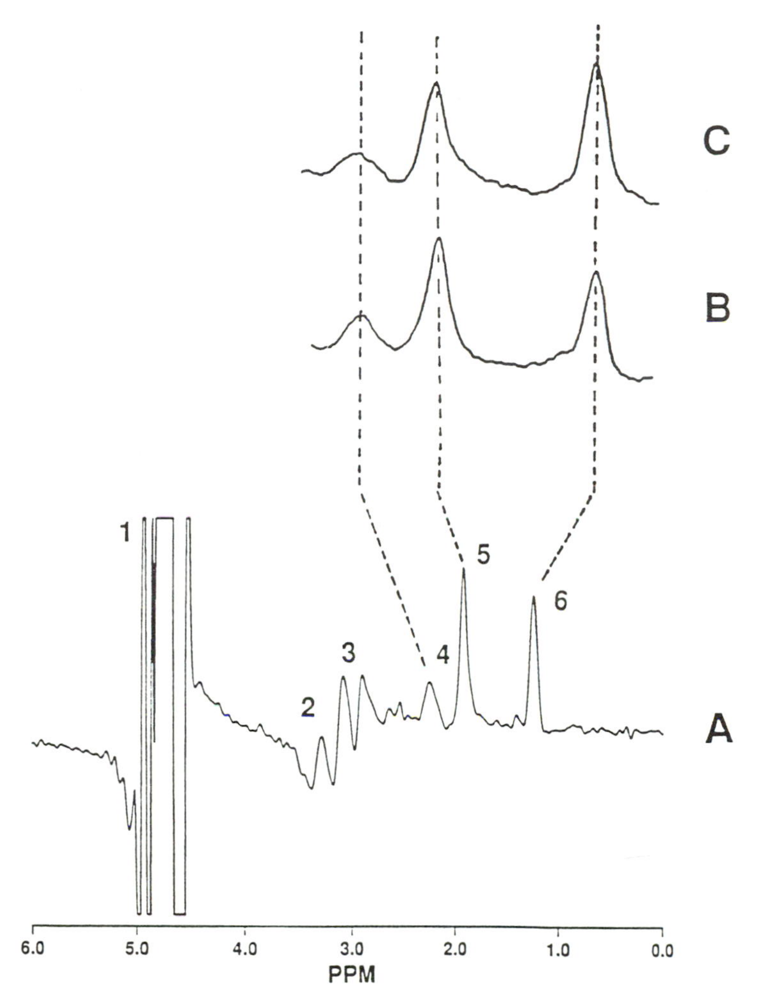

Figure 4: Representative ^1H neonatal brain slice spectra. Reprinted with permission from Espanol *et al.* (1992). Resonance peak assignments are as follows: 1, residual water peak; 2, phosphocholine; 3, PCr/Cr; 4, glutamate/glutamine; 5, N-acetyl-aspartate (NAA); 6, lactate. In trace A (bottom) one has the control spectrum. In the expanded portion shown in trace B (middle), 5% CO_2 is administered in 95% O_2. Finally, in expanded trace C (top), lactate starts to increase if 70% CO_2 is administered in 30% O_2. A comparable increase was seen if 70% N_2 is used instead. Thus the threshold for slice hypoxia can be determined in such preparations.

Intracellular pH and energy during excitotoxicity

The primary role of $^{31}P/^{1}H$ NMR spectroscopy in brain hypoxia/ischemia/excitotoxicity studies has been to provide information on intracellular energy failure and acidosis. When brain tissue is deprived of oxygen, changes in pHi, ATP, and PCr can be detected as rapidly as NMR spectra can be obtained - typically in a few minutes. However, during hypoxia/ischemia numerous pathophysiological actions are underway - with many occurring more rapidly than energy failure. For example, in the intact organism, electroencephalogram (EEG) activity ceases within 30 seconds after cerebral blood flow is stopped, despite the presence of substantial ATP that, under normal circumstances, would be sufficient for sustaining electrical signaling. At a molecular level injurious brain chemistry is known to occur on a time scale of milliseconds or less. An example includes free radical generation and injury. During reoxygenation after ischemia, or at low oxygen concentrations, highly reactive superoxide anions ($O_2^{-\bullet}$) can be produced, and these can diffuse across membranes (as the $HO_2^{\bullet}$ radical). Additionally, even when superoxide is degraded by superoxide dismutase to hydrogen peroxide, reactive hydroxyl radicals can be produced from H_2O_2 via ferrous or cuprous reduction (Cramer and Knaff, 1991). One might naively speculate that sustained, near-normal ATP levels after an insult might be sufficient to permit the restoration of brain cell integrity. However, this possibility is refuted by various phenomena, including excitatory amino acid neurotoxicity, especially one particular form, "delayed neuronal death", which occurs days after oxygen has been restored to ischemic tissue. Neuronal death in excitotoxicity appears to be caused by endogenous increases in brain extracellular glutamate, which can occur with or without ischemia (Choi, 1992). Glutamate excitotoxicity is believed to be a prominent mechanism in neurodegenerative disorders, including amyotrophic lateral sclerosis and Parkinson's disease (Beal, 1992). Earlier studies of intracellular, molecular mechanisms of ischemic brain injury also placed emphasis on calcium-induced processes, such as activation of arachidonic acid cascades. Such biochemical processes, along with free-radical mechanisms, are naturally accommodated by concepts of excitotoxicity, as NMDA-type glutamate receptors regulate calcium ion channels, and also mediate neuronal nitric oxide production (Dugan, 1994).

We recently completed brain slice NMR spectroscopy studies during toxicity from exogenous glutamate (Espanol *et al.*, 1994b). In these studies slices were kept well oxygenated, so as to simulate somewhat the situation in penumbral tissue. During the first hours of a stroke, it is alleged that there is a "penumbra", a small, potentially salvageable region that surrounds the ischemic tissue. In the ischemic core, cell death is certain. However, in the "penumbra", it is alleged that blood flow is sufficient for tissue rescue, if appropriate interventions are made in a timely fashion. Avoiding penumbral glutamate toxicity appears to be an important concern (Hossmann, 1994).

When high concentrations of glutamate were administered to well-oxygenated brain slices, there was immediate energy failure, and cell injury and death.

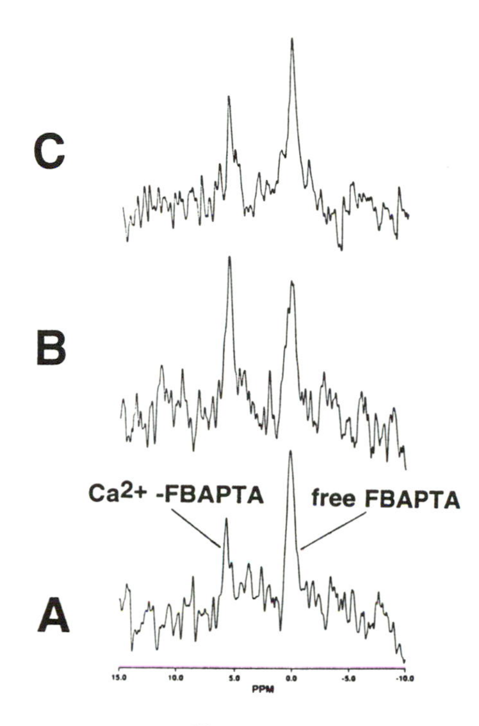

Figure 5: Representative 5 minute ^{19}F spectra in brain slices loaded with 5FBAPTA, as described in Espanol *et al.* (1994a). These data have not been published previously. In each spectrum, the ^{19}F resonance peak at the left (or right) corresponds to 5FBAPTA that is (or is not) bound to calcium. Intracellular calcium concentrations are determined from the ratio of these two signal intensities. The control spectrum is shown at the bottom, while the middle spectrum was obtained after 60 minutes of exposure to 2 mM glutamate. From the substantial increase of left to right signal intensities, it is apparent that intracellular calcium increased. The top spectrum shows recovery after glutamate washout. Note that cells are healthier when there is a relatively smaller peak of Ca^{2+}-bound 5FBAPTA.

However, we reduced the concentration of administered glutamate, and found that 2 mM was the largest concentration where rescue from irreversible metabolic damage could be performed. When 2 mM glutamate was administered alone for 60 minutes, this glutamate concentration caused an irreversible depletion of PCr to ~50% of control, and of ATP to ~80%. Spectra from a typical study of glutamate toxicity are exhibited in Figure 2. However, if there was concurrent treatment with 150 μM dizocilpine, an antagonist specific for NMDA-type glutamate receptors, complete metabolic rescue ensued. Two other glutamate receptor antagonists were also studied: kynurenate and NBQX. These are not specific for NMDA-type receptors, and metabolic rescue was minimal. We also studied edema (cell swelling) in our slices, using histological investigations of Nissl stained sections and wet-weight/dry-weight quantitations. There was excellent correlation with the NMR energy state in all cases (Espanol *et al.*, 1994c). When there was energy recovery, cell swelling and edema water were minimal. As energy recovery was reduced, cell swelling and brain water were increased. Furthermore, we carried out a ^{19}F NMR survey of intracellular calcium levels, as demonstrated in Figure 5, and showed that these increased during glutamate exposure (Espanol *et al.*, 1994a). Thus the studies demonstrated a fidelity between NMR spectroscopy measures and tissue condition. However, NBQX, which was minimally effective in our excitotoxicity slice studies, turns out to be more protective *in vivo* than dizocilpine during brain ischemia. We have not yet studied NBQX during slice ischemia (Xue *et al.*, 1994). Nevertheless, the *in vivo* NBQX studies show that it is important to compare slice results with *in vivo* results. If slice studies are to provide supplemental insights that cannot be obtained *in vivo* , rapport must be established between slice outcome measures and *in vivo* outcome measures.

Hypoxic protection by Fructose-1,6-Biphosphate (FBP)

Although most current studies about cellular mechanisms of hypoxic/ischemic cerebral protection are clearly focused on notions regarding excitotoxicity, free radical injury, and nitric oxide associated phenomena, a serendipitous empirical approach, pursued by various investigators, has not yet blended with standard models. In certain *in vivo* and *in vitro* animal studies, investigators found that pretreatment with fructose-1,6-bisphosphate (FBP) provided hemodynamic and brain metabolic protection during ischemia and hypoxia (Farias *et al.*, 1990; Kelleher *et al.*, 1994). It has been postulated that FBP somehow protects by sustaining ATP levels. However, metabolic mechanisms of FBP's protection have not been elucidated. In fact, in some studies to the contrary, FBP appeared to provide no protection at all. Advocates of FBP claim that in these studies, FBP was either administered after ischemia, or the ischemia was particularly severe, or FBP was below a threshold dose. We recently conducted *ex vivo* NMR spectroscopy studies in respiring, neonatal slices to determine the response of intracellular energy and histological patterns during and after thirty minutes of hypoxia (PO_2 <15 mm Hg) (Litt *et al.*, 1994). Our aim was to establish whether or not there is direct, cellular, metabolic protection by FBP, given either as pretreatment or post-treatment. Nine studies were performed, with n=3 for each of the three groups (no treatment, FBP pretreatment, FBP post-treatment). The FBP concentration, selected from successful *in vitro* studies by others, was 2 mM. Additionally, slices were removed for Nissl stain histology at different times in the protocol: post-decapitation, at the start of NMR studies, at the start of hypoxia, at the end of hypoxia, at the end of recovery. The no treatment and post-treatment groups were identical with respect to any measure, including histological grading. In these two groups, ATP and PCr rapidly fell to NMR-undetectable levels. In the group pretreated with FBP, ATP levels were unchanged throughout the entire study, although PCr decreased and became undetectable, albeit more slowly than in the no treatment group. In all groups pHi decreased during hypoxia to ~6.4 from 7.1, and then returned to control values during subsequent hyperoxia. However, the lactate/NAA ratio remained unchanged in the group pretreated with FBP. Nissl stained sections showed shrunken nuclei and cell swelling in the no treatment and post-treatment groups, but were minimally different from control in the group pretreated with FBP. Our slice studies validated the hypothesis that acute energy failure during hypoxia, normally of rapid onset, is substantially ameliorated by FBP pretreatment. However, it was not apparent why hypoxia-induced lactate increases were absent after FBP treatment. Although a metabolic modulation appears likely, these studies raised numerous questions. Fortunately, additional NMR spectroscopy studies can probably elucidate what is happening, especially if ^{13}C studies are performed and concentrations are determined for intermediate metabolites in glycolysis and other pools (Ben-Yoseph *et al.*, 1993; Badar-Goffer *et al.*, 1990; Kauppinen *et al.*, 1994). Because a low level of oxygen was detected during hypoxia it is possible that our test-tube insult was not as severe as in stroke. Nevertheless, dramatic early protection of ATP levels by FBP was easily detected. This might turn out to be a finding of substantial clinical importance.

Studies with adult rat brain slices

Once mastered, carrying out the biological preparation for NMR spectroscopy studies of neonatal brain slices is straightforward, taking no more than a few hours. It is considerably more difficult to obtain high integrity brain slices from older animals, where the cranium is thicker and more vascularized, and the brain has a shorter revival time and a longer harvesting time. However, if brain slice studies are to have relevance to adult stroke, the study of older brains is essential. Exerting substantial efforts, we recently demonstrated that it is possible to obtain an adult brain slice preparation where metabolic integrity is high, and ^{31}P NMR spectra have identical pHi and energy as is found *in vivo* (Espanol *et al.*, 1994d). Hypothermic cerebral metabolic protection, a well-studied phenomenon often used for obtaining slices for electrophysiology studies, was induced prior to slice harvesting. First, young adult Wistar rats (250-300 gms) underwent *in vivo* surface cooling while anesthetized with isoflurane, a general anesthetic agent that permitted spontaneous ventilation.

Even at low doses, isoflurane impairs thermoregulation and lowers the value for critical cerebral blood flow. The adult rats then underwent surface cooling with ice until their core temperatures decreased to ~30°C. While anesthetized and cold, a thoracotomy was performed, and a transcardiac, intra-aortic injection of 4 °C heparinized saline was used to further cool the brain to 19 °C and provide substantial metabolic protection. As in neonatal studies, harvested slices were allowed to recover and rewarm. However only 40 slices (from 10 adult rats) were needed to obtain an equivalent brain tissue aggregate. Because the harvesting of four slices from a single adult rat took ~25 minutes, slice harvesting took ~4 hours instead of 1 hour. After harvesting and recovery, NMR studies were carried out as before. Figure 6 shows typical ^{31}P brain slice spectra from adult brain

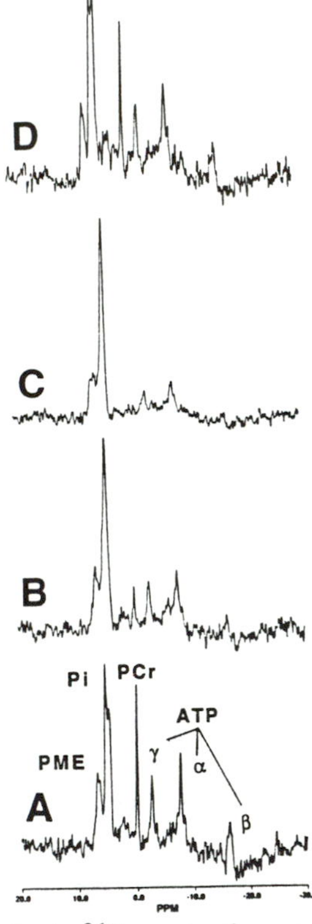

Figure 6: Representative 5 minute ^{31}P spectra from a study using adult brain slices. The methods are outlined in the text, and in Espanol *et al.* (1994d). The bottom spectrum is the control, while the one immediately above it is after 10 minutes of hypoxia. Above that is a spectrum from the end of a 20 minute period of hypoxia. The top spectrum is after 120 minutes of hyperoxic recovery. The ratio of metabolites in the control and recovery spectra are the same as in brain spectra obtained *in vivo* in anesthetized rats.

slice study. The high quality of the *ex vivo* slice spectrum is evident. In our perfused, adult brain slice studies of stroke, as in the neonatal excitotoxicity studies, NMR metabolic impairment and histology correlated well with each other. Thus it is possible to obtain respiring adult brain slices of high metabolic integrity for NMR spectroscopy studies. However, a substantial effort is necessary.

Conclusion

We have tried to demonstrate that interesting, relevant NMR spectroscopy studies can be conducted in perfused, respiring brain slices during periods of oxygen deprivation. We believe that such studies are relevant for testing pharmacologic agents designed to provide treatment and protection in stroke. Ultimately, all pharmacologic agents must prove their efficacy in clinical studies having well chosen outcome measures. Nevertheless, there appears to be a place for *ex vivo* NMR studies that can provide important, mechanism-oriented insights which might otherwise be unobtainable.

Acknowledgements and special tribute to Oleg

The authors deeply appreciate the following NIH research grant support: GM34767 (LL), NS22022 (PRW), NS14543 (PHC), NS25372 (PHC), and RR03841 (TLJ). Additionally, a special tribute to Oleg Jardetzky is appropriate. For one of the authors (LL), the motivation for embarking upon *in vivo* NMR spectroscopy research began at the Stanford Magnetic Resonance Laboratory (SMRL) in 1982 under the tutelage of Oleg Jardetzky and Norma Wade-Jardetzky. SMRL was a very active and friendly place. Besides Oleg and Norma, several other SMRL alumni, including contributors to this volume, took time out from their exciting studies in order to be helpful. Due to multiple circumstances, the center of gravity of Dr. Litt's efforts soon migrated north to UCSF, where it took form in the laboratories of Drs. Tom James and Pak Chan, directors respectively of UCSF research programs for *in vivo* NMR spectroscopy, and brain edema and stroke. Despite the institutional change, many of Oleg's original questions have remained as beacons, including general ones such as, "when and how will *in vivo* NMR spectroscopy become clinically useful?" Oleg's encouragement at a crucial, early stage will always be remembered and appreciated.

References

Adler, S. (1990) "Nuclear magnetic resonance and cell pH, with a focus on brain pH". in Pettegrew JW, ed. NMR, Principles and Applications to Biomedical Research. New York: Springer-Verlag, pp 485-505.

Alberti, K.G. and Cuthbert, C. (1982). "The hydrogen ion in normal metabolism: a review". in Ciba Found Symp. pp 1-19.

Badar-Goffer, R., Bachelard, H.S., and Morris, P.G. (1990a). *Biochem. J.* **266**, 133.

Badar-Goffer, R., Ben-Yoseph, O., Dolin, S.J., Morris, P.G., Smith, G.A., and Bachelard, H.S. (1990b). *J. Neurochem.* **55**, 878.

Barrère, B., Peres, M., Gillet, B., Mergui, S., Beloeil, J.C., and Seylaz, J. (1990). *FEBS. Lett.* **264**, 198.

Beal, M.F. (1992). *Ann. Neurol.* **31**, 119.

Ben-Yoseph, O., Badar-Goffer, R., Morris, P.G., and Bachelard, H.S. (1993). *Biochem. J.* **291 (Pt 3)**, 915.

Bodenhausen, G., Freeman, R., Turner, D.L., Morris, G.A., and Niedermeyer, R. (1977). *J. Mag. Res.* **25**, 559.

Bright, G.R., Fisher, G.W., Rogowska, J., and Taylor, D.L. (1987). *J. Cell Biol.* **104**, 1019.

Chacon, E., Reece, J.M., Nieminen, A.L., Zahrebelski, G., Herman, B., and Lemasters, J.J. (1994). *Biophys. J.* **66**, 942.

Chang, L.H., Pereira, B.M., Weinstein, P.R., Keniry, M.A., Murphy, B.J., and Litt, L. (1987). *Magn. Reson. Med.* **4**, 575.

Choi, D.W. (1992). *J. Neurobiol.* **23**, 1261.

Cohen, Y., Chang, L.H., Litt, L., Kim, F., Severinghaus, J.W., and Weinstein, P.R. (1990). *J. Cereb. Blood Flow Metab.* **10**, 277.

Corbett, R.J., and Laptook, A.R. (1990). *J. Neurochem.* **54**, 1208.

Corbett, R.J., Laptook, A.R., Garcia, D., and Ruley, J.I. (1992). *J. Neurochem.* **59**, 216.

Cramer, W.A., and Knaff, D.B. (1991). *Energy Transduction in Biological Membranes--A Textbook of Bioenergetics*. New York: Springer-Verlag, p 46.

Dugan, L.L., and Choi, D.W. (1994). *Ann. Neurol.* **35 Suppl**, S17.

Espanol, M.T., Litt, L., Yang, G.Y., Chang, L.H., Chan, P.H., and James T.L. (1992). *J. Neurochem.*, **59**, 1820.

Espanol, M.T., Litt, L., Xu, Y., Chang, L.H., James, T.L., and Weinstein, P.R. (1994a). *Brain Research* **647**, 172.

Espanol, M.T., Xu, Y., Litt, L., Yang, G.Y., Chang, L.H., and James, T.L. (1994b). *J. Cereb. Blood Flow Metab.* **14**, 269.

Espanol, M.T., Xu, Y., Litt, L., Chang, L.-H., James, T.L. and Weinstein, P.R. (1994c). *Acta. Neurochir.* **60 Supplement**, 58.

Espanol, M.T., Litt, L., Chang, L.-H., Weinstein, P.R., James, T.L., and Chan, P.H. (1994d). *Anesthesiology* **81 (Supplement)**, A819.

Farias, L.A., Smith, E.E., and Markov, A.K. (1990). *Stroke* **21**, 606.

Feinleib, M., Ingster, L., Rosenberg, H., Maurer, J., Singh, G., and Kochanek, K. (1993). *Ann. Epidemiol.* **3**, 458.

Hetherington, H.P., Avison, M.J., and Shulman, R.G. (1985). *Proc. Natl. Acad. Sci. U.S.A.* **82**, 3115.

Hossmann, K.A. (1994). *Brain Pathol.* **4**, 23.

Jacquin, T., Gillet, B., Fortin, G., Pasquier, C., Beloeil, J.C., and Champagnat, J. (1989). *Brain Research* **497**, 296.

Kaku, D.A., Giffard, R.G., and Choi, D.W. (1993). *Science* **260**, 1516.

Katsura, K., Kristian, T., Smith, M.L., and Siesjo, B.K. (1994). *J. Cereb. Blood Flow Metab.* **14**, 243.

Kauppinen, R.A., Pirttila, T.R., Auriola, S.O., and Williams, S.R. (1994). *Biochem. J.* **298**, 121.

Kelleher, J.A., Gregory, G.A., and Chan, P.H. (1994). *Neurochem. Res.* **19**, 209.

Kogure, K., and Hossmann, K.A. (1985) *Molecular Mechanisms of Ischemic Brain Damage*. New York, Elsevier.

Kogure, K., Hossmann, K.A., and Siesjö, B.K. (1993) *Neurobiology of ischemic brain damage*. New York, Elsevier.

Kristian, T., Katsura, K., Gido, G., and Siesjö, B.K. (1994). *Brain Research* **641**, 295.

Litt, L., Gonzalez, M.R., Severinghaus, J.W., Hamilton, W.K., Shuleshko, J., and Murphy B.J. (1985). *J. Cereb. Blood Flow Metab.* **5**, 537.

Litt, L., Xu, Y., Cohen, Y., and James, T.L. (1993). *Magn. Reson. Med.* **29**, 812.

Litt, L., Espanol, M.T., MacDonald, J., Chang, L.-H., Gregory, G.A., and Weinstein, P.R. (1994) "Fructose-1,6-biphosphate protects ATP levels in hypoxic neonatal rat brain slices". Presented at the 20th Annual Meeting of the Society for Neuroscience in Miami, Florida.

Loubinoux, I., Meric, P., Borredon, J., Correze, J.L., Gillet, B., and Beloeil, J.C. (1994). *Brain Research* **643**, 115.

Petroff, O.A., Prichard, J.W., Behar, K.L., Alger, J.R., den Hollander, J.A., and Shulman, R.G. (1985). *Neurology* **35**, 781.

Pirttila, T.R., and Kauppinen, R.A. (1993). *Neuroreport* **5**, 213.

Pirttila, T.R., Hakumaki, J.M., and Kauppinen, R.A. (1993). *J. Neurochem.* **60**, 1274.

Pirttila, T.R., and Kauppinen, R.A. (1994). *J. Neurochem.* **62**, 656.

Schanne, F.A., Gupta, R.K., and Stanton, P.K. (1993). *Biochim. Biophys. Acta* **1158**, 257.

Seo, Y., Murakami, M., Watari, H., Imai, Y., Yoshizaki, K., and Nishikawa, H. (1983). *J. Biochem. (Tokyo)* **94**, 729.

Williams, S.R., Proctor, E., Allen, K., Gadian, D.G., and Crockard, H.A. (1988). *Magn. Reson. Med.* **7**, 425.

Xu, Y., Cohen, Y., Litt, L., Chang, L.H., and James, T.L. (1991). *Stroke* **22**, 1303.

Xue, D., Huang, Z.G., Barnes, K., Lesiuk, H.J., Smith, K.E., and Buchan, A.M. (1994). *J. Cereb. Blood Flow Metab.* **14**, 251.

Zivin, J.A., and Choi, D.W. (1991) *Stroke Therapy Scientific American* **265**, 56.

Index